AF352797

Constructed and Floating Wetlands for Sustainable Water Reclamation

Constructed and Floating Wetlands for Sustainable Water Reclamation

Editors

Muhammad Arslan
Muhammad Afzal
Naser A. Anjum

MDPI • Basel • Beijing • Wuhan • Barcelona • Belgrade • Manchester • Tokyo • Cluj • Tianjin

Editors

Muhammad Arslan
University of Alberta
Canada

Muhammad Afzal
National Institute for
Biotechnology and Genetic
Engineering (NIBGE)
Pakistan

Naser A. Anjum
Aligarh Muslim University
India

Editorial Office
MDPI
St. Alban-Anlage 66
4052 Basel, Switzerland

This is a reprint of articles from the Special Issue published online in the open access journal *Sustainability* (ISSN 2071-1050) (available at: https://www.mdpi.com/journal/sustainability/special_issues/constructed_floating_wetlands).

For citation purposes, cite each article independently as indicated on the article page online and as indicated below:

LastName, A.A.; LastName, B.B.; LastName, C.C. Article Title. *Journal Name* **Year**, *Volume Number*, Page Range.

ISBN 978-3-0365-3214-1 (Hbk)
ISBN 978-3-0365-3215-8 (PDF)

Cover image courtesy of Muhammad Afzal.

Contents

About the Editors . . . . **vii**

Muhammad Arslan, Muhammad Afzal and Naser A. Anjum
Constructed and Floating Wetlands for Sustainable Water Reclamation
Reprinted from: *Sustainability* **2022**, *14*, 1268, doi:10.3390/su14031268 . . . . **1**

Zepei Tang, Jonaé Wood, Dominae Smith, Arjun Thapa and Niroj Aryal
A Review on Constructed Treatment Wetlands for Removal of Pollutants in the
Agricultural Runoff
Reprinted from: *Sustainability* **2021**, *13*, 13578, doi:10.3390/su132413578 . . . . **5**

Munazzam Jawad Shahid, Ameena A. AL-surhanee, Fayza Kouadri, Shafaqat Ali,
Neeha Nawaz, Muhammad Afzal, Muhammad Rizwan, Basharat Ali and Mona H. Soliman
Role of Microorganisms in the Remediation of Wastewater in Floating Treatment Wetlands:
A Review
Reprinted from: *Sustainability* **2020**, *12*, 5559, doi:10.3390/su12145559 . . . . **33**

Shafaqat Ali, Zohaib Abbas, Muhammad Rizwan, Ihsan Elahi Zaheer, İlkay Yavaş, Aydın
Ünay, Mohamed M. Abdel-DAIM, May Bin-Jumah, Mirza Hasanuzzaman and Dimitris
Kalderis
Application of Floating Aquatic Plants in Phytoremediation of Heavy Metals Polluted Water:
A Review
Reprinted from: *Sustainability* **2020**, *12*, 1927, doi:10.3390/su12051927 . . . . **63**

Pietro Denisi, Nicola Biondo, Giuseppe Bombino, Adele Folino, Demetrio Antonio Zema
and Santo Marcello Zimbone
A Combined System Using Lagoons and Constructed Wetlands for Swine
Wastewater Treatment
Reprinted from: *Sustainability* **2021**, *13*, 12390, doi:10.3390/su132212390 . . . . **97**

Neeha Nawaz, Shafaqat Ali, Ghulam Shabir, Muhammad Rizwan,
Muhammad Bilal Shakoor, Munazzam Jawad Shahid, Muhammad Afzal, Muhammad Arslan,
Abeer Hashem, Elsayed Fathi Abd Allah, Mohammed Nasser Alyemeni and Parvaiz Ahmad
Bacterial Augmented Floating Treatment Wetlands for Efficient Treatment of Synthetic Textile
Dye Wastewater
Reprinted from: *Sustainability* **2020**, *12*, 3731, doi:10.3390/su12093731 . . . . **111**

Muhammad Fahid, Shafaqat Ali, Ghulam Shabir, Sajid Rashid Ahmad, Tahira Yasmeen,
Muhammad Afzal, Muhammad Arslan, Afzal Hussain, Abeer Hashem,
Elsayed Fathi Abd Allah, Mohammed Nasser Alyemeni and Parvaiz Ahmad
Cyperus laevigatus L. Enhances Diesel Oil Remediation in Synergism with Bacterial Inoculation
in Floating Treatment Wetlands
Reprinted from: *Sustainability* **2020**, *12*, 2353, doi:10.3390/su12062353 . . . . **129**

Fan Wei, Munazzam Jawad Shahid, Ghalia S. H. Alnusairi, Muhammad Afzal, Aziz Khan,
Mohamed A. El-Esawi, Zohaib Abbas, Kunhua Wei, Ihsan Elahi Zaheer, Muhammad Rizwan
and Shafaqat Ali
Implementation of Floating Treatment Wetlands for Textile Wastewater Management: A Review
Reprinted from: *Sustainability* **2020**, *12*, 5801, doi:10.3390/su12145801 . . . . **145**

Momina Yasin, Muhammad Tauseef, Zaniab Zafar, Moazur Rahman, Ejazul Islam, Samina Iqbal and Muhammad Afzal
Plant-Microbe Synergism in Floating Treatment Wetlands for the Enhanced Removal of Sodium Dodecyl Sulphate from Water
Reprinted from: *Sustainability* **2021**, *13*, 2883, doi:10.3390/su13052883 . . . . **175**

Chu-Wen Yang, Yi-En Chen and Bea-Ven Chang
Microbial Communities Associated with Acetaminophen Biodegradation from Mangrove Sediment
Reprinted from: *Sustainability* **2020**, *12*, 5410, doi:10.3390/su12135410 . . . . **187**

About the Editors

Muhammad Arslan is a postdoctoral fellow at The University of Alberta, Edmonton, Canada. He obtained his doctoral degree in natural sciences from RWTH Aachen University, Germany, while working as a guest scientist at the Helmholtz Center for Environmental Research, Leipzig, Germany. By training, Dr. Arslan is an environmental microbiologist, currently attempting to optimize the bio-/phyto-remediation of industrial wastewaters via constructed wetlands, bio-reactors, and hybrid systems. His scientific interest lies in theoretical and applied aspects of the microbial ecology of these systems, primarily to understand the underlying mechanisms of pollutant degradation and/or transformation, along with the response of bacterial communities to the environmental stressors. His latest research has discovered that antimicrobials (cotrimoxazole) in constructed wetlands can cause in planta dysbiosis. Although the knowledge established to date is in its infancy, such a perturbance of plant endophytic communities could affect the host's health and, hence, compromise the system's remediation performance. He is also studying the evolution of specific microbes that are exposed to recalcitrants (e.g., naphthenic acids) in natural ecosystems.

Muhammad Afzal, Principal Scientist at the National Institute for Biotechnology and Genetic Engineering (NIBGE), Faisalabad (Pakistan), has pioneered the "Floating Treatment Wetland" (FTW) technology in Pakistan at the field scale for the treatment of sewage and industrial wastewater. Since 2014, floating treatment wetland technology and constructed wetlands have been applied at more than 20 sites in Pakistan to improve the quality of 700 million m3 of wastewater. Recently, Dr. Afzal held the runner-up position for the Nature Research Awards for Driving Global Impact. He is the first Asian scientist to win this prestigious award, which includes a cash prize of USD 10,000, in recognition for his work addressing polluted soil and wastewater through self-sustaining and environment-friendly technologies. Previously, Dr. Afzal was awarded a Gold Medal in Earth and Environmental Sciences by the Pakistan Academy of Sciences. Muhammad is an author of 90 research articles, with a total impact factor of 400 and 4800 citations; he published an article in Nature Sustainability as the first and corresponding author. Muhammad is also a member of the Society of Wetland Scientists.

Naser A. Anjum, Assistant Professor at the Department of Botany, Aligarh Muslim University, Aligarh, U.P. (India), received postdoctoral research training at the University of Aveiro (Portugal), Agricultural Biotechnology Research Centre, Academia Sinica (Taiwan), and Aligarh Muslim University, Aligarh (India). Dr. Anjum has contributed to the biology of "plant/crop–environment adaptation". His work has elucidated the physiology and biochemistry of plants (crops and salt-marsh macrophytes), and their adaptation to/tolerance of various abiotic stress factors and emerging pollutants. Dr. Anjum has also contributed to the field of phytoremediation and assessed plant/salt-marsh macrophyte chemical-pollutant-remediation potential and the underlying physiological/(bio)chemical mechanisms. He has been a pioneer worker unveiling nano-graphene-mediated oxidative stress, antioxidant metabolism and the glutathione-redox system, and also the significance of glutathione-independent defense systems in food-crop models. Dr. Anjum received prestigious research awards from the Portuguese Foundation for Science and Technology (FCT) (Portugal), Academia Sinica (Taiwan); Council of Scientific and Industrial Research (CSIR); and Department of Biotechnology (DBT), New Delhi (India). Dr. Anjum has published >100 peer-reviewed articles with >5000 citations, and has an h-index of 38 (SCOPUS). He has also edited 15 books and journal special issues. Dr. Anjum is a Fellow of The Linnean Society of London; Plantae, American Society of Plant Biologists (ASPB), USA; and International Society of Environmental Botanists (ISEB), NBRI, India. Dr. Anjum is also on the Editorial boards of several reputed international journals.

sustainability

Editorial

Constructed and Floating Wetlands for Sustainable Water Reclamation

Muhammad Arslan [1,*], Muhammad Afzal [2,*] and Naser A. Anjum [3,*]

1 Civil and Environmental Engineering Department, University of Alberta, Edmonton, AB T6G 2R3, Canada
2 Soil & Environmental Biotechnology Division, National Institute for Biotechnology and Genetic Engineering (NIBGE), Faisalabad 38000, Pakistan
3 Department of Botany, Aligarh Muslim University, Aligarh 202001, India
* Correspondence: marslan@ualberta.ca (M.A.); afzal@nibge.org (M.A.); naanjum1@myamu.ac.in (N.A.A.)

Citation: Arslan, M.; Afzal, M.; Anjum, N.A. Constructed and Floating Wetlands for Sustainable Water Reclamation. *Sustainability* **2022**, *14*, 1268. https://doi.org/10.3390/su14031268

Received: 5 January 2022
Accepted: 20 January 2022
Published: 24 January 2022

Publisher's Note: MDPI stays neutral with regard to jurisdictional claims in published maps and institutional affiliations.

1. Introduction

Modern urbanized societies are facing serious challenges in the maintenance of their water resources. Anthropogenic activities result in the production of large quantities of wastewater, carrying a wide array of organic and inorganic pollutants. Most of these pollutants could successfully be removed using conventional remediation technologies; nevertheless, passive and minimally invasive treatment schemes are preferred, as per the sustainable United Nations guidelines (UN SDG6). Being a part of the sewage treatment infrastructure, this would alleviate a substantial economic burden on low- and lower-middle-income countries. Phytoremediation—the use of plants for wastewater reclamation—is one such ecotechnology that offers engineered solutions such as constructed wetlands (CWs) and/or variants, i.e., floating treatment wetlands (FTWs). One successful example is the field-scale application of FTWs, which effectively attenuated a large fraction of diverse organic and inorganic contaminants, with as low as US$0.0026/m^3 of wastewater in Pakistan [1].

2. The Concept

CWs are modern variants of *Rieselfeld* (German: sewage trickle fields) systems, which were introduced by German social reformers in 1891. In *Rieselfeld*, effluent is trickled over gravel or water-permeable soil and degraded by microflora within the substrate [2]. Later, in the 1950s, Dr. Käthe Seidel (a German limnologist) developed a hybrid system for the faster treatment of municipal wastewater by introducing vegetation/plants in the filtering bed [3]. In these systems, multiple horizontal and single vertical seepage beds were used, along with gravel as a substrate, which were further vegetated with marsh plants (i.e., lakeshore bulrush, *Schoenoplectus lacustris*). These systems were recognized as *Pflanzenkläranlage* or "plant-based sewage treatment systems"; these inspired the terminology of the "constructed wetland". As of today, several CW and FTW variants have been engineered to harness the synergistic interactions among plants, microbes and substrates for the treatment of various contaminants from the water bodies. At first, the application of CWs and variants (FTWs) was limited to municipal and/or domestic wastewater treatment. However, modern research has expanded the scope of this ecotechnology to treat wastewater of variety of origins, i.e., stormwater, industrial water, landfill leachates, mine wastewater, and polluted river water [1,4].

3. The Appraisal

This Special Issue *'Constructed and Floating Wetlands for Sustainable Water Reclamation'* compiles six research and three review articles showcasing the use of CWs and variants, for the treatment of wastewater in diverse environmental settings such as that of swine, textile, hydrocarbons, pharmaceutical, and agricultural origins.

1

Swine breeding farms are major contributors to the production of swine wastewater (SW) that contains large fractions of urine, feces, antibiotics, pathogens, and residues of undigested food. The chemical oxygen demand and nutrient contents are tremendously high in SW, along with a large proportion of pharmaceuticals [5,6]. Hence, the direct discharge of untreated SW could negatively impact the biotic components of the receiving ecosystem [6]. In conventional settings, pre-treatment of SW is carried out in anaerobic lagoons for the degradation of organic matter, whereas CWs are used for the removal of nutrients [7]. However, in the absence of pre-treatment schemes, the performance of CWs is not efficient for the complete depuration of livestock wastewater. Given this, Denisi et al. (https://doi.org/10.3390/su132212390 (accessed on 20 December 2021)) showed that both aerated and non-aerated lagoons, when combined with CWs (plant: *Typha latifolia* L.), improve the depuration-efficiency of SW at the pilot-scale. This system attenuated ~99% of organic matter and total suspended solids, along with 80–95% removal of total nitrogen. The study could provide a starting point to establish similar treatment systems for the effective treatment of livestock wastewater at an impaired C/N ratio.

The agriculture sector heavily relies on agrochemicals such as pesticides, herbicides, fungicides, and hormones, to achieve higher yields and feed the burgeoning world population [8]. As a result, agricultural runoff carries a large proportion of nutrients, suspended solids, pesticides, veterinary medicines, pathogens, and potentially toxic metals. To this end, Tang et al. (https://doi.org/10.3390/su132413578 (accessed on 20 December 2021)) highlighted CWs as a panacea for the effective treatment of various contaminants in agricultural runoff. This review article proposes CWs as an innovative solution to mitigate the emerging negative environmental impacts of agricultural intensification.

In recent years, self-buoyant hydroponic root mats have received tremendous attention to reclaim wastewater in open systems, i.e., lagoon, pond, lake [9]. The enrichment of specialized microorganisms along with appropriate choice of macrophytes could greatly enhance the remediation potential of FTWs. To this end, Nawaz et al. (https://doi.org/10.3390/su12093731 (accessed on 20 December 2021)) reported that the bioaugmentation of plant-growth-promoting and pollutant-degrading bacteria efficiently removed a variety of pollutants from textile wastewater. Further, the high persistence of inoculated bacteria in the water, root interior, and shoots interior of the wetland plant was positively correlated with the performance of FTWs. The proliferation of rhizospheric and endophytic bacteria efficiently reduced the total dissolved solids, total suspended solids, chemical oxygen demand, biochemical oxygen demand, electric conductivity, color, and toxic metals from the dye-polluted wastewater. Additionally, the plant's growth was improved, and toxicity was alleviated from the textile effluent, which ultimately promoted the plants' ability to tolerate pollutant-induced toxicity. Accordingly, Fahid et al. (https://doi.org/10.3390/su12062353 (accessed on 20 December 2021)) reported that the synergism of plant- and hydrocarbon-degrading bacteria could improve the remediation of diesel oil from the contaminated water in FTWs.

Sodium dodecyl sulfate (SDS) is a commonly found anionic surfactant in detergents and is extensively applied in various sectors [10]. The direct discharge of these wastewaters without pre-treatment may have harmful effects on biotic elements, particularly the aquatic life [11]. The adoption of conventional and unsustainable methods (i.e., coagulation, filtration with coagulation, adsorption, ion exchange, ozonation, reverse osmosis) may achieve sufficient removal of pollutants; nevertheless, these methods are also known to produce secondary pollution by generating toxic sludge [12,13]. Here, Yasin et al. (https://doi.org/10.3390/su13052883 (accessed on 20 December 2021)) inoculated a consortium of rhizospheric and endophytic bacteria in FTWs comprising two wetland plants. The system achieved a significant removal of sodium dodecyl sulfate (97.5%) concentration in the contaminated water. The authors argued that plant–bacteria synergism provided a congenial environment for the survival and proliferation of inoculated bacteria within the plant tissues for necessary catabolic functioning.

Significantly higher concentrations of acetaminophen (N-acetyl-p-aminophenol, ACE) were reported in the influents and effluents of sewage treatment plants [14]. This has raised serious concerns for natural aquatic ecosystems [15]. One example is the disturbance of mangroves ecosystems, which are known to sink pollutants in tropical and subtropical regions [16]. The sediments in mangroves can accumulate high concentrations of nonylphenol, polycyclic aromatic hydrocarbons, sulfonamides, which could be degraded with the microbial action [17,18]. The application of ACE-degrading bacteria and white-rot fungus appeared to be a promising approach for ACE removal from the aquatic environment [19,20]. Yang et al. (https://doi.org/10.3390/su12135410 (accessed on 20 December 2021)) achieved the successful removal of ACE in mangrove sediments by adding microcapsules, ACE-degrading bacteria, and electron acceptors (Na_2SO_4, $NaNO_3$, and $NaHCO_3$). To this end, the best ACE-degradation was reported with the addition of $NaNO_3$. It was further reported that the addition of an electron acceptor could enrich sixteen microbial genera, which are primarily involved in the anaerobic transformation of ACE.

4. Conclusions and the Way Forward

A thorough understanding of different wetland variants and their working principles is crucial in the customized treatment of wastewater of multiple origins. The major outcomes of the discussion in the research and reviews of this Special Issue may provoke future studies on the subject and help governmental bodies and/or industries to cost-effectively treat wastewater and meet discharge standards. Furthermore, by employing these ecotechnologies, the accumulation of toxic chemicals in the food chain can be reduced, and the local population can be protected against the potentially toxic effects of organic and inorganic pollutants. This approach can also be applied to promote the sustainable production of bioenergy crops, in conjunction with the remediation of municipal effluents. The plant biomass may also be used as wood fuel, especially in the villages and towns, which could greatly reduce the cutting of trees for fuel in underprivileged societies. Last but not least, public, farmers, industrialists, traders, exporters, and commercial entrepreneurs could directly benefit from the useful aspects of CWs and FTWs that are highlighted in this Special Issue.

Author Contributions: M.A. (Muhammad Arslan), M.A. (Muhammad Afzal) and N.A.A. contributed equally. All authors have read and agreed to the published version of the manuscript.

Funding: This research received no external funding.

Institutional Review Board Statement: Not applicable.

Informed Consent Statement: Not applicable.

Data Availability Statement: Not applicable.

Conflicts of Interest: The authors declare no conflict of interest.

References

1. Afzal, M.; Arslan, M.; Müller, J.A.; Shabir, G.; Islam, E.; Tahseen, R.; Anwar-ul-Haq, M.; Hashmat, A.J.; Iqbal, S.; Khan, Q.M. Floating treatment wetlands as a suitable option for large-scale wastewater treatment. *Nat. Sustain.* **2019**, *2*, 863–871. [CrossRef]
2. Bjarsch, B. 125 Jahre Berliner Rieselfeld-Geschichte. *Wasser Boden* **1997**, *49*, 45–48.
3. Seidel, K. Die Flechtbinse, *Scirpus lacustris* L. *Die Binnengewasser* **1955**, *21*, 1–21.
4. Shahid, M.J.; Arslan, M.; Ali, S.; Siddique, M.; Afzal, M. Floating wetlands: A sustainable tool for wastewater treatment. *Clean-Soil Air Water* **2018**, *46*, 1800120. [CrossRef]
5. Dordio, A.; Carvalho, A.J.P. Constructed Wetlands with Light Expanded Clay Aggregates for Agricultural Wastewater Treatment. *Sci. Total Environ.* **2013**, *463–464*, 454–461. [CrossRef]
6. Folino, A.; Zema, D.A.; Calabrò, P.S. Environmental and Economic Sustainability of Swine Wastewater Treatments Using Ammonia Stripping and Anaerobic Digestion: A Short Review. *Sustainability* **2020**, *12*, 4971. [CrossRef]
7. Villamar, C.A.; Rivera, D.; Neubauer, M.E.; Vidal, G. Nitrogen and Phosphorus Distribution in a Constructed Wetland Fed with Treated Swine Slurry from an Anaerobic Lagoon. *J. Environ. Sci. Health* **2015**, *50*, 60–71. [CrossRef] [PubMed]
8. Carvalho, F. Agriculture, pesticides, food security and food safety. *Environ. Sci. Policy* **2006**, *9*, 685–692. [CrossRef]

9. Colares, G.S.; Dell'Osbel, N.; Wiesel, P.G.; Oliveira, G.A.; Lemos, P.H.Z.; da Silva, F.P.; Lutterbeck, C.A.; Kist, L.T.; Machado, Ê.L. Floating treatment wetlands: A review and bibliometric analysis. *Sci. Total Environ.* **2020**, *714*, 136776. [CrossRef] [PubMed]

10. Olkowska, E.; Polkowska, Z.; Namiesnik, J. Analytics of surfactants in the environment: Problems and challenges. *Chem. Rev.* **2011**, *111*, 5667–5700. [CrossRef] [PubMed]

11. Arslan, M.; Ullah, I.; Müller, J.A.; Shahid, N.; Afzal, M. Organic Micropollutants in the Environment: Ecotoxicity Potential and Methods for Remediation. Available online: https://link.springer.com/chapter/10.1007/978-3-319-55426-6_5 (accessed on 20 December 2021).

12. Khan, S.; Afzal, M.; Iqbal, S.; Khan, Q.M. Plant-bacteria partnerships for the remediation of hydrocarbon contaminated soils. *Chemosphere* **2013**, *90*, 1317–1332. [CrossRef] [PubMed]

13. Ojo, O.A.; Oso, B.A. Biodegradation of synthetic detergents in wastewater. *Afr. J. Biotechnol.* **2009**, *8*, 229–249.

14. Tran, N.H.; Reinhard, M.; Khan, E.; Chen, H.; Nguyen, V.T.; Li, Y.; Goh, S.G.; Nguyen, Q.B.; Saeidi, N.; Gin, K.Y.H. Emerging contaminants in wastewater, stormwater runoff, and surface water: Application as chemical markers for diffuse sources. *Sci. Total Environ.* **2019**, *676*, 252–267. [CrossRef] [PubMed]

15. Sun, C.; Dudley, S.; McGinnis, M.; Trumble, J.; Gan, J. Acetaminophen detoxification in cucumber plants via induction of glutathione S-transferases. *Sci. Total Environ.* **2019**, *649*, 431–439. [CrossRef] [PubMed]

16. Bernard, D.; Pascaline, H.; Jeremie, J.J. Distribution and origin of hydrocarbons in sediments from lagoons with fringing mangrove communities. *Mar. Pollut. Bull.* **1996**, *32*, 734–739. [CrossRef]

17. Chang, B.V.; Lu, Z.J.; Yuan, S.Y. Anaerobic degradation of nonylphenol in subtropical mangrove sediments. *J. Hazard. Mater.* **2009**, *165*, 162–167. [CrossRef] [PubMed]

18. Yang, C.W.; Tsai, L.L.; Chang, B.V. Fungi extracellular enzyme-containing microcapsules enhance degradation of sulfonamide antibiotics in mangrove sediments. *Environ. Sci. Pollut. Res.* **2017**, *25*, 10069–10079. [CrossRef] [PubMed]

19. De Gusseme, B.; Vanhaecke, L.; Verstraete, W.; Boon, N. Degradation of acetaminophen by Delftia tsuruhatensis and Pseudomonas aeruginosa in a membrane bioreactor. *Water Res.* **2011**, *45*, 1829–1837. [CrossRef] [PubMed]

20. Żur, J.; Wojcieszyńska, D.; Hupert-Kocurek, K.; Marchlewicz, A.; Guzik, U. Paracetamol—Toxicity and microbial utilization. Pseudomonas moorei KB4 as a case study for exploring degradation pathway. *Chemosphere* **2018**, *20*, 192–202. [CrossRef] [PubMed]

sustainability

MDPI

Review

A Review on Constructed Treatment Wetlands for Removal of Pollutants in the Agricultural Runoff

Zepei Tang, Jonaé Wood, Dominae Smith, Arjun Thapa and Niroj Aryal *

Department of Natural Resources and Environmental Design, College of Agriculture and Environmental Sciences, North Carolina Agricultural and Technical State University, 1601 E. Market St., Sockwell Hall Room 120, Greensboro, NC 27411, USA; ztang@ncat.edu (Z.T.); jrwood@aggies.ncat.edu (J.W.); dasmith4@aggies.ncat.edu (D.S.); athapa@aggies.ncat.edu (A.T.)
* Correspondence: naryal@ncat.edu; Tel.: +1-(336)-285-3832

Abstract: Constructed wetland (CW) is a popular sustainable best management practice for treating different wastewaters. While there are many articles on the removal of pollutants from different wastewaters, a comprehensive and critical review on the removal of pollutants other than nutrients that occur in agricultural field runoff and wastewater from animal facilities, including pesticides, insecticides, veterinary medicine, and antimicrobial-resistant genes are currently unavailable. Consequently, this paper summarized recent findings on the occurrence of such pollutants in the agricultural runoff water, their removal by different wetlands (surface flow, subsurface horizontal flow, subsurface vertical flow, and hybrid), and removal mechanisms, and analyzed the factors that affect the removal. The information is then used to highlight the current research gaps and needs for resilient and sustainable treatment systems. Factors, including contaminant property, aeration, type, and design of CWs, hydraulic parameters, substrate medium, and vegetation, impact the removal performance of the CWs. Hydraulic loading of 10–30 cm/d and hydraulic retention of 6–8 days were found to be optimal for the removal of agricultural pollutants from wetlands. The pollutants in agricultural wastewater, excluding nutrients and sediment, and their treatment utilizing different nature-based solutions, such as wetlands, are understudied, implying the need for more of such studies. This study reinforced the notion that wetlands are effective for treating agricultural wastewater (removal >90%) but several research questions remain unanswered. More long-term research in the actual field utilizing environmentally relevant concentrations to seek actual impacts of weather, plants, substrates, hydrology, and other design parameters, such as aeration and layout of wetland cells on the removal of pollutants, are needed.

Keywords: constructed wetlands; agricultural runoff; chemicals of emerging concern; veterinary antibiotics; antibiotic resistant genes

Citation: Tang, Z.; Wood, J.; Smith, D.; Thapa, A.; Aryal, N. A Review on Constructed Treatment Wetlands for Removal of Pollutants in the Agricultural Runoff. *Sustainability* **2021**, *13*, 13578. https://doi.org/10.3390/su132413578

Academic Editors: Muhammad Arslan, Muhammad Afzal and Naser A. Anjum

Received: 12 October 2021
Accepted: 6 December 2021
Published: 8 December 2021

1. Introduction

Agricultural runoff contains excess quantities of diverse pollutants, such as sediments, nutrients, pathogens, veterinary medicines, pesticides, and metals. Modern agriculture heavily relies on agro-chemicals, such as pesticides, herbicides, and hormones, that would grant a greater yield in a shorter period [1]. As the demands for food have increased, so has the intensity of agricultural activities and animal feed operations [2]. As a result, agricultural practices over the past years have included more pesticides and inorganic fertilizers [3,4]. Carvalho and colleagues (1997) reported that North American farmers relied on herbicides 43.3% of the time, while European farmers used it slightly less at 26.3% in 1993 [5]. In 2005, there were more than 800 newly registered pesticides in the European Union [6]. Additionally, approximately two million tons of pesticides were used globally in 2019, with China and the USA being the two major users [7].

These chemicals are perfect for increasing yield but are ecologically detrimental when they leave agricultural ecosystems in runoff water following storms [8]. Studies have shown that only 1% of pesticides applied to crops are effective, the other 99% enter the atmosphere, soils, and bodies of water through non-targeted contamination [9]. In livestock production, animal waste can act as reservoirs for antibiotic resistant genes (ARGs) and antibiotic resistant bacteria (ARBs) [10–12]. Another study found the prevalence of veterinary pharmaceuticals to be higher in soil than in water, indicating likeliness of movement to water resources through agricultural runoff [13].

In the long run, these chemicals have the ability to negatively impact food security and agricultural sustainability [14]. Termed chemicals of emerging concern (CECs), these compounds have long been a threat for water quality. According to the Environmental Protection Agency (EPA), CECs include but are not limited to nanoparticles, pharmaceuticals, personal care products (PCPs), estrogenic compounds, flame-retardants, detergents, and other industrial chemicals. All of these contaminants, many of which have agricultural origin, significantly influence human health and aquatic life [15].

Treatment of diffuse source pollution, such as agricultural runoff, requires a low-cost, passive, and nature-based approach known as an ecological engineering approach. Constructed wetland (CW) is a natural ecological alternative to the conventional methods for treating various types of wastewater, including agricultural runoff [16]. The EPA (1993) defines CWs as engineered systems that are designed and constructed to utilize natural processes [17]. Specially designed CWs could be used to treat wastewater in a system that mimics their natural components. The use of wetland plants to treat wastewater is a technique that was firstly studied in the 1950s by German scientist Dr. Ka the Seidel; since then, the idea has expanded greatly and is a very sustainable way of naturally treating many sources of wastewater [18]. CWs are more beneficial than conventional wastewater treatment methods because they require lower energy and less operational effort, but they are also land intensive [16]. CWs are versatile in their functioning, serving as a tool for water quality improvement, hydrological buffers, reservoirs, and nature development/recreational areas [19]. Through imitation of natural wetland systems, such as marshes with wetland plants, soils, and soil microorganisms, CWs are capable of removing diverse contaminants from different wastewater sources [20].

However, there is still very little known about the biotic and abiotic influences and interactions that allows this treatment of water and soil to take place [17]. While much of the previous reviews focused on how CWs are used to efficiently remove nutrients, such as nitrogen and phosphorus, and sediments from wastewaters [21,22], this paper focuses on the occurrence of pollutants in the agricultural runoff and how this cost-effective green approach [23] can be used to remove pollutants from agricultural runoff for mitigation of the negative environmental impacts of agricultural intensification. Focus pollutants include veterinary medicines, antimicrobial resistant genes, insecticides, herbicides, and pesticides.

2. Approach and Definitions

In this article, we reviewed global literature that focused on CWs used for the treatment of agricultural runoff or wastewater and the characteristics of their design. Scholarly databases were searched using keywords, such as constructed wetlands, agricultural runoff, ARGs, ARBs, pesticides, veterinary antibiotics, chemicals of emerging concern, and their combination to source relevant articles, reports, books, and conference proceedings published in recent years. Both lab-scale and field-scale experiments that studied effective removal rates of contaminants in these systems were considered. The search resulted in over 60 publications that were examined and subsequently summarized in this article directly or indirectly.

CWs are generally classed based on the life form of the dominating large aquatic plant or macrophyte in the system [24] or water-flow regime [25]. Figure 1 shows the classification of CW and their characteristics, which includes flow and flow direction [18,25,26]. Search results were screened based on their relevancy to include CWs that were subsurface horizontal flow (SSHF), subsurface vertical flow (SSVF), surface flow (SF) and hybrid and were used to remove contaminants that were not nutrients (nitrogen (N), phosphorous (P), total nitrogen (TN), total phosphorous (TP), and sediment).

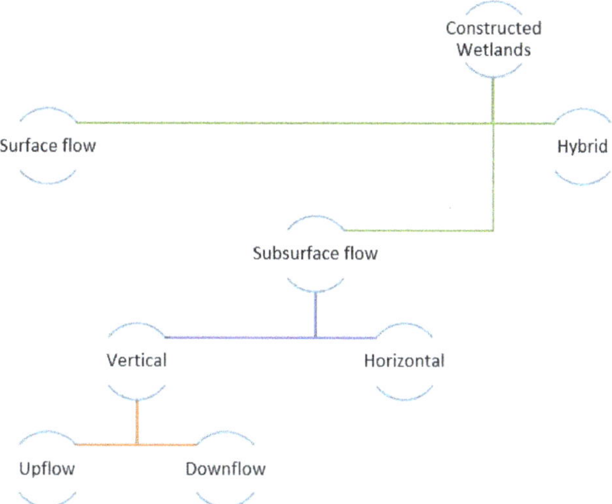

Figure 1. Classification of constructed wetlands.

Hydrological factors dictate the functioning of wetlands as they are directly linked to the ecosystem's biotic and abiotic processes. These processes are what influences both the biological (nutrient availability, microbial community, plant community) and physicochemical (soil pH, water pH, oxidation-reduction potential (ORP)) parameters in CWs [27]. Success of CWs is heavily dependent on the hydraulic residence time (HRT) [28, 29] and the hydraulic loading rate (HLR) [30,31]. Various factors, such as wetland design, scale, size, water depth, HRT, HLR, substrate, experiment duration, source of pollutant, pollutant influent concentration, removal percentage, and major mechanisms responsible for removal of pollutants were tabulated, represented in graphs, or analyzed further.

3. Occurrence of Pollutants in Agricultural Runoff

Diverse pollutants have been measured in agricultural runoff. Pesticides, herbicides, and veterinary pharmaceuticals are present in agricultural runoff and are major threats to water quality health [32]. Concentrations of CECs have been found in quantities in excess of 0.01 mg/L, especially during storm events [33]. The antibiotics found mostly in agricultural runoff from the reviewed articles are mainly tetracyclines, sulfamonomethoxine, enrofloxacin, and trimethoprim, which are either used for disease prevention or as growth promoters in the industry [34–41]. A Chesapeake Bay study found high concentrations of antibiotics (azithromycin (AZI), clarithromycin (CLA), difloxacin (DIF), enrofloxacin (ENR), norfloxacin (NOR), roxithromycin (ROX), and sulfamethoxazole (SMX)), and hormones (mainly estrogen derivatives) due to wastewater effluents and agricultural runoff [33]. Antibiotics in both swine and dairy cattle farm effluents were found at high concentrations in China, which implies frequent application of these antibiotics during the production process [42]. Since China is one of the largest producers of animals in the world, significant consumption and release to the environment are expected.

Animal husbandry is a major source of environmental ARGs and ARBs [12]. ARG dissemination from flowing water normally happens from ground or surface water sources receiving effluents from domestic, municipal, and agricultural sources, such as livestock farms [43,44]. Through horizontal gene transfer, bacteria are able transfer resistance from one organism to the other. Oliver and colleagues studied dairy manure systems and found the presence of bacteria, such as *Enterobacteriaceae* (specifically nontyphoidal *Salmonella*), antibiotic-resistant *Campylobacter*, methicillin- and vancomycin-resistant *Staphylococcus*, and vancomycin-resistant *Enterococcus*, which the Centers for Disease Control and Prevention (CDC) deemed clinically dangerous to be prevalent. Additionally, they also found that some of these bacteria were able to resist up to five antibiotics [45]. A Chinese study (2018) found 18 types of ARGs from swine feedlots in the surrounding environment, namely streams and agricultural soils [46]. Genes dominant in swine manure were found to be those that were resistant to tetracycline (TC), aminoglycoside (AGR), chloramphenicol (CPR), multidrug (MDR), sulfonamide (SNR) and beta-lactam (BLR) [46–59]. SNR genes were also found abundantly in dairy manure storm runoffs [60]. Background bacterial DNA concentrations were indicated by 16S rRNA data as high as 4.10×10^{13} copies/mL [61].

The occurrence of pesticides was also found to be prevalent in agricultural runoff effluent [62–65]. A Mexican study found priority pollutants, such as endosulfan, an insecticide that is authorized for use in the country, to be in excess of 8.656×10^{-3} mg/L in runoff water during storms [66]. Other major contaminants in agricultural runoff include veterinary pharmaceuticals and personal care products (PPCPs), such as naproxen, estrone (and other estrogenic derivatives), which are used mainly for pain suppression or growth enhancement for animals [65–69].

A summary of occurrence of these pollutants in the agricultural wastewater (Figure 2) indicate presence of greater than 1000 mg/L biochemical oxygen demand (BOD), chemical oxygen demand (COD) and total suspended solids (TSS); sub part-per-trillion to 30 part-per-million of antibiotics, hormones, and veterinary pharmaceuticals, and up to 4.1×10^{12} cells per mL of bacteria [42,65,70–72]. Herbicides were found to be more dominant in the agricultural runoff as it had been found as high as 530 mg/L. The prevalence of other contaminants, such as metals and fungicides, were much lower than the other CECs considered [42,66,68,70].

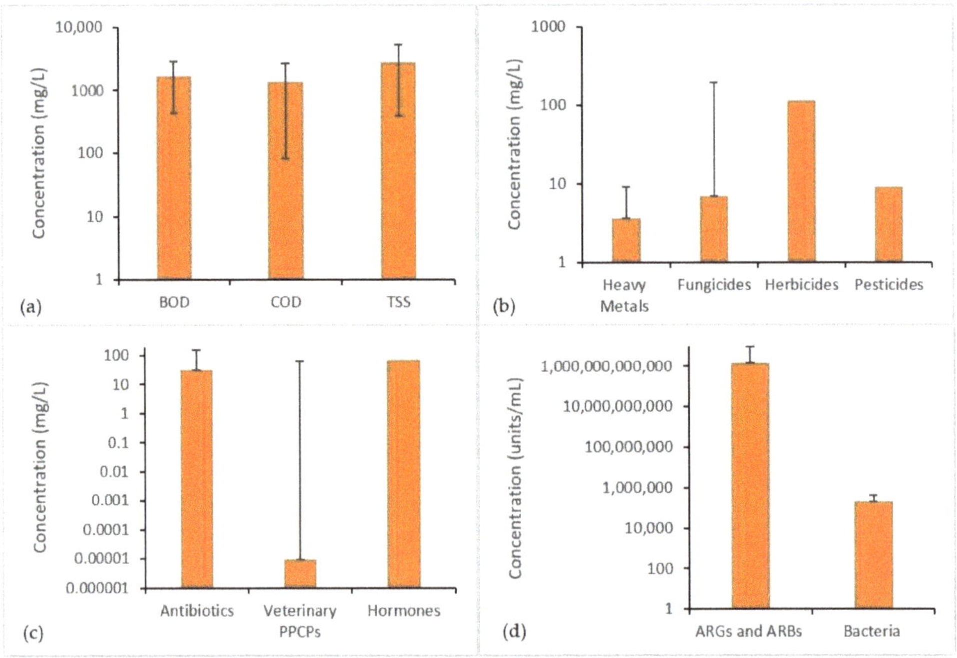

Figure 2. Occurrences of pollutants in the agricultural runoff: (**a**) Total suspended solids, biochemical oxygen demand, and chemical oxygen demand, (**b**) metals, fungicides, herbicides and pesticides, (**c**) antibiotics, hormones and veterinary pharmaceuticals, (**d**) antibiotic resistant bacteria and genes, and bacteria. Data are mean ± standard deviation. Note logarithmic y-axis.

4. Removal of Pollutants by Wetlands and Processes for Their Removal

Many studies have been conducted on the applications of constructed wetlands to remove pollutants from agricultural runoff and wastewater. Based on the study scales, this section has been divided into lab-scale and field-scale for further discussion.

4.1. Lab-Scale

Scientists in Portugal conducted laboratory-scaled microcosm studies to evaluate the removal performance of constructed wetlands for veterinary antibiotics for many years [39,42,73]. In their CW microcosms, multiple layers were set up (from bottom to top) as gravel, lava rock, root bed substrate (which was a mixture of soil and sand to help the vegetation's establishment) and *Phragmites australis* were planted. They used wastewater from swine farms/saline aquaculture facilities as their influent water with antibiotic concentrations spiked-up to 100 μg/L. The results showed that the removal efficiency for vet antibiotics-enrofloxacin (ENR), tetracycline (TET) [39], oxytetracycline (OXY) [73] and ceftiofur (CEF) [42] were over 90% after 9 to 20 weeks treatment period. The major mechanisms for the removal processes were adsorption to the substrate and plant's root (physical process), microbial metabolization and degradation (biological and chemical processes) and plant uptake (biological process) [39,42,73]. Studies conducted using the wide range of pollutants and various influent water types (fresh water and saline water) proved that CW microcosm design was adaptable to various wastewater treatments with satisfying removal efficiencies. Their study in 2020 using the same system even achieved toxic metal removal while maintaining the nutrient levels for agriculture reuse [70]. Another study in 2018 using the same system observed removal of organic micropollutants, such as atrazine, clarithromycin, fluoxetine, and norfluoxetine, from the freshwater aquaculture effluents [74]. Evidence from other studies suggested that such removal was accomplished through microbial degradation [75,76]. Another study conducted in Canada also found that subsurface horizontal flow constructed wetlands could remove 42%, 49% and 49%, respectively, of poultry pharmaceuticals monensin, salinomycin and narasin through sorption onto the soil surface and microbial degradation [77]. This indicates that with successful CW design, we can treat wastewater containing various contaminants in an efficient and economical manner. Such small-scale laboratory studies may not be sufficient for direct field application of constructed wetlands, but they serve as a good role at the proof-of-concept stage for future larger scaled studies.

Besides antibiotics, pesticides, such as chlorpyrifos, have been studied intensively and shown to be highly removable through constructed wetlands [78–82]. Most of the studies showed that biodegradation and adsorption were the primary removal mechanisms of such chemicals from the CW system. In addition, studies have also looked into the removal performance for antibiotic resistance genes [83–85]. According to the study by Song et al. (2018), the accumulation of antibiotics in different layers within the constructed wetland resulted in an abundance of ARGs with a positive correlation relationship [84]. Later studies proved that some CW systems could reduce ARG concentrations as they remove the antibiotic contaminations. Chen et al. 2019 study showed that while antibiotics' major removal mechanism was microbial degradation, ARGs main removal mechanisms were substrate sorption and biological reactions [83]. Another study investigated the comparisons of substrate medium by Du et al. (2020) and observed better removal rates (>95%) for both antibiotics and ARGs when zeolite medium and plant (*Arundo donax*) were used [85].

4.2. Field-Scale

For field-scale studies, it can be further divided into two groups: pilot-scale CW studies and full-scale in situ CW studies. The former one typically had a smaller dimension (within an average volume of 1 m^3 for each CW), while the latter took a greater surface area (typically over 100 m^2) and served as a functional water treatment system for agricultural wastewater or farm runoff. The pilot-scale studies can be seen as a scaled-up version of laboratory-scale studies, as they are in larger volumes and typically operated in greenhouses or open fields receiving more real-world weather conditions than lab studies, but they can still be modified timely during operation to achieve better performance since their scale is still manageable. Therefore, during the pilot-scale study time (ranging from 1 to 16 months), water samples were collected periodically to evaluate the CW performance over time [38,62,63,86–89]. While the full-scale studies were based on fully established CWs that have been operating for several years, therefore, the system typically already reached a steady state for removal performance requiring little manipulation during study time. Compared to the smaller scaled studies that typically collect water samples at the influent and effluent points with multiple and periodic sampling events, the larger scaled studies tend to have more sampling points throughout the system within only one or few sampling events to monitor the removal performance over the entire treatment system [40,41,66,90,91]. Another major difference between pilot-scale and full-scale CW studies was that pilot-scale studies often spiked up the target contamination concentrations even if they already existed in the influent water, but the full-scale field studies treated the existing concentrations and measured field concentrations. Therefore, full-scale studies might show relatively lower removal efficiencies since it is more challenging to achieve high removal performance at lower influent concentrations.

4.2.1. Pilot-Scale

A research group from China studied applying CWs to remove veterinary antibiotics and antibiotic resistant genes from swine wastewater for many years. In their studies, various CW types and their combinations as well as substrate medium and target contamination compounds were investigated. Their 2013 study results found that SSVF-CWs could efficiently remove target antibiotics and ARGs (68–95% and 50–90%, respectively) with the major mechanism being the physical sorption towards the wetland medium [38]. In another study, the results showed that removal performance ranged from high to low in the order of SSVF-Low water level > SSVF-High water level > SSHF > SF (based on average removal rates) indicating that the various design, flow path and water level led to different antibiotic removal rates through impacting the parameters, such as temperature, oxygen transfer, oxidation-reduction potential, sorption sites, etc. [89]. Another key finding from this study was that the seasonality might pose different impacts on different veterinary antibiotics (significant effect on sulfamethazine (SMZ) while no significant effect on TC) and different CW types (significant effect on SF while no significant effect on SSVF) [89]. Another long-term study indicated that high removal rates, ranging from 69.0% to 99.9%, were achieved for the target contaminants in all three treatments with different initial concentrations [86]. In another short-term study, flow direction showed no significant influence since they obtained comparable removal rates, but accumulation of antibiotics and ARGs in the surface soil was observed in down-flow treatments indicating a concern to the local environment due to likeliness of antibiotics enrichment and ARGs abundance [87].

Besides antibiotics, studies have also been conducted on removal of herbicides via application of CW systems, as Gikas et al. demonstrated up to 74% removal of terbuthylazine [92] and 60% removal of S-metolachlor [93] in horizontal subsurface flow CWs. Other researchers also investigated the removal performances of hybrid, SSHF and SF CW systems using various substrate and vegetations [62,63,88]. In a 2019 study, different combinations of CW units (SSHF-SSVF (up-flow); SSHF-SSVF (down-flow); SSFV (down-flow)-SSVF (up-flow)) were run for 84 days to treat antibiotics, ARGs and nutrients from goose wastewater. The researchers reported that the comparable antibiotic removal performance of different combinations of hybrid CWs was probably due to the highly spiked-up influent concentrations (2500 μg/L for tilmicosin (TMS) and 30 μg/L for doxycycline (DOC)), which likely concealed the differences on effluent concentrations among different treatments [63]. This may indicate the importance of conducting full-scale field studies receiving much lower antibiotic concentrations to simulate the real-world scenario, instead of pursuing the high removal efficiency results by dosing up the influent water to an unrealistic level. Besides livestock and poultry, CW has also been applied to treat wastewater from aquaculture. For example, Huang et al. conducted a study using SSHF with different vegetations (single or mixture of *Iris pseudacorus* and *Phragmites australis*) to remove ENR, SMZ and AGRs from wastewater of a local fish farm achieving up to 80% removal performance [88].

4.2.2. Full-Scale

For the full-scale field studies, multiple treatment units either incorporating both traditional water treatment processes (such as filtration, sedimentation, anaerobic digestion, etc.) and constructed wetland treatment processes, or a hybrid system with a series of different CW cells (such as SF, SSHF, SSVF, etc.) were used [40,91]. For assessing treatment performance, an entire system's performance over a long time was monitored [40,91] or the performances at various stages within the system were compared by collecting samples at multiple sampling points [72,90,94]. The application of CW in the field could be an entire system, or sometimes just one unit in addition to the traditional treatment units. For example, the Chen et al. (2012) study compared a traditional swine wastewater treatment system (A) with another system (B) containing additional aquatic vegetation ponds (serving as SF CWs) as a final polishing unit [40]. The results showed that biological activities had a significant impact on the degradation of target contaminants but less impact on the dissipation of contaminants at low concentrations [40]. One common challenge for field studies compared to the pilot-scale or lab-scale studies is there is no perfect "control treatment" to refer to. Therefore, background/influent concentration data for such field studies are extremely important, as they can serve for the comparison of pre and post CW treatment. As an example, Locke et al. (2011) simulated a runoff study which sampled before and after the flushing events for comparisons of the removal rate [91]. Their results indicated that CW could help to protect the downstream water quality through degradation and sorption of the pollutants and retention caused by adsorption and/or uptake by vegetation even after the flushing event throughout the entire 21-day study period [91]. For the integrated/hybrid CW system using multiple treatment units with different designs, the concentrations of target contaminants typically showed a decreasing trend along water flow through various stages. For example, in the study by Chen et al. (2015), which utilized the field CW system containing six units in series and receiving rural wastewater, the antibiotic concentrations decreased continuously along the treatment train as each unit's effluent concentrations were lower than that of the previous unit [90]. This indicated that with careful design and reasonable arrangements, multiple CW treatment units could run in series to achieve better overall removal performance. More complicated systems, such as in the Abdel-Mohsein et al. study (2011), which applied various CW types in series and operated in parallel at each stage with three different treatments in a rotational mode, proved to further enhance the retention time and achieve remarkable removal efficiencies of antibiotic-resistant bacteria with zero residues in the effluent water [72]. Besides these studies looking into the performance at different stages of CW system, some full-scale field studies investigated the removal ability of a single established CW system for various contaminants. For example, Choi et al. (2016) monitored a mature CW system receiving livestock wastewater without any spikes and their results showed various removal rates for the eight antibiotics present in the wastewater [41]. Therefore, it is important to consider whether CW is suitable for the target pollutants based on its chemistry and properties. Conversely, unsatisfying operation performance may occur if the target pollutants are out of the scope from the designed CW's treatment ability.

5. Factors Impacting CW Performance

Based on the literature research, 34 relevant studies have been reviewed for the removal of CECs by CWs. The information has been summarized in Table 1 and various parameters and their impacts on CW performance are discussed in the following section in random order.

Table 1. Removal of pollutants in agricultural runoff wastewater by constructed wetlands.

References	Wetland Type	Scale	Size	Year of Construction	Substrate	Vegetation/ Layering	Water Depth	HLR	HRT	Run Time	Pollutant Studied	Influent Concentration	Removal %	Major Removal Processes	Factors/Conditions Studied/Wastewater Type/Country
Abdel-Mohsein et al., 2020 [72]	Hybrid	Field	Total surface area 111 m², depth 0.7 m Depth of VSF is 0.65 and HSF 0.15 m	2009	Sand and gravel	Coarse sand, small gravel and large gravel for both and concrete is additional for Vertical CW	NA	1.8 cm/d	6–7 d	90 days	ampicillin-resistant bacteria	NA	100%	Combination of physical (sedimentation, mechanical filtration, adsorption to organic matter), chemical (oxidation, exposure to biocides), and biological removal means (predation, competition for nutriments, lytic activity)	Full-scale hybridized CWs, with five stages (first four stages as SSVF and last one as SSHF), each stage has three units in parallel with three cells in rotation for each unit, received raw dairy wastewater, in Japan
											gentamicin-resistant bacteria		100%		
											kanamycin-resistant bacteria		100%		
											streptomycin-resistant bacteria		100%		
											vancomycin-resistant bacteria		99.9% (3.2 log 10 removal)		
Agudelo et al., 2010 [78]	SSHF	Lab	Length 1 m, width 0.6 m, height 0.6 m	NA	Gravel, igneous rock	Phragmites australis, igneous rock	0.2 m	1.1 cm/d	NA	180 days	chlorpyrifos	209.7 µg/L	96.2%	Mineralization process, adsorption into plants and gravel, and biological decomposition.	The wetlands were performed at pilot scale and treated with synthetic wastewater. Four wetlands were designed for this experiment at the university research campus of the University of Antioquia, Columbia.
												305.5 µg/L			
												425.6 µg/L			
Borges et al., 2009 [95]	SSHF	Lab	Length 24 m, width 1 m, height 0.35 m	NA	Fine gravel	Typha latifolia, fine gravel	NA	3.2 cm/d, 2.5 cm/d, 1.9 cm/d	3.8 d	77 days	ametryn	NA	39%	Biodegradation, plant uptake and desorption process	Four constructed wetland cells were used with one being a controlled cell. The wetlands were given ametryn-contaminated water. The country of origin for this experiment was Brazil.
Boto et al., 2016 [73]	SSVF	Lab	Length 0.4 m, width 0.3 m, height 0.3 m	2015	Gravel, lava rock, roots bed substrate (mixture of sand and rhizosediments (1:2))	Phragmites australis, roots bed substrate, lava rock and gravel	0.16 m	NA	7 d	63 days	enrofloxacin	100 µg/L	>99%	Substrate adsorption, microbial biodegradation, and plant uptake	Batch mode CWs with three treatments and triplicate for each treatment, received saline aquaculture wastewater, in Portugal
											oxytetracycline		>99%		

Table 1. *Cont.*

References	Wetland Type	Scale	Size	Year of Construction	Substrate	Vegetation/ Layering	Water Depth	HLR	HRT	Run Time	Pollutant Studied	Influent Concentration	Removal %	Major Removal Processes	Factors/Conditions Studied/Wastewater Type/Country
Carvalho et al., 2013 [39]	SSVF	Lab	Length 0.4 m, width 0.3 m, height 0.3 m	2012	Gravel, lava rock, roots bed substrate	*Phragmites australis*, roots bed substrate, lava rock and gravel	0.16 m	NA	7 d	91 days	enrofloxacin	100 µg/L	98%	Adsorption and/or microbial degradation in the microcosms' substrate	Batch mode CWs with three treatments and triplicate for each treatment, received swine farm wastewater, in Portugal
											tetracycline		94%		
Chen et al., 2012 [40]	SF	Field	Surface area 2000 m^3	NA	NA	*Inter alia*, the family *Lemnaceae*; *Eichhornia crassipes*; and *Alternanthera philoxeroides*	NA	NA	20 d	350 days	tetracycline	41.6 µg/L	27.0–97.1%	Sorption, biological degradation, photolysis, and phytoremediation	Aquatic vegetation ponds (SF) applied as the last unit of a treatment system, received swine farm wastewater, in China
											oxytetracycline	23.8 µg/L	94.1–100.0%		
											chlortetracycline	13.7 µg/L	82.8–90.2%		
											doxycycline	685.6 µg/L	57.1–74.3%		
											sulfadiazine	98.8 µg/L	NA		
Chen et al., 2015 [90]	Hybrid	Field	Total surface area 981 m^2	2012	Chaff and soil	*Pontederia cordata* and *M. verticillatum L.*	NA	7 cm/d	1.5 d	NA	leucomycin	0.12 µg/L	100%	Adsorption onto medium, photodegradation, biodegradation (especially anaerobic degradation)	CW system with five units in series (consisting of SF, SSVF, SSHF), received both domestic sewage (70%) and livestock (swine) sewage (30%), in China
											ofloxacin	0.193 µg/L	100%		
											lincomycin	0.061 µg/L	78.0%		
											sulfamethazine	0.054 µg/L	95.1%		
											ARG *sul1*	2.64×10^6 copies/mL	97.2%		
											ARG *sul2*	1.14×10^6 copies/mL	95.4%		
											ARG *tetM*	1.47×10^6 copies/mL	99.8%		
											ARG *tetO*	1.02×10^6 copies/mL	98.9%		

Table 1. *Cont.*

References	Wetland Type	Scale	Size	Year of Construction	Substrate	Vegetation/ Layering	Water Depth	HLR	HRT	Run Time	Pollutant Studied	Influent Concentration	Removal %	Major Removal Processes	Factors/Conditions Studied/Wastewater Type/Country
Chen et al., 2019 [83]	Hybrid	Lab	Length 0.8 m, width 0.6 m, height 0.8 m	NA	zeolite	Iris tectorum maxim, zeolite	0.77 m	40 cm/d	NA	240 days	sulfamonomethoxine			Microbial degradation, substrate adsorption, and plant uptake	Eight mesocosm scale were used for the hybrid system. Four were HSSF and four were VSSF. The wetlands were constructed in the campus of Guangzhou Institute of Geochemistry. Guangzhou, Guangdong, China. Raw domestic sewage was used to treat the wetlands. The hybrid systems treated antibiotic spiked wastewater for two weeks before the first sampling.
											sulfamethazine		87.8% to 99.1% ARG		
											sulfameter	4.05– 6.24 μg/L			
											trimethoprim				
											norfloxacin				
											ofloxacin				
											enrofloxacin				
											erythromycin-H_2O				
											roxithromycin		87.4% to 97.3% antibiotics		
											oxytetracycline				
											lincomycin				
											sul 1 and sul 2				
											tetG and tetO				
											ermB				
											qnrS and qnrD				
											cmlA and floR				
Choi et al., 2016 [41]	Hybrid	Field	Total surface area 4492 m², total storage volume 4006 m³	2007	NA	Phragmites australis (PA) and Miscanthus sacchariflorus (MS)	0.89 m	44 cm/d	2 d	240 days	sulfamethoxazole	10.03– $11,583.33$ μg/L	49.43%	Biodegradation (uptake into wetland soil and plants by microorganisms) and direct adsorption into soil and plants	CW system with six units in series (consisting of SF, SSVF, SSHF), received secondary piggery wastewater and stormwater runoff, in Korea
											sulfathiazole	1263.33– $57,833.33$ μg/L	81.86%		
											sulfamethazine	1055– $30,033.33$ μg/L	85.00%		
											trimethoprim	1.76– 673.33 μg/L	2.32%		
											tetracycline	8.41– 69.5 μg/L	NA		
											oxytetracycline	12.33– 48.83 μg/L	NA		
											chlortetracycline	4300– $16,100$ μg/L	29.47%		
											enrofloxacin	34.26– 262.16 μg/L	27.26%		
Du et al., 2020 [85]	SSVF	Lab	Length 0.4 m, width 0.2 m, height 0.4 m	2016	Clinoptilolite zeolite, quartz sand	Arundo donax	N/A	9.3 cm/d	7 d	365 days	sulfamethoxazole	6×10^{-4} μg/L		Substrate adsorption	Four treatments, each with two separated cells (a down flow cell followed by an up flow cell), received swine wastewater, in China
											sulfamethazine	10 ng/L 0.01 μg/L	56.1–68.8%		
											sulfadiazine	22.1 ng/L 0.022 μg/L			
											tetracycline	0.539 μg/L	85.9–96.4%		
											oxytetracycline	0.233 μg/L			
											antibiotic resistant genes	N/A	71.7–95.3%		

Table 1. *Cont.*

References	Wetland Type	Scale	Size	Year of Construction	Substrate	Vegetation/ Layering	Water Depth	HLR	HRT	Run Time	Pollutant Studied	Influent Concentration	Removal %	Major Removal Processes	Factors/Conditions Studied/Wastewater Type/Country
Feng et al., 2021 [58]	SSVF	Lab	Height 0.65 m, diameter 0.2 m	N/A	Gravel and biochar with or without air	Gravel, biochar, and *Iris pseudacorus* with or without air	N/A	N/A	3d	180 days	tetA	5.2×10^3–7.03×10^4 copies /mL.	26.2–99.3%	Biodegradation	CWs with four treatments and triplicate for each treatment, received synthetic wastewater and swine wastewater, in China
											tetM	2.7×10^5–7.15×10^6 copies /mL.			
											tetO	9.87×10^4–4.82×10^6 copies /mL.			
											tetW	1.51×10^5–9.42×10^5 copies /mL.			
George et al., 2003 [98]	SSHF	Field	Length 4.9 m, width 1.2 m or 2.4 m height 0.1–0.3 m	1992	Quartz gravel	S. Validus, quartz gravel	0.3 m	4 cm/d	2.2 d	70 days	simazine	750 µg/L and 1400 µg/L	59%	Plant absorption and microbial degradation	Twelve cells were used with half containing vegetation and half with no vegetation. The water used in this experiment was a synthetic mix of water and pesticides. This experiment took place in Baxter, Tennessee, USA.
													59%		
								4 cm/d	2 d				78%		
													54%		
								2 cm/d	4.9 d				96%		
													78%		
								2 cm/d	3.8 d				53%		
													63%		
								1 cm/d	12.2 d				51%		
													79%		
								1 cm/d	7.3 d				57%		
													96%		
								2 cm/d	7.3 d				70%		
													96%		
							0.46 m	4 cm/d	3.5 d		metolachlor	3866 to 300 µg/L.	62%		
													34%		
								4 cm/d	3.1 d				90%		
													68%		
								2 cm/d	7.7 d				96%		
													93%		
								2 cm/d	6.1 d				76%		
													60%		
								1 cm/d	19.3 d				46%		
													89%		
								1 cm/d	11.6 d				75%		
													95%		
								2 cm/d	19.3 d				87%		
													97%		

Table 1. *Cont.*

References	Wetland Type	Scale	Size	Year of Construction	Substrate	Vegetation/ Layering	Water Depth	HLR	HRT	Run Time	Pollutant Studied	Influent Concentration	Removal %	Major Removal Processes	Factors/Conditions Studied/Wastewater Type/Country
Gikas et al., 2017 [92]	SSHF	lab	Length 3 m, width 0.75 m, height 1 m	2003	Medium gravel	*Phragmites australis,* gravel ——————— *T. latifolia,* gravel	0.55 m	2.4 cm/d, 1.8 cm/d	6 d, 8 d	420 days	terbuthylazine	NA	73.7% ——— 58.4%	Uptake through plants	Two constructed wetlands were used each containing a different wetland plant. The wetlands were treated with water enriched with terbuthylazine. The experiment took place at the laboratory of Ecol. Eng. and technology, Department of Environmental Engineering, Democritus University of Thrace.
Gikas et al., 2018 [93]	SSHF	lab	Length 3 m, width 0.75 m, depth 1 m	2003	Medium gravel (carbonate rock)	*Phragmites australis,* gravel ——————— *T. latifolia,* gravel	1 m	2.4 cm/d, 1.8 cm/d	6 d, 8 d	420 days	S-metolachlor	NA	68.9%, 47.8%, 40.8%	Plant uptake and sorption on substrate. Biodegradation	Three constructed wetlands were used with one being left as the control. Water enriched with S-metolachlor was used to treat the wetlands. This experiment was conducted at Department of Environmental Engineering, Democritus University of Thrace.

Table 1. *Cont.*

References	Wetland Type	Scale	Size	Year of Construction	Substrate	Vegetation/ Layering	Water Depth	HLR	HRT	Run Time	Pollutant Studied	Influent Concentration	Removal %	Major Removal Processes	Factors/Conditions Studied/Wastewater Type/Country
Goritio et al., 2018 [74]	SSVF	Lab	Length 0.4 m, width 0.3 m, height 0.3 m	2017	Gravel, lava rock, root bed substrate	Phragmites australis, root bed substrate, lava rock, gravel	0.22 m	N/A	7 d	30 days	Alachlor	0.1 µg/L	>87% <87% for EHMC	Absorption into soil, plant uptake, microbial degradation	. Six microcosms assembled with three spiked and three not spiked each treating 2 L of aquaculture effluent. Aquaculture effluent added at the beginning of each week, microcosms were weekly drained and refilled with spiked or non-spiked aquaculture effluents. This experiment took place in Portugal.
											atrazine				
											chlorfenvinphos				
											isoproturon				
											PFOS				
											simazine				
											azithromycin				
											clarithromycin				
											erythromycin				
											diclofenac				
											methiocarb				
											acetamiprid				
											clothianidin				
											thiacloprid				
											thiamethoxam				
											EHMC				
											atorvastatin				
											carbamazepine				
											cephalexin				
											ceftiofur				
											citalopram				
											clindamycin				
											clofibric acid				
											diphenhydramine				
											enrofloxacin				
											fluoxetine				
											ketoprofen				
											metoprolol				
											norfluoxetine				
											ofloxacin				
											propranolol				
											tramadol				
											trimethoprim				
											venlafaxine				
											warfarin				

Table 1. *Cont.*

References	Wetland Type	Scale	Size	Year of Construction	Substrate	Vegetation/ Layering	Water Depth	HLR	HRT	Run Time	Pollutant Studied	Influent Concentration	Removal %	Major Removal Processes	Factors/Conditions Studied/Wastewater Type/Country
Hsieh et al., 2015 [94]	SF	Field	Length 3.4 m, width 1.4 m, height 1.5 m	NA	Soil, gravel	Cyperus, phragmite, vetiveria, gravel, soil	SF 1: 0.8 m / SF 2: 0.8 m / SF 3: 0.8 m	NA	2.2 d	540 days	chloramphenicol	0.59 ± 0.464 µg/L	98.2%	Removal process must be additionally studied though it is shown wetland was successful in removing antibiotics and other chemicals.	Three free water surface constructed wetlands were used along with a lotus pond and filter bed to achieve purification. The wetlands are field scale and usually treat wastewater on a regular basis which flow through each cell. This experiment was conducted in Taiwan.
											oxolinic acid	NA	100%		
											chlortetracycline	NA	NA		
											oxytetracycline	0.218 ± 0.170 µg/L	97%		
											tetracycline	NA	NA		
											enrofloxacin	NA	NA		
											ciprofloxacin	0.018 ± 0.009 µg/L	100%		
											sulfamerazine	NA	NA		
											sulfamonomethoxine	0.093 ± 0.020 µg/L	87%		
											sulfadimethoxine	0.0843 ± 0.0507 µg/L	90.6%		
											sulfamethazine	NA	NA		
											malachite green	NA	NA		
											leucomalachite green	NA	NA		
											nonylphenol di-ethoxylate	0.173 ± 0.275 µg/L	85.2%		
											nonylphenol mono-ethoxylates	0.291 ± 0.457 µg/L	76.5%		
											nonylphenol	1.65 ± 1.81 µg/L	89.7%		
											octylphenol	1.12 ± 3.02 µg/L	85.1%		
											bisphenol A	0.932 ± 0.684 µg/L	88.8%		
											17b-estradiol	0.189 ± 0.274 µg/L	95.2%		
											estriol	0.156 ± 0.140 µg/L	76.6%		
											17a-ethynylestradiol	0.025 ± 0.039 µg/L	31.8%		
Huang et al., 2015 [96]	SSVF	Field	Height 0.8 m, diameter 0.4 m	2012	Oyster shell, bricks, and red soil	Phragmites australis, oyster shell and red soil	0.7 m	4 cm/d	5 d	480 days	oxytetracycline	14, 64, 164 µg/L	92.7–99.9%	Substrate absorption, plant uptake, microbial degradation, hydrolysis and photodecomposition	Three outdoor VSSF (up flow) CWs served as three treatments, received swine wastewater, in China
											tetracycline	5.56 µg/L	69.0–99.7%		
											chlortetracycline	4.32 µg/L	88.4–98.3%		
											target antibiotic resistant genes	NA	45.4–99.9%		

Table 1. *Cont.*

References	Wetland Type	Scale	Size	Year of Construction	Substrate	Vegetation/ Layering	Water Depth	HLR	HRT	Run Time	Pollutant Studied	Influent Concentration	Removal %	Major Removal Processes	Factors/Conditions Studied/Wastewater Type/Country
Huang et al., 2017 [57]	SSVF	Field	height 0.6 m, diameter 0.25 m	2015	Brick particle or oyster shell, red soil, and humus soil (2:1)	*Phragmites australis*, brick particle/oyster shell, red soil, and humus soil	0.55 m	5.1 cm/d	1.6–5.8 d	90 days	oxytetracycline	250 µg/L	>33%	Substrate adsorption, plant uptake, microbial degradation, hydrolysis and photodegradation	Four CWs served as four treatments (different substrates and flow directions), received swine wastewater, in China
											difloxacin	250 µg/L	>33%		
											tetracycline resistance genes	NA	>33.2 to 99.6%		
		Lab	Height 0.4 m, diameter 0.03 m		Brick and oyster shells	Brick and oyster shells	N/A	2 cm/d	N/A	20 days	oxytetracycline and difloxacin	250 µg/L	N/A		
Huang et al., 2019a [61]	Hybrid	Field	SSHF: length 0.7 m, width 0.43 m, height 0.9 m; SSVF: diameter 0.62 m, height 0.9 m	2018	Gravel, ceramsite, zeolite, red soil	*Phragmites australis*, red soil, ceramsite, zeolite, gravel	0.9 m	300 cm/d	6 d	84 days	tilmicosin	2500 µg/L	100%	Adsorption and degradation of antibiotics	Six CW cells with three treatments, each treatment with two cells in series (SSHF-SSVF(U); SSHF-SSVF(D); SSFV(D)-SSVF(U)), received goose wastewater, in China
											doxycycline	30 µg/L	98–99%		
											intI1, ermB, ermC	NA	>90%		
											tet genes except tetX	NA	>50%		
Huang et al., 2019b [88]	SSHF	Field	Length 1.5 m, width 0.4 m, depth 0.8 m	2014	Gravel and zeolite	*Iris pseudacorus* and/or *Phragmites australis* (50 plants/m²), zeolite, gravel	0.6 m	8.4 cm/d	3 d	120 days	enrofloxacin	0.026–0.067 µg/L	75.6–81.1%	Adsorption for ENR; degradation, transformation and anaerobic fermentation for SMZ	Four SSHF cells served as four treatments (different plant species and planting patterns), received aquaculture farm wastewater, in China
											sulfamethoxazole	0.064–0.211 µg/L	54.3–68.7%		
											antibiotic resistant genes	NA	36.5–58.2%		
Hussain, 2011 [77]	SSHF	Lab	Length 6.1 m, width 1.5 m, height 0.75 m	2011	Sandy soil	Phalaris arundinaceae, Typha Latifolia, Sandy Soil	4.17 m cubed	NA	4 d	30 days	monensin, salinomycin, narasin	10,50,100, and 500 µg/L	Monensin 42%, salinomycin 49%, narasin 49%	Sorption into soil, biodegradation, microbial dissipation	SSHF constructed wetland performance was compared to the performance of free water surface performance wetlands. A mixture of contaminated wastewater was used to treat the wetlands. Three CW units were assembled for the experiment. Water was supplied through an inflow manifold. This experiment took place in Canada.
Liu et al., 2013 [88]	SSVF	Field	Height 0.7 m, surface area 1 m²	NA	Volcanic rocks/zeolite	Red soil; volcanic rocks (CW1)/zeolite (CW2); gravel	0.7 m	3 cm/d	1.25 d	120 days	ciprofloxacin HCl	All at 40 µg/L	82% and 85%	Sorption to wetland medium	Two SSVF CWs served as two treatments (different substrate medium), received swine farm wastewater, in China
											oxytetracycline HCl		91% and 95%		
											sulfamethazine		68% and 73%		
					Volcanic rock						tet gene and 16S rRNA	N/A	50%		
					Zeolite						tet gene and 16S rRNA	N/A	90%		

Table 1. *Cont.*

References	Wetland Type	Scale	Size	Year of Construction	Substrate	Vegetation/ Layering	Water Depth	HLR	HRT	Run Time	Pollutant Studied	Influent Concentration	Removal %	Major Removal Processes	Factors/Conditions Studied/Wastewater Type/Country
Liu et al., 2014 [89]	Hybrid	Field	SF and SSVF: height 0.8 m, surface area 6 m²; SSHF: height 0.8 m, surface area 12 m²	NA	Oyster shell, red soil	Phragmites australis (16 stems/m²), red soil and oyster shell (optional)	SF: 0.3 m; SSVF: 0.1–0.4 m; SSHF: 0.4 m	SF and SSVF: 2 cm/d; SSHF: 4 cm/d	SF: 15.5 d; SSVF: 7.3–14.2 d; SSHF: 16.4 d	428 days	sulfamethazine tetracycline	25–35 µg/L 3.5–6 µg/L	40% for SF; 59% for SSHF; 87% for SSVF-L; 70% for SSVF-H 92% for SF; 92% for SSHF; 99% for SSVF-L; 98% for SSVF-H	Mainly dependent on the physicochemical process (adsorption)	Four pilot-scale CWs served as four treatments (SF, SSHF, SSVF-L, SSVF-H), received swine farm wastewater, in China
Locke et al., 2011 [91]	Hybrid	Field	Length 180 m, width 30 m (average), depth 0.45 m (average)	2003	Sediment	Alligatorweed [Alternanthera philoxeroides (Mart.) Griseb.], and two grass plants, junglerice [Echinochloa colona (L.) Link] and barnyard-grass [Echinochloa crus-galli (L.) Beauv.]	0.3–1 m	2667 cm/d	NA	21 days	atrazine fluometuron	Both at 10,650 µg/L	89% 81%	Adsorption/uptake by plants	A full-scale CW consisted of one sediment trap and two treatment cells in series, received simulated agricultural runoff, in the United States
Lyu et al., 2017 [97]	NA	Lab	Height 0.2 m and diameter 0.2 m	2014–2015	Gravel, geotextile, sand	Juncus effusus, typha latifolia, berula erecta, phragmites and iris pseudacorus, gravel, sand geotextile, gravel	About 0.18 m	1.7, 3.4, 6.9, 13.8 cm/d	2, 1, 0.5, 0.25 days	57 days	tebuconazole	10 to 100 µg/L	99.8%	Plant uptake (biodegradation and metabolization inside plant) substrate sorption	In total 36 mesocosm scale constructed wetlands were used. Water used was artificially spiked with tebuconazole.
Moore et al., 2001 [79]	NA	Lab Field	10 m width, 66 m length, 0.24 m depth 36 m wide, 134 m length	NA 1991	Sandy loam Silty loam	Juncus effusus, leersia, ludwigia, sandy loam Typha capensis, Juncus kraussii, Cyperus dives, silty loam	0.24 m 0.3–1 m	NA	NA	84days	chlorpyrifos	147 µg/L 73 µg/L 147 µg/L 733 µg/L 1.3 µg/L	>83%	Sorption by plants and sediments.	Eight CW cells were used, 4 experimental, 1 controlled and 3 as simulated rainfall. Wetlands ere set up at the University of Mississippi Field Station in Mississippi, USA. Chlorpyrifos of various concentration were used to treat the wetlands
Papaevangelou et al., 2017 [98]	SSHF	Lab	3 m length, 0.75 m width, 1 m depth	2003	Fine gravel Cobbles	Phragmites australis, gravel Phragmites australis cobble	1 m	2.4 cm/d, 1.8 cm/d	6 d, 8 d	1 year	boscalid	NA	75.1% 72.5%	Adsorption of organic matter in substrate material and plant uptake	Two constructed wetlands were used in this experiment each containing different substrate media. Water enriched with boscalid was used for the experiment. This experiment was performed at Department of Environmental Engineering, Democritus University of Thrace.

Table 1. *Cont.*

References	Wetland Type	Scale	Size	Year of Construction	Substrate	Vegetation/ Layering	Water Depth	HLR	HRT	Run Time	Pollutant Studied	Influent Concentration	Removal %	Major Removal Processes	Factors/Conditions Studied/Wastewater Type/Country
Santos et al., 2019 [42]	SSVF	Lab	Length 0.4 m, width 0.3 m, height 0.3 m	2014	Gravel, lava rock, root bed substrate	*Phragmites australis* (Cav.) Trin. ex Steud, root bed substrate, lava rock and gravel	0.3 m	NA	7 d	140 days	enrofloxacin	both at 100 µg/L	>90%	Substrate adsorption, microbial biodegradation, and plant uptake	Batch mode CWs with four treatments and triplicate for each treatment, received swine farm wastewater, in Portugal
											ceftiofur		>90%		
											antibiotic resistant bacteria	N/A	>90%		
Sherrard et al., 2003 [80]	NA	Lab	Length 1.85 m, width 0.63 m, height 0.63 m	NA	Sand/organic matter	C. Dubia, P. Promelas, organic mixture/sand	About 0.3 m	NA	3 d	Performed on consecutive weeks	chlorpyrifos	0.90 µg/L, 19.9 µg/L, 19.4 µg/L	98%	Plant uptake	Four experiments were carried out to assess the impact of pesticide removal in constructed wetlands. Well water spiked with pesticide were used for this experiment. This experiment took place in the
											chlorothalonil	148 µg/L, 326 µg/L, 296 µg/L	100%		
Souza et al., 2017 [81]	SSHF	Lab	0.35 m height, 0.5 m length, 2 m width	NA	Pea gravel, biofilm (sewage)	Cyon spp., M. aquatica, P. punctatum, biofilm mixture, pea gravel.	0.35 m	NA	1 d, 2 d, 4 d, 6 d, 8 d	30 days	chlorpyrifos	1000 µg/L	98.6%	Adsorption due to the biofilm and plants. Biodegradation	Four wetlands were used with different vegetation. One was a controlled. Water used in this experiment was spiked with chlorpyrifos. The experiment took place at the Department of Agricultural Engineering, Federal University of Vicosa.
Tang et al., 2018 [82]	SSVF	Lab	Diameter 0.24–0.27 m, height 0.3 m	NA	Gravel	C. Alternifolius, C. Indica, I. pseudacorus, J. effusus, T. Orientalis, gravel media	0.15 m	NA	7 d	42 days	chlorpyrifos	50 µg/L and 500 µg/L	94–98%	Sorption and biodegradation. Plants enhance removal through biodegradation	Constructed wetlands that were planted performed better in removing pesticides than the control groups which had no plants. Synthetic wastewater was used in the experiment which was conducted in China.
Wu et al., 2016 [99]	SSHF	Lab	Length 1.2 m, width 0.4 m, height 0.8 m	NA	Ceramsite	C. Indica, Ceramsite	0.8 m	20 cm/d	NA	365 days	triazophos	100 µg/L	96%	High urease activity of the substrate	Four constructed wetlands were used with one a control. Relationships between pollutant and microbial communities were discussed. Different concentrations of triazophos and raw water were pumped into the wetland systems.
												1000 µg/L	97%		
												5000 µg/L	75%		

Table 1. *Cont.*

References	Wetland Type	Scale	Size	Year of Construction	Substrate	Vegetation/ Layering	Water Depth	HLR	HRT	Run Time	Pollutant Studied	Influent Con- centration	Removal %	Major Removal Processes	Factors/Conditions Stud- ied/Wastewater Type/Country
Xian et al., 2010 [62]	SF	Field	Length 0.5 m, width 0.4 m, height 0.4 m	2009	HDPE foam plates	Ryegrass (*Dryan, Tachimasari* and *Waseyutaka*)	0.3 m	NA	NA	35 days	sulfadiazine	100 μg/L	98.7–99.2%	Sorption, abiotic transformation and biotic transformation	Three SF CWs served as three treatments (different plant species), received raw swine wastewater, in China
											sulfamethazine	100 μg/L	88.8–91.8%		
											sulfamethoxazole	10 μg/L	99.0–99.5%		
Zhu et al., (2020) [56]	Hybrid	field	Surface area 600 m^2	2004	N/A	N/A	N/A	N/A	7 d	N/A	antibiotic resistant genes	N/A	62%	Microbial degradation and physical adsorption	CW applied as the last unit of a treatment system, received swine farm wastewater, in China

Note: SF—surface flow; SSHF—subsurface horizontal flow; SSVF—subsurface vertical flow; HLR—hydraulic loading rate; HRT—hydraulic retention time; NA—not applicable.

5.1. Target Contaminant Property

Based on the various physicochemical properties, such as pKa, molecular weight, solubility, and functional groups, different contaminants showed different levels of removal by CW systems. From the study by Choi et al. (2016), the major removal mechanism was the adsorption to soil, which was favored for compounds with lower molecular weights and higher pKa values [41]. Gorito et al. (2018) also suggested that high removal of azithromycin through sorption onto the soil and plant uptake were likely due to its high octanol–water coefficient (K_{ow}) and pK_a values [74]. In addition, contaminants with low solubility and high soil adsorption coefficient ($K_{oc} > 1000$) would have better sorption and retention in soils. For example, Gikas et al. (2018) found poor adsorption of selected pesticides due to moderate solubility and low K_{oc} [92]. This was also supported by a pesticide study conducted by Agudelo et al. (2010) as target contaminant chlorpyrifos with low solubility and high adsorption coefficient showed great sorption into the soil substrate or the humic colloids suspended in the water [78]. Vystavna et al. (2017) indicated that compounds, such as propranolol, tend to accumulate in sediments due to its hydrophobicity, therefore, the utilization of porous filter materials with high sorption ability could improve the removal percentages for such compounds [100]. Functional group and structure could also impact pollutant removal mechanism, as the Lyu et al. (2018) study showed that hydrolysis was negligible for tebuconazole removal due to its chemical properties [97]. Overall, to achieve optimal removal performance by CW systems, one should consider the physical and chemical properties of the target contaminant during the design of the CWs as those properties are likely to impact their removal mechanisms.

5.2. Aeration

Depending on the type of CW system and the specific spot within a CW unit, aerobic or anoxic conditions may exist favoring certain pollutant removal processes. The removal of antibiotics, such as monensin, salinomycin and naracin, through microbial degradation was most active at the water/air interface or within the root zone under aerobic conditions [77]. Other works have also shown aerobic conditions to support the removal of veterinary pharmaceuticals from wastewaters [101]. On the contrary, the biodegradation of chloroacetanilide herbicides might be favored under anoxic conditions, as Elsayed et al. (2015) found that bacterial communities were most abundant and active at anoxic rhizosphere zone and anaerobic degradation accounted for the most dissipation of chloroacetanilides [102]. Besides the natural established aerobic/anaerobic conditions, some studies also introduced artificial aeration to promote the removal rates. For example, Chen et al. (2019) compared four different hybrid CW systems with/without the addition of aeration from an air blower and their results showed enhanced ARGs removal rates in both VSSF and HSSF with additional aeration [83]. Similar findings were also observed in a Feng et al. (2021) study, as they also noticed improved target ARGs removal with aerated treatments [58]. This indicated that for future applications, aeration units should be considered in the CW system design to improve the ARGs removal efficiencies. Alternatively, better designs to enhance aeration naturally in the CWs will likely enhance removal of ARGs.

5.3. Types and Design of CWs

Out of the 32 studies listed in Table 1 with identifiable CW type/design, 3 of them (around 9%) were surface flow (SF), 10 of them (around 31%) were subsurface horizontal flow (SSHF), 11 of them (around 35%) were subsurface vertical flow (SSVF), and 8 of them (around 25%) were hybrid systems containing more than one type (SF/SSHF/SSVF). In general, SSHF and SSVF are more widely applied in single CW type studies comparing to SF. This is due to how SSHF CW provides an anoxic system which promotes denitrification and other anoxic microbial processes; whereas SSVF CW provides an aerobic system which supports nitrification and other aerobic microbial processes [72,102]. In addition, SSVF CW can also remove organic compounds and suspended solids effectively [72]. In order to achieve better removal efficiency for various pollutants, a lot of studies applied hybrid system containing SSVF and SSHF [72,90,102]. For example, Huang et al. (2019a) showed that all three two-stage CW systems removed over 98% of the antibiotics without significant differences among treatments [63]. SSFV (down-flow) and SSVF (up-flow) had a better performance for ARGs and nutrients (especially for N) removal due to its establishment of anaerobic ammonium oxidation condition and limitation of bacterial growth [63].

Besides CW types, studies have also investigated the impacts of different flow directions (up-flow vs. down-flow) [85,87] as well as the water level (high-level vs. low-level) [89]. Their results showed that the configuration of down-flow SSVF followed by up-flow SSVF provided best pollutant removal performance, however, they also expressed concern about accumulation of ARGs in the surface soil for down-flow SSVFs [85,87]. Meanwhile, Liu et al. (2014) reported relatively higher removal efficiencies for SSVF with low water level [89] and this was supported by Lyu et al. (2018) as they found significantly higher tebuconazole removal in unsaturated CWs than saturated CWs [97]. For larger field scale studies consisting of multiple CW units, various configurations are utilized. The most common one was connecting CW units in series [40,41,90,91] but there were more complicated setups in some studies. For example, George et al. (2003) first had eight cells connected in parallel as the first stage and the remaining six cells connected in parallel as the second stage with series connection between stages [96]. Another study conducted in Japan had five stages connected in series and three treatments connected in parallel for each stage; and within each treatment there were three cells operating in rotational mode [72,102]. Such sophisticated design could not only provide better removal performance but also allow the avoidance of cross-contamination between different cells and provide the chances to perform operation and maintenance on a specific cell without disturbing the entire system.

5.4. Hydraulic Parameters (HLR and HRT)

Compared to wastewater treatment plants, CWs typically need lower HLR and longer HRT to achieve the similar level of removal performances. The hydraulic parameters are very important to consider during CW system design since lower HLR/longer HRT may provide better treatment but require much larger land area, while higher HLR/shorter HRT may occupy a smaller footprint but face low treatment efficiency and frequent clogging events and need more operation and maintenance inputs. Based on the study data listed in Table 1, Figures 3 and 4 were plotted to show the relationships between HLR/HRT and target contamination removal percentage. It is apparent from Figure 3 that removal efficiency had a positive correlation with HRT, meaning a greater removal rate with the longer retention time. After 7 days, an average removal efficiency of 90% was achieved, which is in agreement with the findings from previous research that a hydraulic retention time of 6–7 days was adequate for the removal of pollutants [93,103]. Meanwhile, Figure 4 showed that with the increase in HLR, removal efficiency would first increase, reach to a steady level (>90%), and later start to decrease. That is to say, the ideal HLR should be 10–30 cm/d for best pollutant removal performance, as greater or lesser HLR would both result in reduced removal efficiency. This was supported by findings from Lyu et al. (2017) as they reported decreasing removal rates over increased HLR [97]. Therefore, choosing the appropriate HRT/HLR for CW system has great impacts on the system performance. Furthermore, in some studies, hydraulic retention times were adjusted based on seasons, with them being longer in warmer seasons (8 days) and shorter in colder seasons (6 days) to address the water requirement variations due to evapotranspiration [92,93,98].

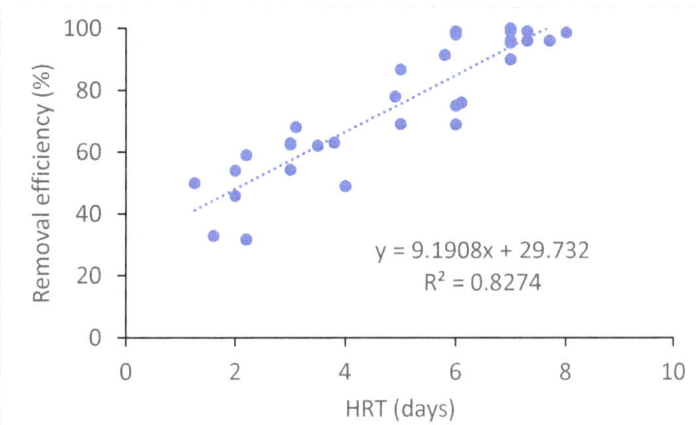

Figure 3. Relationship between HRT and removal efficiency in CWs.

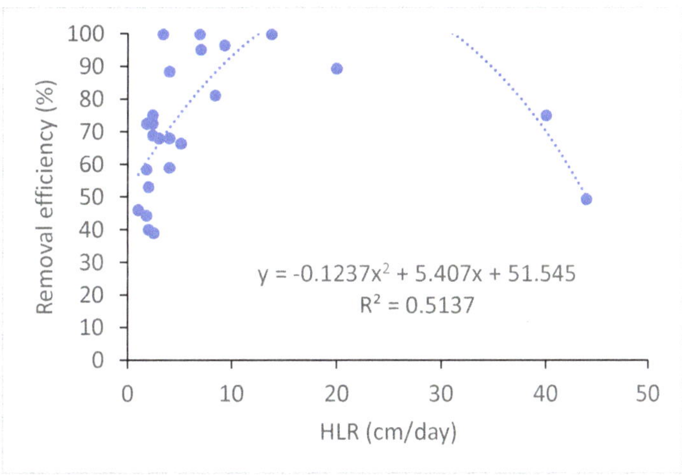

Figure 4. Relationship between HLR and removal efficiency in CWs.

5.5. Substrate Medium

Since adsorption is one of the major mechanisms for pollutant removal in CW systems, the substrate medium's physical and chemical properties would have a huge impact on the removal performances. The Papaevangelou et al. study (2017) compared two substrates (fine gravel and cobbles) from the same riverbed with various sizes and found better removal performance of fine gravel for target pollutant-boscalid (fungicide) in the preliminary tests but no significant difference in performance of the substrates over long-term field study [98]. A lot of previous research have compared the removal efficiency of specific contaminants with various substrate medium. For example, Liu et al. (2013) showed that compared to volcanic rock, zeolite had a lower point of zero charge (PZC) indicating a higher affinity to cationic form of antibiotics at neutral pH levels and had smaller pore size indicating greater sorption sites; therefore, zeolite showed better removal efficiencies for selected antibiotics and ARGs [38]. Similar observations were made by Du et al. (2020) as zeolite medium had a better removal performance for both antibiotics and ARZs compared to quartz sand medium [85]. In the Huang et al. study (2017), brick-based substrate achieved better antibiotic removal performance compared to oyster shell-based substrate due to two major reasons: (1) greater porosity and average pore size that provided more surface areas for sorption processes; (2) higher iron oxides contents in brick that provided better adsorption capacity [87]. Besides zeolite and brick material that were widely applied in CW substrates, medium with high organic matter content was also investigated since it could potentially increase pollutant removal through interactions with organic functional groups (such as phenolic and carboxyl groups), hydrogen bonding, and ion exchange [104]. For example, Feng et al. (2021) compared biochar and gravel based CWs and found that while treatment with only biochar-based substrate had no significant improvement in target contaminants removal, treatment with both biochar-based substrate and aeration showed much higher removal rates [58]. They also measured abundance of ARGs in the substrate indicating the accumulation of antibiotics in the substrate and proliferation of ARGs during the long-term operation [58]. That is to say, appropriate operation methods need to be taken to address the potential risks of ARGs development in such substrates. Therefore, to achieve better elimination of ARGs in practical approaches, the suitable selection of CW substrate medium is an important decision.

5.6. Vegetation

Vegetation is another key component in CW systems as plants cannot only directly uptake pollutants, but also modify the surrounding environment, for example, by transporting oxygen into a rhizosphere to enhance the diversity and biomass of microorganisms, microbial degradation, and sequestration [104]. Several studies have compared treatment with plants versus without plants and most of them showed higher removal performance for herbicides [92,93,96] and pesticides [82,97] with plants as treatment, while one study presented no significant difference with plant treatment for veterinary antibiotics [39]. Research has also been performed to compare the removal performances of various plant species. For example, Lyu et al. (2018) compared five plant species (*Typha latifolia*, *Phragmites australis*, *Iris pseudacorus*, *Juncus effusus* and *Berula erecta*) for pesticide removal and found *Berula* to contribute to significantly higher removal efficiency compared to the rest four plant species [97]. Moreover, Tang et al. study (2019) indicated that *Canna indica*, *Cyperus alternifolius* and *Iris pseudacorus* had better removal performance for pesticides than *Juncus effusus* and *Typha orientalis* [82]. However, Souza et al.'s study (2017) showed no significant differences in pesticide removal among *Polygonum punctatum*, *Cynodon spp.* and *Mentha aquatica* [81].

Another study also confirmed that vegetation type impacts antibiotic removal efficiencies in surface flow CWs. The authors compared three varieties of ryegrass (*Dryan*, *Tachimasari* and *Waseyutaka*) to treat three antibiotics (sulfadiazine, sulfamethazine, and sulfamethoxazole) and found that *Dryan* outcompeted the other two types of plants due to its highest removal rates for both nutrients and antibiotics [62]. Gikas et al. (2018) compared treatments planted with *Phragmites australis* and *Typha latifolia*, with an unplanted control and the results showed that *Phragmites australis* had the highest removal capacity for both herbicide (S-metolachlor) [96] and pesticide (terbuthylazine) [92].

Besides the plant species, studies have also shown that various planting patterns may impact the removal performance. Huang et al. (2019) compared CW treatments with single plant species and mixed plant species and found that CWs with single plant type performed better in reducing antibiotic and ARG concentrations [88]. These findings imply that different plant species and planting patterns should be applied to achieve best performance depending on the target contaminant. Furthermore, studies indicated that after a certain time of exposure to the pollutant, the plant would uptake the pollutants with more concentrations in the root part than in the shoot part [62]. Harvesting the vegetations planted in the CW reduced the concentration of antibiotics in the soil, implying plant harvest as an effective procedure to maintain sustainable efficient removal performance [86].

6. Research Bottlenecks and Prospects

As stated in previous sections and presented in Table 1, numerous studies and reviews have been performed either on the topics of constructed wetlands pollutant removal performance or chemicals of emerging concern (such as pesticides, herbicides, veterinary antibiotics, etc.), but fewer studies have been focused on the overlapping research area of these two topics, which is using constructed wetlands to remove CECs. Among those studies, even fewer are related to agricultural runoff, since most of them studied treating domestic sewage or effluent from wastewater treatment plants. Even those on agricultural runoff, the studies are dominated by nutrients and sediment. Therefore, most future research needs to be performed on the application of CWs to remove CECs from agricultural runoffs. In addition, compared to livestock and poultry wastewater treatment applications, even fewer data were collected and reported from aquacultural wastewater and farm runoff either due to irrigation or precipitation. That is to say, more studies need to be conducted in these specific areas to safeguard our water resources, environmental, and human health.

From the scale's perspective, the majority of current studies are mainly in lab scale or pilot field scale, with only a few papers reporting the data from full-scale field studies. Small-scale studies in a controlled environment in the laboratory or greenhouse setting are valuable to serve as the first step attempt to address the research questions, but eventually large-scale studies fitting the real-world scenario are still needed for future applications. The designed CW system needs to be tested under real field conditions with fluctuating temperatures, flow rate, redox state, etc. to prove its durability. Nowadays, climate change has resulted in more extreme weather conditions happening more frequently; therefore, future research should also take consideration of the impacts of extreme weather, such as flooding and drought, on designed CW systems. To assist the optimization of design parameters, predication models coupled with remote sensing data could be built for simulating various conditions and potential extreme weather events. With the screening feedback from such models, suggestions could be provided for future application development.

In addition, after a period of operation, the CWs could accumulate the CECs within the system and lead to the development of ARGs into the local environment by self-developing and transferring to other microorganisms [101]. Especially for the down-flow SSVF CWs, the enrichment of pollutants and ARGs in the surface soil could become problematic in the long term [86,87]. Therefore, it is necessary to investigate methods to periodically remove and safely treat the accumulated contaminations from the CW system in order to maintain a sustainable high removal performance in the long term. Currently, very few papers have reported such operation and maintenance practices for CW applications.

For the theoretical investigation part, it is broadly accepted that the CW removes pollutants through a variety of processes, including adsorption to the substrate and soil, plant uptake, and biological degradation. The physiochemical sorption process has been well studied based on parameters, such as solubility (S), sorption coefficient (K_d), octanol–water coefficient (K_{ow}), oxidation-reduction potential (ORP), pH and pKa, with a lot of studies reporting certain correlations between the above parameters and removal efficiencies. However, most of the current studies failed to provide detailed explanations on biological processes and their role in the pollutant removal [40,41,86,89]. Therefore, further research is also needed for understanding the mechanisms of microbial biodegradation and plant uptake of CECs within the CW systems. For example, more research can explore various microorganisms' functions under aerobic/anaerobic conditions and compare contaminant uptake at different plants parts (root/stem/leave/shoot/etc.). The identification of optimal conditions for biodegradation and extraction of plant tissues with highest accumulation could be beneficial for future CW system applications by providing suggestions of ideal set-up conditions as well as operation protocols, such as harvesting the plant parts with greatest pollutant accumulations to maintain a high removal rate throughout the entire treatment period.

Current studies have also reported contradictory results of ARG occurrence and removal within the CW systems as some of the studies showed significant removal of ARGs with CWs since they arrest and inhibit the growth of bacteria, while others reported increases of ARGs due to the exposure and adaptations to accumulated contaminants in the substrate/soil. Therefore, future research is also needed to determine the internal complicated processes and mechanisms underlying various conditions of ARG sequestration and removal within the CW system. Based on these results, application suggestions of CW could be provided to avoid ARG accumulation during operation. In addition, further studies could be performed on the evaluation of potential impacts of ARG accumulation within the CW system, such as whether accumulated ARGs are going to change the structure of microorganisms within the system and the system performance; or whether the accumulation may lead to increase in effluent ARG concentrations. If severe impacts are noticed from such accumulation, future research on appropriate approaches to prevent the ARG accumulations will be needed.

With successful CW design, we can treat wastewater containing various contaminants in an efficient and economical manner. However, there are several ways we can improve the performance removal of the pollutants by CWs. For example, finding ways to promote aeration in the CWs can enhance aerobic biodegradation. Additionally, selection of substrate medium is key to achieving better elimination of ARGs. Studies also showed hybrid setup to perform differently based on the order of SF or SSF. Moreover, plant species affect the performance of the CWs. Consequently, screening of plants and plant selection is important for improving the removal efficiency. Another potential method is to improve the design of CWs, for example, CWs in series to boost performance.

Because nature-based systems need time to establish and function, real field studies over longer period without spiking concentrations are needed. As short-term studies with spiked concentrations may not represent the true removal efficiency, real field studies conducted for a long time are required. In addition, sampling strategies, for example, before vs. after in long-term study rather than treatment vs. control, may be needed to represent the efficiency of removal.

7. Conclusions

The paper reviewed recent findings on the applications of wetlands for the treatment of agricultural wastewater that contained pesticides and herbicides, veterinary medicines, antimicrobial resistant bacteria, or ARGs. By the volume of the search results, it can be concluded that these topics are understudied but are gaining major attention lately, likely due to concerns with ARGs. Wetlands are nature-based treatment systems, which are capable of treating many pollutants in the agricultural wastewater simultaneously by utilizing several physico-chemical and biological mechanisms. For example, adsorption to the substrate and plant's root (physical process), microbial metabolization and degradation (biological and chemical processes), and plant uptake (biological process) were found to be responsible for removal of veterinary medicines. While a major removal mechanism for antibiotics was microbial degradation, substrate sorption was a major mechanism for ARGs. The major parameters, such as target contaminants' property, aeration condition, types and designs of CWs, hydraulic parameters, substrate medium and vegetation that impact the CW system's removal performances, were also discussed to provide suggestions for successful future designs. Since CWs are adaptable to various wastewater treatments with satisfying removal efficiencies, CWs can be a key tool to fight against current and emerging environmental problems, especially when resilient and climate smart solutions are needed more than ever.

Author Contributions: Conceptualization, N.A. and Z.T.; writing—original draft preparation, J.W., Z.T., D.S. and A.T.; writing—review and editing, N.A. and Z.T.; supervision and project administration, N.A.; funding acquisition, N.A. All authors have read and agreed to the published version of the manuscript.

Funding: This work was supported by Grant project number NC.X333-5-21-130-1 and Capacity Building Grants Program grant no. 2020-38821-31114 from the USDA National Institute of Food and Agriculture. Any opinions, findings, conclusions, or recommendations expressed in this publication are those of the author(s) and do not necessarily reflect the view of the U.S. Department of Agriculture.

Institutional Review Board Statement: Not applicable.

Informed Consent Statement: Not applicable.

Conflicts of Interest: The authors declare no conflict of interest.

References

1. Willis, G.H.; McDowell, L.L. Pesticides in agricultural runoff and their effects on downstream water quality. *Environ. Toxicol. Chem.* **1982**, *1*, 267–279. [CrossRef]
2. Vymazal, J.; Březinová, T. The use of constructed wetlands for removal of pesticides from agricultural runoff and drainage: A review. *Environ. Int.* **2015**, *75*, 11–20. [CrossRef]
3. García-Galán, M.J.; Matamoros, V.; Uggetti, E.; Díez-Montero, R.; García, J. Removal and environmental risk assessment of contaminants of emerging concern from irrigation waters in a semi-closed microalgae photobioreactor. *Environ. Res.* **2021**, *194*, 110278. [CrossRef]
4. Popp, J.; Pető, K.; Nagy, J. Pesticide productivity and food security. A review. *Agron. Sustain. Dev.* **2013**, *33*, 243–255. [CrossRef]
5. Arora, K.; Mickelson, S.K.; Helmers, M.J.; Baker, J. Review of pesticide retention processes occurring in buffer strips receiving agricultural runoff. *JAWRA J. Am. Water Resour. Assoc.* **2010**, *46*, 618–647. [CrossRef]
6. Carvalho, F.; Fowler, S.; Villeneuve, J.; Horvat, M. Pesticide residues in the marine environment and analytical quality assurance of the results. In *Environmental Behaviour of Crop Protection Chemicals, Proceedings of an FAO-IAEA International Symposium, Vienna, Austria, 7–11 April 1997*; International Atomic Energy Agency: Vienna, Austria, 1997.
7. Carvalho, F. Agriculture, pesticides, food security and food safety. *Environ. Sci. Policy* **2006**, *9*, 685–692. [CrossRef]
8. Sharma, A.; Kumar, V.; Shahzad, B.; Tanveer, M.; Sidhu, G.P.S.; Handa, N.; Kohli, S.K.; Yadav, P.; Bali, A.S.; Parihar, R.D. Worldwide pesticide usage and its impacts on ecosystem. *SN Appl. Sci.* **2019**, *1*, 1446. [CrossRef]
9. Devi, M.; Singh, S. Wellbeing, Measuring impacts of fertilizers and pesticides on the agriculture production. *Indian J. Health Wellbeing* **2018**, *9*, 895–899.
10. EPA. Chemicals of Emerging Concern in the Columbia River. Available online: https://www.epa.gov/columbiariver/chemicals-emerging-concern-columbia-river (accessed on 8 September 2021).
11. Srivastava, P.R. Pesticides: Past, Present and Future. In Proceedings of the 27th Training on Managing Plant Microbe Interactions for the Management of Soil-borne Plant Pathogens, Pantnagar, India, 22 January–11 February 2013; Indian Council of Agricultural Research: New Delhi, India; pp. 165–176.
12. He, Y.; Yuan, Q.; Mathieu, J.; Stadler, L.; Senehi, N.; Sun, R.; Alvarez, P.J.J. Antibiotic resistance genes from livestock waste: Occurrence, dissemination, and treatment. *npj Clean Water* **2020**, *3*, 4. [CrossRef]
13. Jacobs, K.B. Recovery of Antibiotic Resistance Genes from Agricultural Runoff. Master's Thesis, Virginia Polytechnic Institute and State University, Blacksburg, Virginia, 19 September 2017.
14. Dolliver, H.; Gupta, S. Antibiotic losses in leaching and surface runoff from manure-amended agricultural land. *J. Environ. Qual.* **2008**, *37*, 1227–1237. [CrossRef] [PubMed]
15. Kim, Y.; Lee, K.B.; Choi, K. Effect of runoff discharge on the environmental levels of 13 veterinary antibiotics: A case study of Han River and Kyungahn Stream, South Korea. *Mar. Pollut. Bull.* **2016**, *107*, 347–354. [CrossRef]
16. Wu, H.; Zhang, J.; Ngo, H.H.; Guo, W.; Hu, Z.; Liang, S.; Fan, J.; Liu, H. A review on the sustainability of constructed wetlands for wastewater treatment: Design and operation. *Bioresour. Technol.* **2015**, *175*, 594–601. [CrossRef]
17. Stottmeister, U.; Wießner, A.; Kuschk, P.; Kappelmeyer, U.; Kästner, M.; Bederski, O.; Müller, R.; Moormann, H. Effects of plants and microorganisms in constructed wetlands for wastewater treatment. *Biotechnol. Adv.* **2003**, *22*, 93–117. [CrossRef]
18. Rousseau, D.P.; Vanrolleghem, P.A.; De Pauw, N. Model-based design of horizontal subsurface flow constructed treatment wetlands: A review. *Water Res.* **2004**, *38*, 1484–1493. [CrossRef]
19. Vymazal, J. Constructed wetlands for wastewater treatment: Five decades of experience. *Environ. Sci. Technol.* **2011**, *45*, 61–69. [CrossRef]
20. EPA. Constructed Wetlands for Wastewater Treatment and Wildlife Habitat: 17 Case Studies. Available online: https://www.epa.gov/sites/default/files/2018-07/documents/constructed_wetlands_for_wastewater_treatment_and_wildife_habitat_17_case_studies_epa832-r-93-005.pdf (accessed on 8 September 2021).
21. Pericherla, S.; Karnena, M.K.; Vara, S. A review on impacts of agricultural runoff on freshwater resources. *Int. J. Emerg. Technol.* **2020**, *11*, 829–833.
22. Xia, Y.; Zhang, M.; Tsang, D.C.; Geng, N.; Lu, D.; Zhu, L.; Igalavithana, A.D.; Dissanayake, P.D.; Rinklebe, J.; Yang, X. Recent advances in control technologies for non-point source pollution with nitrogen and phosphorous from agricultural runoff: Current practices and future prospects. *Appl. Biol. Chem.* **2020**, *63*, 8. [CrossRef]
23. Sundaravadivel, M.; Vigneswaran, S. Constructed wetlands for wastewater treatment. *Crit. Rev. Environ. Sci.* **2001**, *31*, 351–409. [CrossRef]
24. Kivaisi, A.K. The potential for constructed wetlands for wastewater treatment and reuse in developing countries: A review. *Ecol. Eng.* **2001**, *16*, 545–560. [CrossRef]
25. Vymazal, J. Removal of nutrients in various types of constructed wetlands. *Sci. Total Environ.* **2007**, *380*, 48–65. [CrossRef] [PubMed]
26. Haberl, R.; Perfler, R.; Mayer, H. Constructed wetlands in Europe. *Water Sci. Technol.* **1995**, *32*, 305–315. [CrossRef]
27. Scholz, M.; Lee, B. Constructed wetlands: A review. *Int. J. Environ. Sci.* **2005**, *62*, 421–447. [CrossRef]

28. Akratos, C.S.; Tsihrintzis, V.A. Effect of temperature, HRT, vegetation and porous media on removal efficiency of pilot-scale horizontal subsurface flow constructed wetlands. *Ecol. Eng.* **2007**, *29*, 173–191. [CrossRef]
29. Zurita, F.; De Anda, J.; Belmont, M.A. Treatment of domestic wastewater and production of commercial flowers in vertical and horizontal subsurface flow constructed wetlands. *Ecol. Eng.* **2009**, *35*, 861–869. [CrossRef]
30. Tarimo, I.A.; Mbwette, T. Municipal Wastewater Management: Use of Horizontal Subsurface Flow Constructed Wetland (HSSFCW) for Aquaculture and Agriculture. In *Encyclopedia of Environmental Management*, 1st ed.; Jorgensen, S.K., Ed.; CRC Press: Boca Raton, FL, USA, 2012; pp. 1–8. [CrossRef]
31. Caselles-Osorio, A.; García, J. Impact of different feeding strategies and plant presence on the performance of shallow horizontal subsurface-flow constructed wetlands. *Sci. Total Environ.* **2007**, *378*, 253–262. [CrossRef]
32. Old, G.; Naden, P.; Granger, S.; Bilotta, G.; Brazier, R.; Macleod, C.; Krueger, T.; Bol, R.; Hawkins, J.; Haygarth, P. A novel application of natural fluorescence to understand the sources and transport pathways of pollutants from livestock farming in small headwater catchments. *Sci. Total Environ.* **2012**, *417*, 169–182. [CrossRef]
33. He, K.; Hain, E.; Timm, A.; Tarnowski, M.; Blaney, L. Occurrence of antibiotics, estrogenic hormones, and UV-filters in water, sediment, and oyster tissue from the Chesapeake Bay. *Sci. Total Environ.* **2019**, *650*, 3101–3109. [CrossRef]
34. Fairbairn, D.J.; Karpuzcu, M.E.; Arnold, W.A.; Barber, B.L.; Kaufenberg, E.F.; Koskinen, W.C.; Novak, P.J.; Rice, P.J.; Swackhamer, D.L. Sources and transport of contaminants of emerging concern: A two-year study of occurrence and spatiotemporal variation in a mixed land use watershed. *Sci. Total Environ.* **2016**, *551*, 605–613. [CrossRef]
35. Tian, Z.; Wark, D.A.; Bogue, K.; James, C.A. Suspect and non-target screening of contaminants of emerging concern in streams in agricultural watersheds. *Sci. Total Environ.* **2021**, *795*, 148826. [CrossRef]
36. Moeder, M.; Carranza-Diaz, O.; López-Angulo, G.; Vega-Aviña, R.; Chávez-Durán, F.A.; Jomaa, S.; Winkler, U.; Schrader, S.; Reemtsma, T.; Delgado-Vargas, F. Potential of vegetated ditches to manage organic pollutants derived from agricultural runoff and domestic sewage: A case study in Sinaloa (Mexico). *Sci. Total Environ.* **2017**, *598*, 1106–1115. [CrossRef]
37. Zhou, L.J.; Ying, G.G.; Liu, S.; Zhang, R.Q.; Lai, H.J.; Chen, Z.F.; Pan, C.G. Excretion masses and environmental occurrence of antibiotics in typical swine and dairy cattle farms in China. *Sci. Total Environ.* **2013**, *444*, 183–195. [CrossRef] [PubMed]
38. Liu, L.; Liu, C.; Zheng, J.; Huang, X.; Wang, Z.; Liu, Y.; Zhu, G. Elimination of veterinary antibiotics and antibiotic resistance genes from swine wastewater in the vertical flow constructed wetlands. *Chemosphere* **2013**, *91*, 1088–1093. [CrossRef]
39. Carvalho, P.N.; Araújo, J.L.; Mucha, A.P.; Basto, M.C.P.; Almeida, C.M.R. Potential of constructed wetlands microcosms for the removal of veterinary pharmaceuticals from livestock wastewater. *Bioresour. Technol.* **2013**, *134*, 412–416. [CrossRef]
40. Chen, Y.; Zhang, H.; Luo, Y.; Song, J. Occurrence and dissipation of veterinary antibiotics in two typical swine wastewater treatment systems in east China. *Environ. Monit. Assess.* **2012**, *184*, 2205–2217. [CrossRef] [PubMed]
41. Choi, Y.J.; Zoh, K.D.; Kim, L.Y. Removal characteristics and mechanism of antibiotics using constructed wetlands. *Ecol. Eng.* **2016**, *91*, 85–92. [CrossRef]
42. Santos, F.; Almeida, C.M.; Ribeiro, I.; Mucha, A.P. Potential of constructed wetland for the removal of antibiotics and antibiotic resistant bacteria from livestock wastewater. *Ecol. Eng.* **2019**, *129*, 45–53. [CrossRef]
43. Zainab, S.M.; Junaid, M.; Xu, N.; Malik, R.M. Antibiotics and antibiotic resistant genes (ARGs) in groundwater: A global review on dissemination, sources, interactions, environmental and human health risk. *Water Res.* **2020**, *187*, 116455. [CrossRef]
44. Liu, L.; Xin, Y.; Huang, X.; Liu, C. Response of antibiotic resistance genes in constructed wetlands during treatment of livestock wastewater with different exogenous inducers: Antibiotic and antibiotic-resistant bacteria. *Bioresour. Technol.* **2020**, *314*, 123779. [CrossRef]
45. Oliver, J.P.; Gooch, C.A.; Lansing, S.; Schueler, J.; Hurst, J.J.; Sassoubre, L.; Crossette, E.M.; Aga, D.S. Invited review: Fate of antibiotic residues, antibiotic-resistant bacteria, and antibiotic resistance genes in US dairy manure management systems. *Int. J. Dairy Sci.* **2020**, *103*, 1051–1071. [CrossRef]
46. Fang, H.; Han, L.; Zhang, H.; Long, Z.; Cai, L.; Yu, Y. Dissemination of antibiotic resistance genes and human pathogenic bacteria from a pig feedlot to the surrounding stream and agricultural soils. *J. Hazard. Mater.* **2018**, *357*, 53–62. [CrossRef]
47. Sui, Q.; Zhang, J.; Chen, M.; Tong, J.; Wang, R.; Wei, Y. Distribution of antibiotic resistance genes (ARGs) in anaerobic digestion and land application of swine wastewater. *Environ. Pollut.* **2016**, *213*, 751–759. [CrossRef]
48. Cheng, D.; Ngo, H.H.; Guo, W.; Chang, S.W.; Nguyen, D.D.; Liu, Y.; Wei, Q.; Wei, D. A critical review on antibiotics and hormones in swine wastewater: Water pollution problems and control approaches. *J. Hazard. Mater.* **2020**, *387*, 121682. [CrossRef]
49. Cheng, D.; Ngo, H.H.; Guo, W.; Chang, S.W.; Nguyen, D.D.; Liu, Y.; Zhang, X.; Shan, X.; Liu, Y. Contribution of antibiotics to the fate of antibiotic resistance genes in anaerobic treatment processes of swine wastewater: A review. *Bioresour. Technol.* **2020**, *299*, 122654. [CrossRef]
50. Hall, M.C.; Duerschner, J.; Gilley, J.E.; Schmidt, A.M.; Bartelt-Hunt, S.L.; Snow, D.D.; Eskridge, K.M.; Li, X. Antibiotic resistance genes in swine manure slurry as affected by pit additives and facility disinfectants. *Sci. Total Environ.* **2021**, *761*, 143287. [CrossRef]
51. Hall, M.C.; Mware, N.A.; Gilley, J.E.; Bartelt-Hunt, S.L.; Snow, D.D.; Schmidt, A.M.; Eskridge, K.M.; Li, X. Influence of setback distance on antibiotics and antibiotic resistance genes in runoff and soil following the land application of swine manure slurry. *Environ. Sci. Technol.* **2020**, *54*, 4800–4809. [CrossRef] [PubMed]
52. Neher, T.P.; Ma, L.; Moorman, T.B.; Howe, A.; Soupir, M.L. Seasonal variations in export of antibiotic resistance genes and bacteria in runoff from an agricultural watershed in Iowa. *Sci. Total Environ.* **2020**, *738*, 140224. [CrossRef]

53. Christofilopoulos, S.; Kaliakatsos, A.; Triantafyllou, K.; Gounaki, I.; Venieri, D.; Kalogerakis, N. Evaluation of a constructed wetland for wastewater treatment: Addressing emerging organic contaminants and antibiotic resistant bacteria. *New Biotechnol.* **2019**, *52*, 94–103. [CrossRef] [PubMed]
54. Ben, W.; Wang, J.; Pan, X.; Qiang, Z. Dissemination of antibiotic resistance genes and their potential removal by on-farm treatment processes in nine swine feedlots in Shandong Province, China. *Chemosphere* **2017**, *167*, 262–268. [CrossRef] [PubMed]
55. Tao, C.W.; Hsu, B.M.; Ji, W.T.; Hsu, T.K.; Kao, P.M.; Hsu, C.P.; Shen, S.M.; Shen, T.Y.; Wan, T.J.; Huang, Y.L. Evaluation of five antibiotic resistance genes in wastewater treatment systems of swine farms by real-time PCR. *Sci. Total Environ.* **2014**, *496*, 116–121. [CrossRef]
56. Zhu, N.; Jin, H.; Ye, X.; Liu, W.; Li, D.; Shah, G.M.; Zhu, Y. Fate and driving factors of antibiotic resistance genes in an integrated swine wastewater treatment system: From wastewater to soil. *Sci. Total Environ.* **2020**, *721*, 137654. [CrossRef]
57. Yuan, Q.B.; Zhai, Y.F.; Mao, B.Y.; Schwarz, C.; Hu, N. Fates of antibiotic resistance genes in a distributed swine wastewater treatment plant. *Water Environ. Res.* **2019**, *91*, 1565–1575. [CrossRef]
58. Feng, L.; Wu, H.; Zhang, J.; Brix, H. Simultaneous elimination of antibiotics resistance genes and dissolved organic matter in treatment wetlands: Characteristics and associated relationship. *Chem. Eng. Sci.* **2021**, *415*, 128966. [CrossRef]
59. Ávila, C.; García-Galán, M.J.; Borrego, C.M.; Rodríguez-Mozaz, S.; García, J.; Barceló, D. New insights on the combined removal of antibiotics and ARGs in urban wastewater through the use of two configurations of vertical subsurface flow constructed wetlands. *Sci. Total Environ.* **2021**, *755 Pt 2*, 142554. [CrossRef] [PubMed]
60. Jacobs, K.; Wind, L.; Krometis, L.A.H.; Hession, W.C.; Pruden, A. Fecal indicator bacteria and antibiotic resistance genes in storm runoff from dairy manure and compost-amended vegetable plots. *J. Environ. Qual.* **2019**, *48*, 1038–1046. [CrossRef] [PubMed]
61. Lan, L.; Kong, X.; Sun, H.; Li, C.; Liu, D. High removal efficiency of antibiotic resistance genes in swine wastewater via nanofiltration and reverse osmosis processes. *J. Environ. Manag.* **2019**, *231*, 439–445. [CrossRef] [PubMed]
62. Xian, Q.; Hu, L.; Chen, H.; Chang, Z.; Zou, H. Removal of nutrients and veterinary antibiotics from swine wastewater by a constructed macrophyte floating bed system. *J. Environ. Manag.* **2010**, *91*, 2657–2661. [CrossRef]
63. Huang, X.; Luo, Y.; Liu, Z.; Zhang, C.; Zhong, H.; Xue, J.; Wang, Q.; Zhu, Z.; Wang, C. Influence of Two-Stage Combinations of Constructed Wetlands on the Removal of Antibiotics, Antibiotic Resistance Genes and Nutrients from Goose Wastewater. *Int. J. Environ. Res. Public Health* **2019**, *16*, 4030. [CrossRef]
64. Pereira, W.E.; Domagalski, J.L.; Hostettler, F.D.; Brown, L.R.; Rapp, J.B. Occurrence and accumulation of pesticides and organic contaminants in river sediment, water and clam tissues from the San Joaquin River and tributaries, California. *Environ. Toxicol. Chem.* **1996**, *15*, 172–180. [CrossRef]
65. Joseph, S.M.; Ketheesan, B. Microalgae based wastewater treatment for the removal of emerging contaminants: A review of challenges and opportunities. *Case Stud. Chem. Environ. Eng.* **2020**, *2*, 100046. [CrossRef]
66. Srikanth, K.; Sukesh, K.; Rao, A.R.; Pavan, G.; Ravishankar, G.A. Emerging Contaminants Effect on Aquatic Ecosystem: Human Health Risks-A Review. *Agric. Res. Technol.* **2019**, *19*, 556104. [CrossRef]
67. Matamoros, V.; Caiola, N.; Rosales, V.; Hernández, O.; Ibáñez, C. The role of rice fields and constructed wetlands as a source and a sink of pesticides and contaminants of emerging concern: Full-scale evaluation. *Ecol. Eng.* **2020**, *156*, 105971. [CrossRef]
68. Saini, A.; Solomon, S.S. A critical review of contaminants and its analytical management. *Int. J. Res. Anal. Rev.* **2019**, *6*, 386–393.
69. Fairbairn, D.J.; Karpuzcu, M.E.; Arnold, W.A.; Barber, B.L.; Kaufenberg, E.F.; Koskinen, W.C.; Novak, P.J.; Rice, P.J.; Swackhamer, D.L. Sediment–water distribution of contaminants of emerging concern in a mixed use watershed. *Sci. Total Environ.* **2015**, *505*, 896–904. [CrossRef]
70. Dias, S.; Mucha, A.P.; Duarte Crespo, R.; Rodrigues, P.; Almeida, C.M.R. Livestock Wastewater Treatment in Constructed Wetlands for Agriculture Reuse. *Int. J. Environ. Res. Public Health* **2020**, *17*, 8592. [CrossRef]
71. Tada, C.; Ikeda, N.; Nakamura, S.; Oishi, R.; Chigira, J.; Yano, T.; Nakano, K.; Nakai, Y. Animal wastewater treatment using constructed wetland. *J. Integr. Field Sci.* **2011**, *8*, 41–47.
72. Abdel-Mohsein, H.S.; Feng, M.; Fukuda, Y.; Tada, C. Remarkable Removal of Antibiotic-Resistant Bacteria During Dairy Wastewater Treatment Using Hybrid Full-scale Constructed Wetland. *Water Air Soil Pollut.* **2020**, *231*, 397. [CrossRef]
73. Bôto, M.; Almeida, C.M.R.; Mucha, A.P. Potential of Constructed Wetlands for Removal of Antibiotics from Saline Aquaculture Effluents. *Water* **2016**, *8*, 465. [CrossRef]
74. Gorito, A.M.; Ribeiro, A.R.; Gomes, C.R.; Almeida, C.M.R.; Silva, A.M.T. Constructed wetland microcosms for the removal of organic micropollutants from freshwater aquaculture effluents. *Sci. Total Environ.* **2018**, *644*, 1171–1180. [CrossRef]
75. Radke, M.; Ulrich, H.; Wurm, C.; Kunkel, U. Dynamics and attention of acidic pharmaceuticals along a river stretch. *Environ. Sci. Technol.* **2010**, *44*, 2968–2974. [CrossRef]
76. Matamoros, V.; Arias, C.; Brix, H.; Bayona, J.M. Removal of pharmaceuticals and personal care products (PPCPs) from urban wastewaters in pilot vertical flow constructed wetland and a sand filter. *Environ. Sci. Technol.* **2007**, *41*, 8171–8177. [CrossRef]
77. Hussain, S.A. Removal of Poultry Pharmaceuticals by Constructed Wetlands. Doctor of Philosophy Dissertation, McGill University, Montreal, QC, Canada, July 2011.
78. Agudelo, R.M.; Peñuela, G.; Aguirre, N.J.; Morató, J.; Jaramillo, M.L. Simultaneous removal of chlorpyrifos and dissolved organic carbon using horizontal sub-surface flow pilot wetlands. *Ecol. Eng.* **2010**, *36*, 1401–1408. [CrossRef]
79. Moore, M.T.; Schulz, R.; Cooper, C.M.; Smith, S.; Rodgers, J.H. Mitigation of chlorpyrifos runoff using constructed wetlands. *Chemosphere* **2002**, *46*, 827–835. [CrossRef]

80. Sherrard, R.M.; Bearr, J.S.; Murray-Gulde, C.L.; Rodgers, J.H.; Shah, Y.T. Feasibility of constructed wetlands for removing chlorothalonil and chlorpyrifos from aqueous mixtures. *Environ. Pollut.* **2004**, *127*, 385–394. [CrossRef]

81. de Souza, T.D.; Borges, A.C.; de Matos, A.T.; Mounteer, A.H.; de Queiroz, M.E.L.R. Removal of chlorpyrifos insecticide in constructed wetlands with different plant species. *Rev. Bras. Eng. Agric. Ambient.* **2017**, *21*, 878–883. [CrossRef]

82. Tang, X.Y.; Yang, Y.; McBride, M.B.; Tao, R.; Dai, Y.N.; Zhang, X.M. Removal of chlorpyrifos in recirculating vertical flow constructed wetlands with five wetland plant species. *Chemosphere* **2019**, *216*, 195–202. [CrossRef]

83. Chen, J.; Deng, W.J.; Liu, Y.S.; Hu, L.X.; He, L.Y.; Zhao, J.L.; Wang, T.T.; Ying, G.G. Fate and removal of antibiotics and antibiotic resistance genes in hybrid constructed wetlands. *Environ. Pollut.* **2019**, *249*, 894–903. [CrossRef]

84. Song, H.L.; Zhang, S.; Guo, J.; Yang, Y.L.; Zhang, L.M.; Li, H.; Yang, X.L.; Liu, X. Vertical up-flow constructed wetlands exhibited efficient antibiotic removal but induced antibiotic resistance genes in effluent. *Chemosphere* **2018**, *203*, 434–441. [CrossRef]

85. Du, L.; Zhao, Y.; Wang, C.; Zhang, H.; Chen, Q.; Zhang, X.; Zhang, L.; Wu, J.; Wu, Z.; Zhou, Q. Removal performance of antibiotics and antibiotic resistance genes in swine wastewater by integrated vertical-flow constructed wetlands with zeolite substrate. *Sci. Total Environ.* **2020**, *721*, 137765. [CrossRef]

86. Huang, X.; Liu, C.; Li, K.; Su, J.; Zhu, G.; Liu, L. Performance of vertical up-flow constructed wetlands on swine wastewater containing tetracyclines and tet genes. *Water Res.* **2015**, *70*, 109–117. [CrossRef]

87. Huang, X.; Zheng, J.; Liu, C.; Liu, L.; Liu, Y.; Fan, H. Removal of antibiotics and resistance genes from swine wastewater using vertical flow constructed wetlands: Effect of hydraulic flow direction and substrate type. *Chem. Eng. Sci.* **2017**, *308*, 692–699. [CrossRef]

88. Huang, X.; Ye, G.; Yi, N.; Lu, L.; Zhang, L.; Yang, L.; Xiao, L.; Liu, J. Effect of plant physiological characteristics on the removal of conventional and emerging pollutants from aquaculture wastewater by constructed wetlands. *Ecol. Eng.* **2019**, *135*, 45–53. [CrossRef]

89. Liu, L.; Liu, Y.H.; Wang, Z.; Liu, C.X.; Huang, X.; Zhu, G.F. Behavior of tetracycline and sulfamethazine with corresponding resistance genes from swine wastewater in pilot-scale constructed wetlands. *J. Hazard. Mater.* **2014**, *278*, 304–310. [CrossRef] [PubMed]

90. Chen, J.; Liu, Y.S.; Su, H.C.; Ying, G.G.; Liu, F.; Liu, S.S.; He, L.Y.; Chen, Z.F.; Yang, Y.Q.; Chen, F.R. Removal of antibiotics and antibiotic resistance genes in rural wastewater by an integrated constructed wetland. *Environ. Sci. Pollut. Res.* **2015**, *22*, 1794–1803. [CrossRef] [PubMed]

91. Locke, M.A.; Weaver, M.A.; Zablotowicz, R.M.; Steinriede, R.W.; Bryson, C.T.; Cullum, R.F. Constructed wetlands as a component of the agricultural landscape: Mitigation of herbicides in simulated runoff from upland drainage areas. *Chemosphere* **2011**, *83*, 1532–1538. [CrossRef]

92. Gikas, G.D.; Pérez-Villanueva, M.; Tsioras, M.; Alexoudis, C.; Pérez-Rojas, G.; Masís-Mora, M.; Lizano-Fallas, V.; Rodríguez-Rodríguez, C.E.; Vryzas, Z.; Tsihrintzis, V.A. Low-cost approaches for the removal of terbuthylazine from agricultural wastewater: Constructed wetlands and biopurification system. *Chem. Eng. Sci.* **2018**, *335*, 647–656. [CrossRef]

93. Gikas, G.D.; Vryzas, Z.; Tsihrintzis, V.A. S-metolachlor herbicide removal in pilot-scale horizontal subsurface flow constructed wetlands. *Chem. Eng. Sci.* **2018**, *339*, 108–116. [CrossRef]

94. Hsieh, C.Y.; Liaw, E.T.; Fan, K.M. Removal of veterinary antibiotics, alkylphenolic compounds, and estrogens from the Wuluo constructed wetland in southern Taiwan. *J. Environ. Sci. Health Part A* **2015**, *50*, 151–160. [CrossRef]

95. Borges, A.C.; Calijuri, M.; Matos, A.; Lopes, M.; Queiroz, R. Horizontal subsurface flow constructed wetlands for mitigation of ametryn-contaminated water. *Water Sa* **2009**, *35*, 441–446. [CrossRef]

96. George, D.; Stearman, G.K.; Carlson, K.; Lansford, S. Simazine and METOLACHLOR removal by subsurface Flow Constructed Wetlands. *Water Environ. Res.* **2003**, *75*, 101–112. [CrossRef]

97. Lyu, T.; Zhang, L.; Xu, X.; Arias, C.A.; Brix, H.; Carvalho, P.N. Removal of the pesticide tebuconazole in constructed wetlands: Design comparison, influencing factors and modelling. *Environ. Pollut.* **2018**, *233*, 71–80. [CrossRef]

98. Papaevangelou, V.A.; Gikas, G.D.; Vryzas, Z.; Tsihrintzis, V.A. Treatment of agricultural equipment rinsing water containing a fungicide in pilot-scale horizontal subsurface flow constructed wetlands. *Ecol. Eng.* **2017**, *101*, 193–200. [CrossRef]

99. Wu, J.; Feng, Y.; Dai, Y.; Cui, N.; Anderson, B.; Cheng, S. Biological mechanisms associated with triazophos (TAP) removal by horizontal subsurface flow constructed WETLANDS (HSFCW). *Sci. Total Environ.* **2016**, *553*, 13–19. [CrossRef]

100. Vystavna, Y.; Frkpva, Z.; Marchand, L.; Vergeles, Y.; Stolberg, F. Removal efficiency of pharmaceuticals in a full scale constructed wetland in east Ukraine. *Ecol. Eng.* **2017**, *108*, 50–58. [CrossRef]

101. Senarathna, D.D.T.T.D.; Abeysooriya, K.H.D.N.; Vithushana, T.; Dissanayake, D.M.N.A. Veterinary pharmaceuticals in aquaculture wastewater as emerging contaminant substances in aquatic environment and potential treatment methods. *MOJ Ecol. Environ. Sci.* **2021**, *6*, 98–102. [CrossRef]

102. Elsayed, O.F.; Maillard, E.; Vulleumier, S.; Millet, M.; Imfeld, G. Degradation of chloroacetanilide herbicides and bacterial community composition in lab-scale wetlands. *Chem. Eng. Sci.* **2018**, *339*, 108–116. [CrossRef]

103. Zhang, D.; Gersberg, R.M.; Ng, W.J.; Tan, S.K. Removal of pharmaceuticals and personal care products in aquatic plant-based systems: A review. *Environ. Pollut.* **2013**, *184*, 620–639. [CrossRef]

104. Guan, Y.; Wang, B.; Gao, Y.; Liu, W.; Zhao, X.; Huang, X.; Yu, J. Occurrence and Fate of Antibiotics in the Aqueous Environment and Their Removal by Constructed Wetlands in China: A review. *Pedosphere* **2017**, *27*, 42–51. [CrossRef]

 *sustainability*

Review

Role of Microorganisms in the Remediation of Wastewater in Floating Treatment Wetlands: A Review

Munazzam Jawad Shahid [1], **Ameena A. AL-surhanee** [2], **Fayza Kouadri** [3], **Shafaqat Ali** [1,4,*], **Neeha Nawaz** [1], **Muhammad Afzal** [5], **Muhammad Rizwan** [1], **Basharat Ali** [6] and **Mona H. Soliman** [7]

[1] Department of Environmental Sciences and Engineering, Government College University, Faisalabad 38000, Pakistan; munazzam01@gmail.com (M.J.S.); neehanawaz66@gmail.com (N.N.); mrazi1532@yahoo.com (M.R.)

[2] Biology Department, College of Science, Jouf University, Sakaka 2014, Saudi Arabia; amaserhani@ju.edu.sa

[3] Biology Department, Faculty of Science, Taibah University, AL-Madina AL-Munawarah 344, Saudi Arabia; fayzakouadri@yahoo.com

[4] Department of Biological Sciences and Technology, China Medical University, Taichung 40402, Taiwan

[5] Soil and Environmental Biotechnology Division, National Institute of Biotechnology and Genetic Engineering, Faisalabad 38000, Pakistan; manibge@yahoo.com

[6] Department of Agronomy, University of Agriculture, Faisalabad 38040, Pakistan; basharat2018@yahoo.com

[7] Botany and Microbiology Department, Faculty of Science, Cairo University, Giza 12613, Egypt; monahsh3344@gmail.com

[*] Correspondence: shafaqataligill@gcuf.edu.pk

Received: 6 June 2020; Accepted: 29 June 2020; Published: 10 July 2020

Abstract: This article provides useful information for understanding the specific role of microbes in the pollutant removal process in floating treatment wetlands (FTWs). The current literature is collected and organized to provide an insight into the specific role of microbes toward plants and pollutants. Several aspects are discussed, such as important components of FTWs, common bacterial species, rhizospheric and endophytes bacteria, and their specific role in the pollutant removal process. The roots of plants release oxygen and exudates, which act as a substrate for microbial growth. The bacteria attach themselves to the roots and form biofilms to get nutrients from the plants. Along the plants, the microbial community also influences the performance of FTWs. The bacterial community contributes to the removal of nitrogen, phosphorus, toxic metals, hydrocarbon, and organic compounds. Plant–microbe interaction breaks down complex compounds into simple nutrients, mobilizes metal ions, and increases the uptake of pollutants by plants. The inoculation of the roots of plants with acclimatized microbes may improve the phytoremediation potential of FTWs. The bacteria also encourage plant growth and the bioavailability of toxic pollutants and can alleviate metal toxicity.

Keywords: floating treatment wetlands; water; plants; microbes; pollutants

---

1. Introduction

Constructed wetlands (CWs) are purposely designed and constructed systems, based on the physical, chemical, and biological principles and processes of natural wetlands [1]. The vegetation, soil, and microorganisms are the main components of a CW that contribute to pollutant removal processes from wastewater. The associated environmental and economic benefits have established CWs as a viable option for wastewater treatment [2]. These have been widely applied in the treatment of various types of wastewater, such as municipal, agricultural runoff, storm runoff, and industrial [3–8]. Floating treatment wetland (FTW) is a novel technology, based on a floating vegetated system, that has unique abilities to remediate wastewater [9,10]. In FTWs, plants are supported by a buoyant mat or raft that floats on the surface of the water [11]. The roots of the plants develop below the floating mat, extending down the water column, and develop an extensive root system beneath the water level [10,12,13]. The development of a widespread and dense root system is necessary for the effective performance of FTWs [14]. FTWs move freely and thus cover a wider area of water than the emergent root system. In a FTW system, the rhizomes and dense root structure develop a special hydraulic flow in the water zone between the mat and the bottom of the water body, and the floating roots act as a filter [15]. This leads to an effective removal of pollutants from the water due to the availability of the increased surface area of roots for adsorption and absorption [16]. The roots and rhizomes provide a habitat for microbial growth and development. The roots and attached biofilms perform different physical and biochemical processes for the removal of pollutants from the contaminated water [17,18]. In FTWs, pollutants are removed by three main processes, namely adsorption, sedimentation, and biodegradation [19].

The benefits associated with FTWs have made it a promising ecological remediation technology in the field of wastewater treatment. These benefits include economic and convenient construction, no digging/earth moving or extra land acquisition, easy operation and maintenance, floating mats that are adjustable with a change in the water level, and excellent treatment performance [10,20,21]. Furthermore, the planted vegetation provides economic and ecological benefits such as the use of vegetation as fodder, providing a habitat for wildlife/aquatic animals, and enhancing the aesthetic value of the pond [10,22]. Globally, FTWs are being applied to remediate various types of wastewater, such as eutrophic water, sewage and domestic, storm water runoff, and industrial [23–29].

Microbes have a fundamental role in the remediation of polluted water by FTWs. The bacteria attached to the roots form biofilms through a repeated proliferation process [30]. The oxygen and exudates released by the plants create a substrate for microbial growth and colonization on the root beneath the water level [31]. Thus, along the vegetation, the performance of FTWs also depends upon the metabolism of the microbial community in water, attached to the roots and floating mats [32–34]. The application of plants in combination with microorganisms in FTWs is an effective and sustainable approach for the treatment of wastewater [35]. The plant–microbe interaction enhances the efficacy of FTWs [36]. Although the plant–bacteria interaction plays an essential role in the removal of contaminants from aquatic ecosystem, the interaction of the plant with bacteria in the FTWs is not well explored [37].

This paper discusses this important component of FTWs and provides a detailed overview of the specific role of microorganisms in FTWs. We have summarized the important species of bacteria that colonize the roots of plants. Furthermore, the specific role of rhizospheric bacteria, endophytes, and algae in the pollutant removal process in FTWs has been elaborated.

2. Mechanism of FTWs

In FTWs, pollutants are removed from the wastewater by different mechanisms induced by plants, microbes, and their mutualistic relationships. The presence of a vegetated floating mat in a water body boosts the pollutant removal efficiency of the system by modifying the physicochemical properties of the water [38,39]. The physical characteristics of the plant's roots and the nutrient uptake are interdependent/interlinked. The type of medium in which the roots exit and the nutrients present in the medium specify the root's physical characteristics [9,40]. In general, the roots of plants filter the particulates present in the water. Nutrients are taken up by the plant's roots and accumulated in them, as well as in the parts of the plant above the mat [14]. Most organic pollutants are degraded by microorganisms present on the roots. However, some of the organic pollutants are taken by the plants. The organic pollutants can either be accumulated in the biomass of vegetation or degraded by endophytic bacteria present inside the plants [41,42].

The plants in FTWs contribute to the pollutant removal process by entrapping pollutant particles in the roots [11,43,44]. The roots of plants act as physical filters, and remove suspended particulate matter from the water. For an effective removal, there should be dense roots, so that they can act as a physical filter and a bio-sorbent [15].

The bioactive substances released by the roots have a unique role in the removal of nutrients. These substances balance pH, and increase the humic content in the water, which results in the adsorption and/or precipitation of pollutants in the form of insoluble material [15,21]. The neutral pH induced by the vegetation helps in the settlement of dissolved particulate pollutants [24]. Moreover, these substances alter the physicochemical condition of water, and increase metal and nutrient removal and the sorption characteristics of biofilms [45,46]. For example, plants may remove phosphorus by direct uptake, but the key mechanisms of phosphorus removal are sorption, settlement at the bottom, and physical entrapment in the roots [47]. The FTWs also inhibit the growth of algal communities by removing nutrients from the water, thus reducing their population [48].

Roots act as a suitable surface for the formation of biofilms, which enhance the degradation of organic pollutants and removal of nutrients from wastewater [11]. Root exudates aid in the retention of microbes on the roots by providing them with nutrients [49]. The roots also provide oxygen to rhizospheric bacteria for aerobic degradation of organic matter. The biodegradation of organic matter into simple nutrients occurs when it comes in contact with the biofilm [50,51]. Plants remove these nutrients through direct uptake [52]. Trapping in the biofilm of the roots of macrophytes is an essential mechanism for particulate matter removal. Furthermore, roots let microbial colonies assimilate the carbon compounds and help in the reduction in biological oxygen demand and chemical oxygen demand [26]. Floating wetlands can work under both aerobic and anaerobic conditions. However, the nutrient removal under aerobic conditions is higher than under anaerobic conditions [53]. Other organic compounds are degraded by heterotrophic microorganisms either aerobically or anaerobically, depending upon the oxygen level in water [54].

3. Important Components of FTWs

FTW is composed of plants that are vegetated in a floating mat. Different types of material are used as floating mats. The detail of these important components is described below (Figure 1).

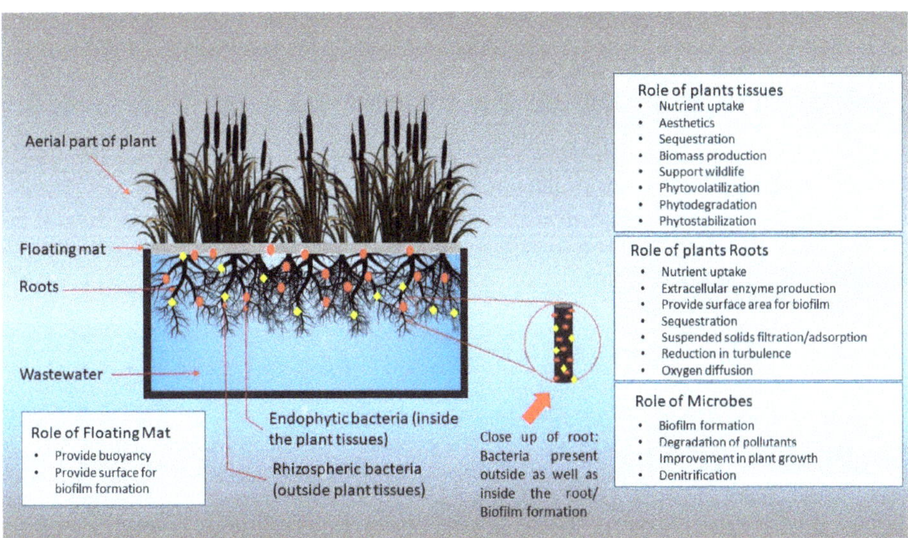

Figure 1. Schematic representation of floating treatment wetland and pollutant removal process.

3.1. Growth Media

Different types of growth media have been used to provide support to the plants growing on the floating mat. This growth media can be coconut fiber, peat, soil, bamboo crush, sand, peat rice straw, and compost [55]. The selection of growth media also influences the pollutant removal process. For instance, the use of rice straw as growth media improved the total nitrogen removal process by the formation of thick biofilms, boosting the nitrification/denitrification process [56].

3.2. Buoyancy

In FTWs, different materials have been applied with different natural buoyancies. These floating materials serve as a platform to fix the plants. The floating mats are made up of different materials such bamboo sticks, polyester fibers, plastic and foaming sheets [57–59]. The floating material should be hydrophobic, nutrient absorbent, bacterial adhesive, and with no desorption [15].

Some patent floating mats are also available commercially, such as Beemat®, and Bioheaven®, made up of buoyant material with holes for plantation. The wrapped plastic tubes and pipes manufactured from polyethylene (PE), polypropylene (PP) and polyvinyl chloride (PVC), and PS (polystyrene) foams are most commonly used for the construction of floating frames and rafts [38]. A natural buoyant material, bamboo, has been found to be a cheap and cost-effective material for the construction of floating rafts [60].

3.3. Plants

The selection of plant species has a great influence on the pollutant removal process. The selection of plants depends upon their local availability, the nature of pollutants, and the climate zone. The plants mostly used to develop FTWs are of *Canna*, *Typha*, *Phragmites*, and *Cyperus* genera. They have been widely applied in FTWs for the remediation of different types of wastewater [30,56,61–66]. Some species of the *Poaceae* family (*Lollium* sp., *Zizania* sp., and *Chrysopogon* sp.) have been successfully applied in Italy, China, Singapore, and Thailand to develop FTWs. Some plant species are suitable for particular regions and have efficiently removed nutrients and other pollutants in a specific climate. Some other plants such as *Phragmites*, *Carex*, *Acorus*, and *Juncus* were also successfully applied in FTWs, and these effectively adapted in several locations. The selection of macrophytes to develop FTWs is very important for pollutant removal as well as for ecosystem sustainability. The selected plants should be native, easily available, non-invasive species, perennial, able to thrive in a hydroponic environment with an extensive root system and aerenchyma [67]. The application of invasive species in FTWs may result in damage to the ecosystem, and the ultimate cost of habitat restoration may suppress the benefits gained by pollutant removal. [68]. The characteristics that make these macrophytes ideal for FTWs are their robust growth tall shoot length, extensive root system, and large aerenchyma in their roots and rhizomes. Plants with relatively thin fibrous roots have a better performance in total nitrogen removal, and plants with high total root biomass have a better performance in NH^+-N removal [69]. The root development depends upon various factors such as species, age, type of plant and concentration of nutrients, trophic status of water, nature of pollutants, redox conditions, and use of supporting mats and growth media. A high nutrient load at an earlier plant stage can be harmful to plants and can damage the root system [70].

Similarly, the high load of toxicants can also hinder the growth of the root by permanently damaging young plants. The root development of *P. australis* was constrained up to 40-cm deep after 3 years of plantation due to the toxic effects of digestate liquid fraction. On the other hand, *Typha latifolia* and *Juncus maritimus* did not establish themselves due to the high pollutant load [71].

3.4. Bacterial Biofilm

Bacteria have a unique ability to form biofilms, also known as epiphytic microbes. Biofilm formation begins with the attachment of free-floating microbes to gas–liquid and solid–liquid interfaces.

These biofilms have a key role in the assimilation of the biogeochemical cycles and the dynamics of an ecosystem process [72]. In the aquatic ecosystem, aquatic plants are an essential substrate for the establishment, growth, and development of biofilms. Aquatic plants release oxygen, essential for aerobic bacteria attached to roots, and stimulate the nitrogen cycle in the roots' surroundings [73,74]. Biofilms are composed of an extracellular matrix comprised of polysaccharide biopolymers, proteins, and DNA that hold the cell together [75]. The structural integrity of biofilms is obtained by secreted proteins, various types of exopolysaccharides and cell surface adhesions [76]. The development and maintenance of these biofilms rely on small molecules such as homoserine lactones, antibiotics, and secondary metabolites, such as the *Staphylococcus aureus* matrix, provide proteins for the synthesis of biofilm. The extracellular matrix also facilitates the formation of adhesive protein found anchored to the cell wall of *S. aureus*, holding the cells together within the biofilm by interaction with other proteins [77,78]. The extracellular DNA also strengthens the structural integrity of the biofilms. For example, *Pseudomonas aeruginosa* contains a significant amount of DNA to provide stability to biofilms [79]. The nature of biofilms and associated matrices depends upon the types of substrates, medium, and growth conditions. *Bacillus subtilis*, a Gram-positive bacterium, can make biofilms via production of two different polymers: polysaccharide extracellular polymeric substances and poly-d-glutamate. Both of these polymers contribute to biofilm formation; however, the contribution of each polymer is determined by strain and prevailing conditions [80]. The plants can also modify the function and structure of the microbial community in their rhizosphere [81]. The biodiversity and species of bacteria determine the functions of the biofilms. The biofilm-forming bacteria have been reported as diverse and host specific. The secretion of macrophytes and growth status can determine the bacterial composition of biofilms in the aquatic ecosystem [82]. Moreover, the bacterial community of biofilms was found to be different than those in the surrounding water column [37].

4. Microorganisms

Microbial communities have an essential role in the organic and inorganic pollutant removal process and plant growth promotion in FTWs (Figure 2); however, little has been explored about specific microbial species in roots and their functions in pollutant removal processes from water [83,84]. Some bacteria, such as rhizospheric bacteria, are essential for vigorous plant growth [85]. The bulk soil is the main source of these microbial populations. However, the rhizospheric bacterial population is different from the soil bacterial community [86–88]. Similarly, in FTWs, the microbes can be categorized into biofilm-forming bacteria and water column bacteria.

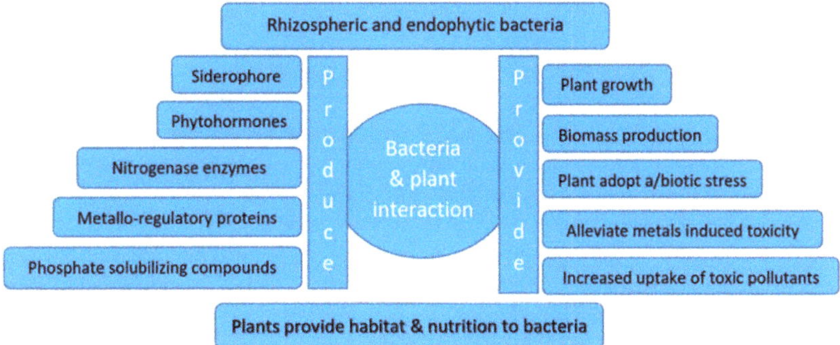

Figure 2. Role of rhizospheric and endophytic bacteria in plant growth promotion and pollutant removal processes.

In FTWs, the microbial communities mostly originate from ambient water. The amelioration and scrapping specific to the plants' roots perform a central part in the formation of specific rhizosphere microbial communities.

Actinobacteria was found to be a dominant group in the water of FTW systems; however, Proteobacteria was mainly found in the roots and biofilm samples [89]. In Proteobacteria, Alphaproteobacteria was found to be abundant in the rhizoplane of plants vegetated in FTWs, and biofilms were mostly composed of Gammaproteobacteria. The second largest phylum in water and plant root samples was *Cyanobacteria*, but it was not found in biofilm samples. In a comparison of the microbial communities in the roots of *Canna* and *Juncus*, it was found that different plants host different types of microbes in their roots. This difference reveals that plant roots secrete specific exudates and compounds, which attract specific microbial communities [89]. The plant rhizoplane in the water column attracts microbes and develops large microbial mass manifests in the shape of a thick, slimy coat on plant roots.

The presence of autotrophic microbial populations may also depend upon the presence of sunlight, although, in most cases, the floating mat covers the water surface to minimize the availability of sunlight. However, some amount of sunlight may be available under the water to support the Cyanobacterial community. However, the relative abundance of Cyanobacteria in plant root and water samples was found to be similar. In the roots of FTW plants, the genera of Cyanobacteria (*Anabaena* and *Nostochopsis*) that forms a heterocyst was abundantly observed. This indicates the ability of Cyanobacteria to associate with the roots of floating macrophytes and survive in available light conditions. In floating macrophytes, the rhizoplane was found to be enriched with sulfate-reducing bacteria [90]. In FTWs, even in aerobic conditions, anaerobic zones were found in the rhizoplane of the aquatic plants. These anaerobic microorganisms belong to sulfate-reducing bacteria and *Clostridium*. In FTWs, different sulfur oxidizers and sulfate reducers are essential to make out the sulfur cycle, yield, and depletion of hydrogen sulfide within the plant rhizoplane [70]. The sulfur-oxidizing bacteria are essential to protect the plants by the detoxification of reduced sulfides such as hydrogen sulfide.

The FTWs are efficient for nitrogen removal through denitrification by the microbial process. The nitrifiers are augmented in the aquatic root system of FTWs and responsible for ammonia oxidation. The *Nitrosomonas* and *Nitrosovibrio* (*Nitrosospira*) were found only on the plant roots of FTWs plants. The presence of *Rhizobium*, *Bradyrhizobium*, *Azorhizobium* and *Azovibrio* contributes toward nitrogen fixation within the FTWs. Several methanotrophs and methylotrophs were also found on plant roots in the FTWs [91]. These methanotrophs and methylotrophs were also abundant in the rhizosphere of terrestrial plants, and these were not specific to the aquatic plants. However, these bacteria have a key role in the rhizoplane of FTWs plants, predominantly under reduced oxygen levels [92].

Proteobacteria were found in the various rhizosphere systems [91,93–95]. The comparison between FTW plants and terrestrial plants' rhizosphere microbial communities revealed a distinctive mutualistic association of aquatic microbes with aquatic plants. *Bacillus*, a soil bacterial group, was absent in the rhizoplane of FTWs macrophytes. Similarly, Acidobacteria, the major bacterial group in the terrestrial plant, was not found in the rhizoplane of an aquatic plant [94,96]. Cyanobacteria were different in the plant's rhizosphere compared to the aquatic plant's rhizoplane [91,93,96].

Pseudomonas has the distinctive capability to degrade several polymers, which are difficult to demean by any other group of bacteria [97]. *Pseudomonas* has a dominant role in the degradation of polyethylene in combination with physical degradation [97]. *Pseudomonas* was found abundantly (95.5%) in a sample of floating foam from FTWs. The development of biofilms on floating mats involves a distinctive mechanism that is different from the formation of biofilm on plant roots and in water samples [97].

Ammonia oxidizing archaea (AOA) and bacteria can attach to the suspended roots in an autotrophic water environment [98]. The ammonia-oxidizing archaea and bacteria were found only on the roots as biofilms. The predominant ammonia oxidizers were ammonia-oxidizing bacteria (AOB) on the rhizoplane of macrophytes. The *Nitrosomonas europaea* and *Nitrosomonas ureae* were well adapted to NH_4^+-N rich environments. However, in the terrestrial ecosystem, *Nitrosospira* was found predominantly in AOB communities [98,99].

In a study on three aquatic plants, *N. peltatum, M. verticillatum, and T. japonica*, the dominant phylum detected was Proteobacteria, ranging from 37% to 83%, followed by Bacteroidetes (8–38%). The other phyla found in root biofilms were Chloroflexi, Firmicutes, and Verrucomicrobia at low frequencies. The dominant bacteria in the phylum Proteobacteria were Alphaproteobacteria, followed by Betaproteobacteria and Gammaproteobacteria. The other bacteria detected at a low frequency were Epsilonproteobacteria and Deltaproteobacteria [74].

The class Epsilonproteobacteria was found to be higher in number in vegetated sediment samples compared to un-vegetated sediments and biofilms [74]. The difference in microbial composition and epiphytic biomass may be the effect of the difference in plant exudates such as polyphenols and allopathically active compounds [100]. The plants can increase the quantity and diversity of bacterial biofilms in the aquatic ecosystem, which ultimately can promote the remediation potential of associated macrophytes [72].

Epiphytic bacterial communities are diverse and host specific. A similar phenomenon was also found in other terrestrial and aquatic plants [82,101]. The biofilms attached to roots exhibit particular niches. The difference in bacterial communities is attributed to the different growth environments such as the difference in water flow, the availability of light, and nutrients conditions [37]. Additionally, plant roots, water characteristics, sediment properties, and aquatic animals also influence the nutrient availability, types, and suitability of the environment for the bacteria. The epiphytic bacteria diversity and species richness were generally greater on roots than those on stems and leaves. Similarly, the bacterial species in vegetated sediments were more diverse than in un-vegetated sediments [74].

Similarly, the bacterial population linked with sea grassroots was different from the adjacent bulk sediment [102]. Thus, the roots of the plant may alter the bacterial community in the surrounding environment. This difference may be due to the influence of root rhizospheric zones on organic matter accumulation, chemical exudates, and oxygen concentration [22,103].

Similarly, the biofilm and sediment's microbial communities were found to be dissimilar from one another. In biofilms, the percentage of class Alphaproteobacteria was higher than in sediments. The class Epsilonproteobacteria and Deltaproteobacteria were mostly detected only in sediment. The parallel findings have been stated by other researchers who investigated the bacterial composition in the sediments of two lakes in China [104].

4.1. Role of Endophytes

The microorganisms residing in the roots of plants and soil also have a major contribution to the uptake of metals from the contaminated media. These microorganisms boost the breakdown of complex organic and inorganic compounds into simple nutrients, mobilize metal ions, and increase the bioavailability to plants [105–108]. These bacteria, such as rhizobacteria, stimulate the growth of plants and biomass production, and enhance plants' uptake of toxic pollutants, and the their ability to alleviate metal-induced toxicity [109,110]. Endophytic bacteria reside within different tissues of the plant [111,112], increasing the ability of plants to cope with different biotic and abiotic stresses [113]. Broadly, endophytes perform three major roles in the plant which are its protection from biotic stress, relieving abiotic stress, and supporting it by providing nutrients such as the increasing availability of nitrogen, phosphorus, and other essential elements [114]. The prior inoculation of plants with endophytes can reduce the chances of bacterial, fungal, and viral diseases, and even the damage caused by insects and nematodes [113,115]. The relationship of endophytes with host plants may be either as obligate endophytes and or facultative endophytes [112]. In stress conditions, endophytes may help the plant to relieve stress by the combined action of multiple mechanisms [116]. Direct mechanisms include siderophore production [117], antimicrobial metabolites [118], phosphate-solubilizing compounds [119], nitrogen-fixing abilities [120], and phytohormones [42,121,122]. The indirect methods include bioremediation and biocontrol [123]. It is established that certain endophytic bacteria initiate a system known as induced systematic resistance in their host. This system is effective against different types of pathogenic bacteria, by preventing the induced bacteria from causing any visible disease symptoms in the host plant [113,124]. It is well reported that endophytes stimulate the degradation of xenobiotics and their supplementary compounds by expressing required catabolic genes. The endophytic bacteria have evolved various types of mechanisms to nullify the effect of toxic heavy metals and contaminants, such as the efflux of metal ions, the transformation of pollutants into less toxic forms, and the sequestration of metal ions on the surface of the cell [125]. Endophytes can also mitigate metal stress by promoting photosynthesis, anti-oxidative enzyme activities, modifying translocation, and the storage of heavy metal ions. The inoculation of maize with *Gaeumannomyces cylindrosporus* significantly improved the yield and productivity of maize under lead stress [126]. Similarly, *Pseudomonas aeruginosa* inoculation increases the cadmium tolerance (Cd) of plants and enhances the accumulation and translocation of Cd in inoculated plants [127].

The high concentration of toxic pollutants may cause toxicity to macrophytes, thus decreasing the efficiency of macrophytes to remediate pollutants. The endophytes may overcome this challenge. Endophytes possess plant growth-promoting (PGP) traits and degradation genes that assists the plant in handling with several environmental stresses. The endophytes contribute to the decontamination of mixed contaminants by degradation and heighten the metal translocation by the mutualistic relation of plants and endophytes [128,129]. A few studies have highlighted the application of endophytes in the macrophytes of FTWs for the treatment of sewage effluent, textile effluent, polluted river water and potentially toxic metals [25,130,131]. The major advantage of using endophytes to improve xenobiotic remediation is that it is easier to genetically modify the microorganisms for maximum pollutant degradation than the plants. Furthermore, the efficiency of the remediation process can be easily tracked by the estimation of the abundance and expression of pollutant catabolic genes in soil and plant tissues. The unique environment of plants facilitates the endophytic bacteria to make large population sizes due to the minimal competition. The pollutant is degraded by endophyte bacteria in planta, and eliminates the toxic effect on the plant [113,132].

The application of endophytes in a FTWs system, vegetated with *P. australis,* improved the remediation potential of the plant and successfully removed the toxic metals such as iron, nickel, manganese, lead and chromium from the polluted river water. These inoculated endophytes were tracked in the root/shoot interior of *P. australis,* proving their potential role in pollutant removal [131]. The specific strains of endophytic bacteria inoculated to *T. domingensis* enhanced the remediation of textile effluent [133]. Similarly, the inoculation of *Leptochloa fusca* with a consortium of three endophyte bacteria strains in CWs boosted the efficiency of plants to remediate tannery effluent. This endophytic inoculation also enhanced the growth of *L. fusca*, increased the removal of pollutants and decreased the toxicity of treated wastewater [49].

4.2. Role of Rhizospheric Bacteria

The rhizospheric bacteria in FTWs have a prominent role in the degradation of organic matter, [134,135], and the translocation of potentially toxic metals [81,136,137]. This bacterial population differs qualitatively and quantitatively from those found in the bulk soil [138–140]. The microbial species in soil biota may pathologically infect the roots and rhizosphere biota [141,142]. The plant roots secrete exudates and metabolites, which chemotactically attract bacteria [143]. The rhizospheric bacteria of macrophytes in wetlands have a prominent role in the removal of pollutants [144]. The roots of the plants actually control the microbial colonies in the rhizosphere with the exchange of oxygen, CO_2, nutrients, and bio-chemicals [145,146]. The iron and ammonia can be oxidized by the oxygen released from the roots [81,147]. The roots' microbial populations also have an impact on the emission of methane, as well as other gases from the wetland system [148,149]. The enzymes and organic acids released by rhizophytes modify the nutrients and make them available to roots [135].

The roots of wetland plants secrete bioactive chemicals, which favor the development of microbial communities on roots [150]. The roots can also oxidize and reduce the sulfide present in their rhizosphere by regulating oxygen concentration, redox potential, and the release of low-nitrogen exudates such as sugar [151].

5. Role of Bacteria in Pollutant Removal Process

5.1. Nitrogen Fixation

The nitrogen fixation by microbes is a critical natural source of reactive nitrogen in the wetland ecosystem [152]. The oxygen and organic matter supply from the roots favor the enrichment of nitrogen-metabolizing microorganisms in the rhizosphere [40,153]. In the rhizosphere of wetland plants, bacteria transform the nitrogen by ammonification, nitrification, denitrification, uptake, and the anaerobic oxidation of ammonia by nitrate and nitrogen fixation [154]. The metabolic energy required for this process is obtained from the oxidation of organic matter and lithotrophy. In wetland plants, most of the nitrogen metabolism occurs at or near the roots [155,156]. The roots either take up the produced ammonia or they oxidize it into nitrites and nitrates. That oxidized nitrogen diffuses to the roots or to denitrifiers, which reduces the nitrate to N_2 gas in the absence of oxygen [157]. Microbes perform an N-fixation of non-reactive N_2, and nitrogen is produced [158]. The heterotroph and autotroph prokaryotes contribute toward the production of a large amount of reactive nitrogen by nitrogen fixation [152]. The nitrogen fixation by cyanobacteria in wetlands depends upon the availability of light [152]. The important N-fixing bacterial genera are *Enterobacter*, *Azospirillum*, *Pseudomonas*, *Klebsiella*, and *Vibrio* in wetlands [153,159]. The heterotrophic nitrogen fixer usually makes mutual symbiosis with the roots and exchanges the sugars from the roots for ammonia that bacteria produce [152,160]. The nitrogen fixation process took place several times in the planted area of wetlands relative to the non-planted area, especially in the oxygen-deprived area of wetlands [153,161]. The same bacteria also influence nitrogen fixation and denitrification. Often, these processes take place concurrently near the roots of macrophytes [162]. The nitrogen-fixing bacteria dwell on the roots or in the rhizosphere of most of the aquatic macrophytes such as *P. australis*, *J. effusus*, *J. balticus*, *Sagittaria triflolia*, *Zostera marina* [163–165]. Roots also contribute to nitrogen fixation by reducing nitrogen from their rhizosphere, adjusting the pH level and redox potential [151]. Nitrogen-fixing microorganisms, such as *Azospirillum*, reside in the rhizosphere; these stimulate hormones, such as auxins, to influence the pH and redox potential and boost the nitrogen fixation process [161].

5.2. Degradation of Organic Pollutants

Microbes are known as bio-remediators due to their capability to break down virtually all classes of organic pollutants [166–168]. Microbes degrade the organic pollutants by a process of co-metabolism. In this process, microbes in the rhizospheric zone of aquatic and terrestrial plants degrade the complex carbon-based compounds in order to obtain organic carbon and electron acceptors [169]. In natural water, the biodegradation rate depends upon the microbial population and amount of xenobiotics [170], and the numbers of the microbes are heavily influenced by the macrophyte species [171]. Plants give organic carbon to microbes present in the rhizosphere that assist them to degrade complex organic compounds [172], such as hydrocarbons and aromatic hydrocarbons [173,174]. Bacteria also release indole acetic acid (IAA) to improve plant growth [175]. Many bacteria isolated from aquatic plants also showed pollutant degradation and plant growth-promoting activities [176,177]. The biofilms attached to aquatic plants are capable of degrading organics such as phenolics, amines, and aliphatic aldehydes [178]. Additionally, these biofilms are capable of degrading dissolved organic matter such as polychlorinated biphenyls (PCBs) and atrazine [54,179,180]. The aquatic plant rhizosphere is also enriched with methanotrophs containing a collection of Proteobacteria, which utilize methane for obtaining carbon and energy [181]. Methanotrophs can degrade numerous types of harmful organic complexes [182,183] such as chlorinated ethenes by enzymatic reactions. The *Eichhornia crassipes* can remediate eutrophic water by influencing the production of gaseous nitrogen [184,185].

5.3. Removal of Heavy Metals

The rhizospheric and endophytic bacteria have been reported to play a prominent part in the removal of heavy metals (Table 1). Bacteria promote the removal of metals by their ability to sorb the metallic ion into their cell walls [186]. Metal uptake by plants can be enhanced by bacteria, which increase the bioavailability of metals to plants [187,188]. The microorganisms can accumulate heavy metals with the help of specific metal-binding proteins and peptides such as metallothionein and phytochelatins [189]. The transcription factors of metal-binding proteins facilitate the hormone and redox signaling process upon exposure to toxic metals in the context of toxic metal exposure [190]. Cyanobacteria decrease the metal toxicity by the production of proteins that can bind metals [191]. The genetically modified *Ralstonia eutropha* can reduce the harmful Cd (II) by the production of metallothionein on the surface of the cell [192]. Likewise, *Escherichia coli* regulates the accumulated Cd toxicity by the production of many proteins and peptides [193]. The production of metallo-regulatory protein is a natural resistant method against arsenic (As) and mercury (Hg) in microorganisms [46].

The metal toxicity affects the performance of the phytoremediation process [194]. Microorganisms augment and facilitate plants to make heavy metals and antibiotic-resistant proteins [195]. The antibiotic-resistant proteins can reduce the abiotic and biotic stress induced by metals. Some of the *Bacillus* sp. strains have the ability to devise a mechanism to alleviate the metal stress by an active transport efflux pump [194]. The endophytic bacteria also influence the functional and phenotypic characteristics of the plants in which they reside [196]. Moreover, these bacteria influence the activity of plant antioxidant enzymes and lipid peroxidation, which support the plant resistance system, particularly resisting the oxidative stress in the plants caused by heavy metals [197,198]. Methylation can also be used by a few endophytic bacteria to induce the defense and detoxification of metals. Few gram-negative bacteria possess the specific mercury-resistant *(Mer)* operon gene for the degradation of organic mercurials and reductions in Hg^{+2} [199].

Table 1. Removal of heavy metals by bacteria.

Bacteria	Metal	Reference
Lactobacillus delbrueckii and *Streptococcus thermophillus*	Fe, Zn	[200]
Acinetobacter sp., *Bacillus megaterium* and *Sphingobacterium* sp.	Fe, Mn	[201]
Anoxybacillus flavithermus	Fe, Cu	[202]
Leptothrix, Pseudomonas, Hyphomicrobium and *Planctomyces*	Mn	[203]
Methylobacterium organophilum	Cu, Pb	[204]
Herminiimonas arsenicoxydans	As	[205]
Enterobacter cloaceae	Cd, Cu, Cr	[206]
Acetobacter	Pb, Cu, Mn, Zn, Co	[207]
Chryseomonas luteola	Cd, Co, Cu, Ni	[208]
Ochrobactrum anthropi	Cr, Cu	[209]
Anabaena spiroides	Mn	[210]
Ralstonia solanacearum	Pb	[211]
Proteobacteria and Bacteroidetes	Cu	[212]
Bacillus cereus	Cu	[213]
Bacillus licheniformis	Pb	[214]
Ralstonia solanacearum	Pb	[211]
Enterobacter aerogenes	Cd	[215]
S Pseudomonas azotoformans	Cd, Cu, Pb	[216]

5.4. Metal Biosorption and Bioaccumulation

Generally, bacteria perform metal ion biosorption into their cell wall by two processes, which are passive and active [217]. Passive biosorption takes place in the cell walls of living and dead/inactive bacterial cells, supported by multiple metabolism processes [218]. The reaction between the functional groups (e.g., amine, amide, carbonyl, hydroxyl, sulfonate, etc.) of the cell wall and metal ions causes the adsorption of metal ions to the cell surface [106]. In the metal ion binding process, different mechanisms (e.g., ion exchange, sorption, complexation, chelation and micro-precipitation) may be involved independently or synergistically [219].

On the other hand, in the active biosorption process, metal ions are up taken by living cells. The fate of metals that enter the inside of living cells depends upon the organisms and specific elements. The elements can be bound, stored, precipitated, and sequestered in some specific intracellular organelles and may be transported to a particular structure [106,220].

The endophytic bacteria exhibited outstanding heavy metal bioaccumulation and detoxification abilities [59,221]. The plant–bacteria symbiotic relation improves the phytoremediation potential of plants by the increased uptake of heavy metals due to the secretion of organic acid by bacteria. These organic acids secrete, by bacterial influence, the pH of the system and increase the bioavailability of the metal ions to plants [222]. For example, the application of endophytic bacteria, *Pseudomonas fluorescens* G10 and *Microbacterium* sp. G16, on *Brassica napus* increased the Pb accumulation in plant shoots [223]. *Saccharomyces cerevisiae*, commonly known as baker's yeast, is a successful bio-sorbent for the removal of Zn and Cd due to its ion exchange mechanism [224,225]. Similarly, *Cunninghamella elegans* has been proven an efficient sorbent for the remediation of textile effluent enriched with heavy metals [226].

Bacteria also produce biosurfactants and release them as root exudates. These biosurfactants enhance the bioavailability of metals in the soil and aquatic medium by their interaction and complexation with insoluble metals [227]. On the other hand, the extracellular polymeric substances, mainly composed of proteins, polysaccharides, nucleic acid, and lipids, perform a key part in the complexation of metals and reduce their bioavailability [125]. For example, *Azobacter* sp. formed complexes with chromium and cadmium by the formation of extracellular polymeric substances (EPS) and decreased the uptake of metals by *Triticum aestivum* [228]. The secretion of different metabolites such as siderophores and organic acids (including citric acids, oxalic acid, and acetic acid) influences heavy metals' bioavailability and their translocation in plants [229,230]. In an earlier study, the inoculation of the endophytic bacterium (*Pseudomonas* sp.) improved the plant's growth and increased the nickel (Ni) accumulation in the plant [220].

6. Role of Fungi

Fungi perform a potential role in the remediation of heavy metals by increasing their bioavailability and transformation into less toxic forms [231–233]. Some fungi, such as *Klebsiella oxytoca*, *Allescheriella* sp., *Stachybotrys* sp., *Phlebia* sp. *Pleurotus pulmonarius* and *Botryosphaeria rhodina*, have the capacity to bind metals [234]. Fungal species like *Aspergillus parasitica* and *Cephalosporium aphidicola* can remediate lead-contaminated soil by their biosorption process [235,236]. The fungi *Hymenoscyphus ericae*, *Neocosmospora vasinfecta* and *Verticillum terrestre* showed resistance to Hg and the ability to transform the toxic state of Hg (II) to a non-toxic form [237]. Fungi of the genera *Penicillium*, *Aspergillus*, and *Rhizopus*, have proven efficient in heavy metal removal from polluted water [238,239].

Fungi link closely with the roots in wetland plants and have a significant influence on wetland functioning [240,241]. Root exudates attract fungi toward the rhizosphere. The roots and fungi in wetland plants make multilevel physical, chemical, hormonal, and genetic interactions, which may be species specific [242,243]. The rhizospheric fungi community is different than soil communities. The types and interactions of the fungal community with the rhizosphere may be influenced by plant species, soil characteristics, climate, type of water, and other microorganisms [244]. The plant–fungi association in wetland plants performs different key functions such as the emission of metal-chelating siderophores, denitrification and metal detoxification [245,246]. Bacteria can easily stick to the surface of the substrate compared to algae due to their smaller size [247]. The other reason for the high ratio of attachment of epiphytic bacteria to aquatic plants compared to algae is the specific metabolites released from the plants [184,248].

7. Role of Inoculated Bacteria

It is well established that plant–bacteria synergism is essential to enhance the phytoremediation potential of plants and ultimately FTWs (Table 2) [49,249,250]. The inoculation of FTWs by immobilized denitrifiers greatly improved the nitrogen removal from wastewater [61]. Endophytes can be isolated from and within various plant tissues that include roots, stems, leaves, flower, fruit, and seed [112]. The root is the main source of endophytes, and legume root nodules have a large diversity of endophytes [251]. Some plants have an underground stem, so, in these plants, stem and root endophytes may be similar [252]. Bacterial endophytes that were obtained from the shoot of sugarcane promoted fixation as well as acetylene reduction activities [253]. The inoculation method affects bacterial colonization, and inoculation should be performed appropriately [254]. Nonetheless, no standard method is defined for the inoculation of plant roots in FTWs. The two common methods of inoculation are the inoculation of seeds and the inoculation of soil [252,255,256]. In seed inoculation, the inoculum is introduced into host plants directly when they are in the seed or seedling stage. The soil inoculation is done directly in root media or the pot in which the plant is growing. In FTWs, the roots of the plant are inoculated directly by pouring the inoculum in the water near the root of the plant. For example, Shahid et al. (2019a) prepared the inoculum of five different rhizospheric and endophytic bacterial strains and inoculated the roots of plants by directly adding a specific amount of inoculum into the water [20]. Previously, many attempts have been performed to create an effective partnership between plant and metal-resistant bacteria in order to effectively treat water contaminated with heavy metals [250,257,258]. FTWs vegetated with *Brachia mutica* and inoculated with bacteria were used to treat sewage effluent and it was found that the concentration of heavy metals, including Cd, Fe, Cu, Cr, Mn, Co and Pb, decreased significantly from the effluent. The removal of iron was significant (79 to 85%) [259]. Similarly, in another study, a consortium of hydrocarbon-degrading bacteria was added into the hydrocarbon-enriched water for its remediation by FTWs [260]. The inoculation of these rhizospheric and endophytic bacteria was reported to enhance the degradation of hydrocarbons, and also improved the efficiency of the FTWs.

Table 2. Application of bacteria to enhance phytoremediation potential of floating treatment wetlands.

Bacteria/Bacterial Biofilm	Nature of Bacteria	Plant	Plant–Bacteria Interaction	Summary	Reference
Bacterial Biofilm	—	*Ipomoea aquatic* and *Corbicula fluminea*	—	The removal efficiencies of TN, NH_4^+-N, TP, total organic carbon (TOC), Chl-*a*, total microcystin-LR and extracellular microcystin-LR were 52.7%, 33.7%, 54.5%, 49.2%, 80.2%, 77.4% and 68.0%, respectively.	[261]
Proteobacteria	Nitrosomonadaceae	*Canna Indica* and *Iris pseudacorus*	Bacteria were mainly attached on the fiber filling of floating mat and plant roots	The average removal efficiencies of chemical oxygen demand (COD), TN, NH_3-N and TP for *Canna indica* set-up were 23.1%, 15.3%, 18.1% and 19.4% higher, respectively, than that of the setup with only substrate, and 14.2%, 12.8%, 7.9% and 11.9% higher than *Iris pseudacorus*. FTWs.	[262]
Nitrifying and Denitrifying	Carrying *nirS*, *nirK* and *amoA* genes	Unplanted	Specific microbial communities were visualized with denaturing gradient gel electrophores (DGGE)	COD was efficiently removed in all systems examined (>90% removal). Ammonia was efficiently removed by nitrification. Removal of total dissolved nitrogen was ~50% by day 28	[22]
Biofilms	—	*Carex virgate*, *Cyperus ustulatus*, *Juncus edgariae*, and *Schoenoplectus tabemaemontani*	Biofilm performed a key role in the removal of Cu, P and FSS. Plant roots and biofilm interaction enhanced metal speciation	The presence of a planted floating mat with biofilms improved removal of copper (>six-fold), fine suspended particles (~threefold reduction in turbidity) and dissolved reactive P compared to the control.	[11]
Ammonifying bacterial strains	Engineering bacterial strain	*Cymbidium faberi*	The ammonifying bacteria adhered to plants roots enhanced oxygen supply to microorganism involved in nitrification process and increased capacity of plants roots to absorb ammonia nitrogen.	The organic nitrogen decomposition rate was up to 86.50% by adding the strain agent while it was 75.66% without them in the control test group in FTWs	[263]
Adsorptive biofilm	Natural	*Thalia dealbata*	Combined action of plant and biofilms	The average removal rates for TN, NH_4^+-N, NO_3^--N NO_2^--N, TP and chlorophyll-a in summer–autumn season were 36.9%, 44.8%, 25.6%, 53.2%, 43.3% and 64.5%, respectively, effectively reduced the concentrations of total suspended solids (TSS), *Escherichia coli* and heavy metals.	[55]
Photosynthetic bacteria	—	*Vetiveria zizanioids*	Combined action of plant and inoculated bacteria improved purifying effect of FTWs	Efficiently removed TN and TP	[264]
Biofilm Reactor	Protozoa and Metazoa	*Bambusoideae*	In the batch reactor, COD was mainly removed by the biofilm on the filamentous bamboo	The removal rate of the COD, NH_4^+–N, turbidity, and total bacteria were 11.2–74.3%, 2.2–56.1%, 20–100%	[265]

Table 2. *Cont.*

Bacteria/Bacterial Biofilm	Nature of Bacteria	Plant	Plant–Bacteria Interaction	Summary	Reference
Acinetobacter sp.	Perchlorate reducing bacterium	*Pistia stratiotes*	Phyto-accumulation and rhizo-degradation were key mechanisms involved in perchlorate removal	*Pistia* showed $63.8 \pm 4\%$ (w/v) removal of 5 mg/L level perchlorate in 7 days	[266]
Denitrifying polyphosphate accumulating microorganisms	—	*Festuca arundinacea*	Improved the growth of plant and biomass	The average removal rates were 86.32%, 93.60%, 90.12%, 72.09%, and 84.29%, respectively, for $NH4^+$-N, NO_3^--N, TN, TP, and ortho-P.	[267]
Acinetobacter, Bacillus cereus and *Bacillus licheniformis*	Endophytic bacteria	*Brachiaria mutica*	The inoculated bacteria showed persistence in water as well as successfully colonized the root and shoots of the plants	Maximum reduction in COD, biological oxygen demand (BOD_5), TN, and PO_4 was achieved by the combined use of plants and bacteria.	[259]
Biofilms	Natural	*Juncus effuses* *Carex riparia*	Metals were found in the root biofilm, probably due to microbial respiration activity	Analysis showed Ni concentration in leaves were between 23 and 31 μg/g dry matter, and between 113 and 131 μg/g in roots. Accumulation of Zn was 45-80 μg/g in leaves and 168–210 μg/g in roots.	[14]
Klebsiella sp., *Pseudomonas* sp. and *Acinetobacter* sp.	Endophytic Bacteria	*Typha domingensis*	Possessed pollutant-degrading and plant growth-promoting abilities and successful survival of bacteria was found in plant tissues	The average reduction in COD and BOD_5 was 87% and 87.5%, and significantly removed heavy metals.	[26]
Biofilm	Nitrifying and denitrifying bacteria	*Canna indica*	Improved nitrification and denitrification process and overall high removal of total nitrogen	Significantly higher removal rates of ammonia nitrogen (85.2%), total phosphorus (82.7%), and orthophosphate (82.5%) were observed	[18]
The community was mainly composed of Cyanobacteria, Proteobacteria, Bacteroidetes, Planctomycetes, Firmicutes, Actinobacteria, Chlorobi and Acidobacteria.	Periphyton	—	Improved its nutrient removal capacity	Successfully maintained TN and TP concentration in the river water at less than 2.0 and 0.02 mg L^{-1} respectively	[268]
Dechloromonas, Thiobacillus and *Nitrospira*	Heterotrophic and autotrophic	—	Mixotrophic denitrification occurred in auto and heterotrophic bacteria	About 89.4% of the TN was removed from autotrophic coupled floating wetlands, and 88.5% from heterotrophic enhanced floating wetlands	[39]
Bacillus subtilis, Klebsiella sp., *Acinetobacter Junii* and *Acinetobacter* sp.	Hydrocarbon degrading bacteria	*Brachiara mutica* and *Phragmites australis*	Alkane-degrading gene (*alkB*) abundance confirmed microbial growth in plant's root and shoot and in water.	Reduced oil content (97%), COD (93%), and BOD (97%), in wastewater	[260]
Acinetobacter lwofii, Bacillus cereus, and *Pseudomonas* sp.	Phenol-degrading bacteria	*Typha domingensis*	The inoculated bacteria showed successful colonization and survival in the rhizosphere, root interior and shoot interior of the plant and enhanced plant growth and biomass	Bacterial augmentation enhanced the removal potential significantly, i.e., 0.146 g/m^2/day vs. 0.166 g/m^2/day without bacterial inoculation	[269]

Table 2. *Cont.*

Bacteria/Bacterial Biofilm	Nature of Bacteria	Plant	Plant–Bacteria Interaction	Summary	Reference
Acinetobacter lwofii, Bacillus cereus, and *Pseudomonas* sp.	Phenol degrading bacteria	*Phragmite australis*	Improved plant biomass and high rate of inoculated bacteria survival observed in plant roots, shoot and water	Plant–bacteria synergism significantly improved the phenol degradation and removal. Highest reduction in COD, BOD, and TOC was achieved by bacterial augmentation	[270]
Acinetobacter, Acinetobacter sp., and *Bacillus niabensis*	Hydrocarbons degrading bacteria	*Leptochloa fusca*	Achieved successful degradation of Hexadecane The Inoculated bacteria displayed highest persistence in the roots followed by shoots and then in the wastewater and improved plant growth promoting (PGP) activities	Hydrocarbons degradation was recorded up to 92%, COD was reduced up to 95%, BOD up to 84%, and TDS up to 47% and alleviated the toxicity	[41]
Archaea, anaerobic ammonium oxidation (Anammox) bacteria	Natural	*Oenanthe javanica*	High abundance and diversity of bacteria in planted floating wetland	The average removal rates of NH_4^+-N, NO_3^--N and total nitrogen were 78.3, 44.4 and 49.7% respectively	[44]
Proteobacteria Actinobacteria Cyanobacteria, and *Rhizorhapis*	—	*Eichhornia crassipes*	Bacteria were involved in pollutant degradation and nutrients removal	Suspended solids, TN, TP, NO_3^--N and COD was 86%, 75%, 80%, 95% and 84%, respectively.	[271]
Bacillus subtilis, Klebsiella sp., *Acinetobacter Junii,* and *Acinetobacter* sp.	Hydrocarbon degrading bacteria	*Typha domingensis* and *Leptochloa fusca*	Persistence of bacteria and expression of the alkB gene in the rhizoplane of inoculated plants	Reduction in hydrocarbon (95%), COD (90%), and BOD content (93%)	[272]
Acinetobacter junii, Pseudomonas indoloxydans, and *Rhodococcus* sp.	Rhizospheric and endophytes	*Phragmites australis* and *Typha domingensis*	Removal efficiency was further enhanced by augmentation with bacteria and promoted plant growth	Color, COD and BOD after an 8-day period were 97, 87 and 92%, respectively, 87–99% reduction in heavy metals	[273]
Consortium of five strains namely *Aeromonas salmonicida, Bacillus cerus, Pseudomonas indoloxydans, Pseudomonas gessardii, and Rhodococcus* sp.	Rhizospheric and endophytes	*Phragmites australis* and *Brachia mutica*	Persistence and survival of inoculated bacteria in roots and shoots, and inoculated bacteria improved the plant growth and biomass production	Reduced COD, BOD_5, and TOC up to 85.9%, 83.3%, and 86.6% in 96 h, respectively. TN was reduced from 37.5 to 2.07 mg l^{-1}, N from 33.3 to 1.23 mg l^{-1}, and TP from 2.63 to 0.53 mg l^{-1}. Trace metals were also reduced up to 79.5% for iron, 91.4% for nickel, 91.8% for manganese, 36.14% for lead, and 85.19% for chromium.	[20]
Acinetobacter juniistrain, Rhodococcus sp. strain, and *Pseudomonas indoloxydans*	Dye degrading bacteria	*Phragmites australis*	The inoculated bacteria showed persistence in water, roots and shoots of inoculated plants of FTWs	The COD was reduced to 92%, BOD to 91%, color to 86%, and trace metals to approximately 87% in the treated wastewater.	[274]
Bacillus cerus, Cyperus laevigatus, Aeromonas salmonicida and *Pseudomonas gessardii,*	Rhizospheric and endophytes	*Typha domingensis* and *Leptochloa fusca*	Improved remediation performance of inoculated plants, inoculated bacteria were found in root and shoots of inoculated plants	The TN, NO_3^{-1} and TP contents decreased to 1.77 mg l^{-1}, 0.80 mg l^{-1} and 0.60 mg l^{-1}, respectively. Additionally, the concentration of iron, nickel, manganese, lead, and chromium in the water lowered to 0.41, 0.16, 0.10, 0.25, and 0.08 mg l^{-1},	[131]
These strains were *Ochrobactrum intermedium, Microbacterium oryzae, Pseudomonas, Acinetobacter* sp., *Klebsiella* sp., *Acinetobacter* sp., *P. aeruginosa, Bacillus subtilus,* and *Acinetobacter junii*	Bacteria possessing capabilities of hydrocarbon degradation, rhamnolipid production, and plant growth promotion.	*Phragmites australis, Typha domingensis, Leptochloa fusca,* and *Brachiaria mutica*	Produced biosurfactants and promoted plant growth. Bacteria showed persistent in the rhizoplane, roots and shoots of plants	Reduced COD, BOD, TDS, hydrocarbon content, and heavy metals by 97.4%, 98.9%, 82.4%, 99.1%, and 80%, respectively, within 18 months.	[25]

" __ " no data.

8. Conclusions

Microbes, bacteria and algae are the major components of epiphytic microbes, which colonize the lower surface of floating plants. Bacterial biofilm has a crucial role in the removal of organics, inorganics and metals in FTW systems. The plant species and pollutant concentration in wastewater influence the nature and diversity of bacteria. Furthermore, the availability of nutrients influences the metabolism of bacteria and the pollutant removal efficiency. The rhizosphere and endophytes both have a prominent role in the pollutant removal process. The rhizospheric bacteria mostly remove the pollutants near the root system, whereas the endophytes mostly remove the pollutants inside the roots and shoots. The rhizospheric and endophytic bacterial community also enhances the pollutant removal process by alleviating the pollutant stress, increasing tolerance towards environmental changes, and regulating plant growth by direct and indirect mechanisms. The inoculation of plant roots with specific strains of bacteria also boosts the pollutant removal process.

It is clear from this information that plant–microbe interaction is vital for the pollutant removal process in FTWs. There is a need to conduct further research to gain a better understanding of specific microbe and plant interactions and their beneficial role in the pollutant removal process in the aquatic ecosystem. Environmental factors such as temperature, pH, and the availability of nutrients have a profound effect on the pollutant removal abilities of microorganisms. These factors need further investigation to achieve the optimal performance of microorganisms in FTWs. The nature of pollutants affects the persistence and survival of bacteria and may determine the type of bacterial communities in a wetland system. Bacteria specific to the removal of particular types of pollutants need to be identified and isolated for their future application in FTWs. Bacteria that are easy to culture in the lab with minimal prerequisites, which possess the potential to treat a diverse range of pollutants and can be augmented with diverse macrophytes in FTWs, need to be widely explored for their use in FTWs.

Author Contributions: The paper was written by M.J.S., S.A., N.N. and M.A. The data were collected and coordinated by A.A.A., F.K. and M.H.S. The paper was reviewed and revised by M.R., B.A., and M.A. All authors have read and agreed to the published version of the manuscript.

Funding: The authors are grateful to the Higher Education Commission (HEC) Islamabad, Pakistan, for its support.

Acknowledgments: The authors are grateful to the Higher Education Commission (HEC) Islamabad, Pakistan, for its support.

Conflicts of Interest: The authors declare no conflict of interest.

References

1. Vymazal, J. Constructed wetlands for treatment of industrial wastewaters: A review. *Ecol. Eng.* **2014**, *73*, 724–751. [CrossRef]
2. Stefanakis, A.I. The role of constructed wetlands as green infrastructure for sustainable urban water management. *Sustainability* **2019**, *11*, 6981. [CrossRef]
3. Calheiros, C.S.; Castro, P.M.; Gavina, A.; Pereira, R. Toxicity abatement of wastewaters from tourism units by constructed wetlands. *Water* **2019**, *11*, 2623. [CrossRef]
4. Riva, V.; Mapelli, F.; Syranidou, E.; Crotti, E.; Choukrallah, R.; Kalogerakis, N.; Borin, S. Root bacteria recruited by *Phragmites australis* in constructed wetlands have the potential to enhance azo-dye phytodepuration. *Microorganisms* **2019**, *7*, 384. [CrossRef]
5. Donoso, N.; van Oirschot, D.; Kumar Biswas, J.; Michels, E.; Meers, E. Impact of aeration on the removal of organic matter and nitrogen compounds in constructed wetlands treating the liquid fraction of piggery manure. *Appl. Sci.* **2019**, *9*, 4310. [CrossRef]
6. Kujala, K.; Karlsson, T.; Nieminen, S.; Ronkanen, A.-K. Design parameters for nitrogen removal by constructed wetlands treating mine waters and municipal wastewater under Nordic conditions. *Sci. Total Environ.* **2019**, *662*, 559–570. [CrossRef]

7. Anawar, H.M.; Ahmed, G.; Strezov, V. Long-term Performance and Feasibility of Using Constructed Wetlands for Treatment of Emerging and Nanomaterial Contaminants in Municipal and Industrial Wastewater. In *Emerging and Nanomaterial Contaminants in Wastewater*; Elsevier: Amsterdam, The Netherlands, 2019; pp. 63–81.

8. Smith, E.L.; Kellman, L.; Brenton, P. Restoration of on-farm constructed wetland systems used to treat agricultural wastewater. *J. Agric. Sci.* **2019**, *11*, 1–12. [CrossRef]

9. Jones, T.G.; Willis, N.; Gough, R.; Freeman, C. An experimental use of floating treatment wetlands (FTWs) to reduce phytoplankton growth in freshwaters. *Ecol. Eng.* **2017**, *99*, 316–323. [CrossRef]

10. Headley, T.; Tanner, C. Floating Treatment Wetlands: An Innovative Option for Stormwater Quality Applications. In Proceedings of the 11th International Conference on Wetland Systems for Water Pollution Control, Indore, India, 1–7 November 2008.

11. Tanner, C.C.; Headley, T.R. Components of floating emergent macrophyte treatment wetlands influencing removal of stormwater pollutants. *Ecol. Eng.* **2011**, *37*, 474–486. [CrossRef]

12. Hubbard, R.; Anderson, W.; Newton, G.; Ruter, J.; Wilson, J. Plant growth and elemental uptake by floating vegetation on a single-stage swine wastewater lagoon. *Trans. ASABE* **2011**, *54*, 837–845. [CrossRef]

13. Shahid, M.J.; Arslan, M.; Ali, S.; Siddique, M.; Afzal, M. Floating wetlands: A sustainable tool for wastewater treatment. *CLEAN–Soil Air Water* **2018**, *46*, 1800120. [CrossRef]

14. Ladislas, S.; Gerente, C.; Chazarenc, F.; Brisson, J.; Andres, Y. Floating treatment wetlands for heavy metal removal in highway stormwater ponds. *Ecol. Eng.* **2015**, *80*, 85–91. [CrossRef]

15. Chen, Z.; Cuervo, D.P.; Müller, J.A.; Wiessner, A.; Köser, H.; Vymazal, J.; Kästner, M.; Kuschk, P. Hydroponic root mats for wastewater treatment—a review. *Environ. Sci. Pollut. Res.* **2016**, *23*, 15911–15928. [CrossRef] [PubMed]

16. Kumari, M.; Tripathi, B. Effect of aeration and mixed culture of *Eichhornia crassipes* and *Salvinia natans* on removal of wastewater pollutants. *Ecol. Eng.* **2014**, *62*, 48–53. [CrossRef]

17. Li, W.; Li, Z. In situ nutrient removal from aquaculture wastewater by aquatic vegetable *Ipomoea aquatica* on floating beds. *Water Sci. Technol.* **2009**, *59*, 1937–1943. [CrossRef]

18. Zhang, L.; Zhao, J.; Cui, N.; Dai, Y.; Kong, L.; Wu, J.; Cheng, S. Enhancing the water purification efficiency of a floating treatment wetland using a biofilm carrier. *Environ. Sci. Pollut. Res.* **2016**, *23*, 7437–7443. [CrossRef]

19. Pavlineri, N.; Skoulikidis, N.T.; Tsihrintzis, V.A. Constructed floating wetlands: A review of research, design, operation and management aspects, and data meta-analysis. *Chem. Eng. J.* **2017**, *308*, 1120–1132. [CrossRef]

20. Shahid, M.J.; Arslan, M.; Siddique, M.; Ali, S.; Tahseen, R.; Afzal, M. Potentialities of floating wetlands for the treatment of polluted water of river Ravi, Pakistan. *Ecol. Eng.* **2019**, *133*, 167–176. [CrossRef]

21. Borne, K.E.; Fassman, E.A.; Tanner, C.C. Floating treatment wetland retrofit to improve stormwater pond performance for suspended solids, copper and zinc. *Ecol. Eng.* **2013**, *54*, 173–182. [CrossRef]

22. Faulwetter, J.; Burr, M.D.; Cunningham, A.B.; Stewart, F.M.; Camper, A.K.; Stein, O.R. Floating treatment wetlands for domestic wastewater treatment. *Water Sci. Technol.* **2011**, *64*, 2089–2095. [CrossRef]

23. Chen, Z.; Kuschk, P.; Paschke, H.; Kästner, M.; Müller, J.A.; Köser, H. Treatment of a sulfate-rich groundwater contaminated with perchloroethene in a hydroponic plant root mat filter and a horizontal subsurface flow constructed wetland at pilot-scale. *Chemosphere* **2014**, *117*, 178–184. [CrossRef]

24. Borne, K.E.; Fassman-Beck, E.A.; Tanner, C.C. Floating treatment wetland influences on the fate of metals in road runoff retention ponds. *Water Res.* **2014**, *48*, 430–442. [CrossRef]

25. Afzal, M.; Rehman, K.; Shabir, G.; Tahseen, R.; Ijaz, A.; Hashmat, A.J.; Brix, H. Large-scale remediation of oil-contaminated water using floating treatment wetlands. *npj Clean Water* **2019**, *2*, 3. [CrossRef]

26. Ijaz, A.; Iqbal, Z.; Afzal, M. Remediation of sewage and industrial effluent using bacterially assisted floating treatment wetlands vegetated with *Typha domingensis*. *Water Sci. Technol.* **2016**, *74*, 2192–2201. [CrossRef] [PubMed]

27. Li, X.; Guo, R. Comparison of nitrogen removal in floating treatment wetlands constructed with *Phragmites australis* and acorus calamus in a cold temperate zone. *Water Air Soil Pollut.* **2017**, *228*, 132. [CrossRef]

28. Nawaz, N.; Ali, S.; Shabir, G.; Rizwan, M.; Shakoor, M.B.; Shahid, M.J.; Afzal, M.; Arslan, M.; Hashem, A.; Abd_Allah, E.F. Bacterial augmented floating treatment wetlands for efficient treatment of synthetic textile dye wastewater. *Sustainability* **2020**, *12*, 3731. [CrossRef]

29. Fahid, M.; Ali, S.; Shabir, G.; Rashid Ahmad, S.; Yasmeen, T.; Afzal, M.; Arslan, M.; Hussain, A.; Hashem, A.; Abd Allah, E.F. *Cyperus laevigatus* L. enhances diesel oil remediation in synergism with bacterial inoculation in floating treatment wetlands. *Sustainability* **2020**, *12*, 2353. [CrossRef]

30. Zhang, C.-B.; Liu, W.-L.; Pan, X.-C.; Guan, M.; Liu, S.-Y.; Ge, Y.; Chang, J. Comparison of effects of plant and biofilm bacterial community parameters on removal performances of pollutants in floating island systems. *Ecol. Eng.* **2014**, *73*, 58–63. [CrossRef]

31. Wang, C.-Y.; Sample, D.J.; Bell, C. Vegetation effects on floating treatment wetland nutrient removal and harvesting strategies in urban stormwater ponds. *Sci. Total Environ.* **2014**, *499*, 384–393. [CrossRef]

32. Afzal, M.; Shabir, G.; Tahseen, R.; Islam, E.U.; Iqbal, S.; Khan, Q.M.; Khalid, Z.M. Endophytic Burkholderia sp. strain PsJN improves plant growth and phytoremediation of soil irrigated with textile effluent. *CLEAN–Soil Air Water* **2014**, *42*, 1304–1310. [CrossRef]

33. Arslan, M.; Imran, A.; Khan, Q.M.; Afzal, M. Plant–bacteria partnerships for the remediation of persistent organic pollutants. *Environ. Sci. Pollut. Res.* **2017**, *24*, 4322–4336. [CrossRef] [PubMed]

34. Chang, N.-B.; Islam, K.; Marimon, Z.; Wanielista, M.P. Assessing biological and chemical signatures related to nutrient removal by floating islands in stormwater mesocosms. *Chemosphere* **2012**, *88*, 736–743. [CrossRef]

35. Khan, S.; Afzal, M.; Iqbal, S.; Khan, Q.M. Plant–bacteria partnerships for the remediation of hydrocarbon contaminated soils. *Chemosphere* **2013**, *90*, 1317–1332. [CrossRef]

36. Hussain, I.; Aleti, G.; Naidu, R.; Puschenreiter, M.; Mahmood, Q.; Rahman, M.M.; Wang, F.; Shaheen, S.; Syed, J.H.; Reichenauer, T.G. Microbe and plant assisted-remediation of organic xenobiotics and its enhancement by genetically modified organisms and recombinant technology: A review. *Sci. Total Environ.* **2018**, *628*, 1582–1599. [CrossRef] [PubMed]

37. He, D.; Ren, L.; Wu, Q. Epiphytic bacterial communities on two common submerged macrophytes in Taihu Lake: Diversity and host-specificity. *Chin. J. Oceanol. Limnol.* **2012**, *30*, 237–247. [CrossRef]

38. Headley, T.; Tanner, C. Constructed wetlands with floating emergent macrophytes: An innovative stormwater treatment technology. *Crit. Rev. Environ. Sci. Technol.* **2012**, *42*, 2261–2310. [CrossRef]

39. Gao, L.; Zhou, W.; Huang, J.; He, S.; Yan, Y.; Zhu, W.; Wu, S.; Zhang, X. Nitrogen removal by the enhanced floating treatment wetlands from the secondary effluent. *Bioresour. Technol.* **2017**, *234*, 243–252. [CrossRef]

40. Jun-Xing, Y.; Yong, L.; Zhi-Hong, Y. Root-induced changes of pH, Eh, Fe (II) and fractions of Pb and Zn in rhizosphere soils of four wetland plants with different radial oxygen losses. *Pedosphere* **2012**, *22*, 518–527.

41. Hussain, F.; Tahseen, R.; Arslan, M.; Iqbal, S.; Afzal, M. Removal of hexadecane by hydroponic root mats in partnership with alkane-degrading bacteria: Bacterial augmentation enhances system's performance. *Int. J. Environ. Sci. Technol.* **2019**, *16*, 4611–4620. [CrossRef]

42. Dutta, D.; Puzari, K.C.; Gogoi, R.; Dutta, P. Endophytes: Exploitation as a tool in plant protection. *Braz. Arch. Biol Technol.* **2014**, *57*, 621–629. [CrossRef]

43. White, S.A.; Cousins, M.M. Floating treatment wetland aided remediation of nitrogen and phosphorus from simulated stormwater runoff. *Ecol. Eng.* **2013**, *61*, 207–215. [CrossRef]

44. Wang, P.; Jeelani, N.; Zuo, J.; Zhang, H.; Zhao, D.; Zhu, Z.; Leng, X.; An, S. Nitrogen removal during the cold season by constructed floating wetlands planted with *Oenanthe javanica*. *Mar. Freshw. Res.* **2018**, *69*, 635–647. [CrossRef]

45. Govarthanan, M.; Mythili, R.; Selvankumar, T.; Kamala-Kannan, S.; Rajasekar, A.; Chang, Y.-C. Bioremediation of heavy metals using an endophytic bacterium Paenibacillus sp. RM isolated from the roots of *Tridax procumbens*. *3 Biotech* **2016**, *6*, 242. [CrossRef] [PubMed]

46. Singh, S.; Kang, S.H.; Mulchandani, A.; Chen, W. Bioremediation: Environmental clean-up through pathway engineering. *Curr. Opin. Biotechnol.* **2008**, *19*, 437–444. [CrossRef]

47. Borne, K.E. Floating treatment wetland influences on the fate and removal performance of phosphorus in stormwater retention ponds. *Ecol. Eng.* **2014**, *69*, 76–82. [CrossRef]

48. Urakawa, H.; Dettmar, D.L.; Thomas, S. The uniqueness and biogeochemical cycling of plant root microbial communities in a floating treatment wetland. *Ecol. Eng.* **2017**, *108*, 573–580. [CrossRef]

49. Ashraf, S.; Afzal, M.; Naveed, M.; Shahid, M.; Ahmad Zahir, Z. Endophytic bacteria enhance remediation of tannery effluent in constructed wetlands vegetated with *Leptochloa fusca*. *Int. J. Phytorem.* **2018**, *20*, 121–128. [CrossRef]

50. Ahsan, M.T.; Najam-ul-Haq, M.; Idrees, M.; Ullah, I.; Afzal, M. Bacterial endophytes enhance phytostabilization in soils contaminated with uranium and lead. *Int. J. Phytorem.* **2017**, *19*, 937–946. [CrossRef]

51. Afzal, M.; Yousaf, S.; Reichenauer, T.G.; Sessitsch, A. Ecology of alkane-degrading bacteria and their interaction with the plant. *Mol. Microb. Ecol. Rhizosphere* **2013**, *2*, 975–989.

52. Billore, S.; Sharma, J. Treatment performance of artificial floating reed beds in an experimental mesocosm to improve the water quality of river Kshipra. *Water Sci. Technol.* **2009**, *60*, 2851–2859. [CrossRef]

53. Van de Moortel, A.M.; Meers, E.; De Pauw, N.; Tack, F.M. Effects of vegetation, season and temperature on the removal of pollutants in experimental floating treatment wetlands. *Water Air Soil Pollut.* **2010**, *212*, 281–297. [CrossRef]

54. Tranvik, L.J. Degradation of Dissolved Organic Matter in Humic Waters by Bacteria. In *Aquatic Humic Substances*; Springer: Berlin/Heidelberg, Germany, 1998; pp. 259–283.

55. Zhao, F.; Xi, S.; Yang, X.; Yang, W.; Li, J.; Gu, B.; He, Z. Purifying eutrophic river waters with integrated floating island systems. *Ecol. Eng.* **2012**, *40*, 53–60. [CrossRef]

56. Cao, W.; Zhang, Y. Removal of nitrogen (N) from hypereutrophic waters by ecological floating beds (EFBs) with various substrates. *Ecol. Eng.* **2014**, *62*, 148–152. [CrossRef]

57. Curt, M.; Aguado, P.; Fernandez, J. Nitrogen Absorption by Sparganium Erectum L. and Typha Domingensis (Pers.) Steudel Grown as Floaters. In Proceedings of the International Meeting on Phytodepuration, Lorca, Spain, 20–22 July 2005.

58. Hu, G.-J.; Zhou, M.; Hou, H.-B.; Zhu, X.; Zhang, W.-H. An ecological floating-bed made from dredged lake sludge for purification of eutrophic water. *Ecol. Eng.* **2010**, *36*, 1448–1458. [CrossRef]

59. Shin, C.J.; Nam, J.M.; Kim, J.G. Floating mat as a habitat of *Cicuta virosa*, a vulnerable hydrophyte. *Landsc. Ecol. Eng.* **2015**, *11*, 111–117. [CrossRef]

60. Seo, E.-Y.; Kwon, O.-B.; Choi, S.-I.; Kim, J.-H.; Ahn, T.-S. Installation of an artificial vegetating island in oligomesotrophic Lake Paro, Korea. *Sci. World J.* **2013**, *2013*, 857670. [CrossRef]

61. Sun, L.; Liu, Y.; Jin, H. Nitrogen removal from polluted river by enhanced floating bed grown canna. *Ecol. Eng.* **2009**, *35*, 135–140. [CrossRef]

62. Boonsong, K.; Chansiri, M. Domestic wastewater treatment using vetiver grass cultivated with floating platform technique. *AU J. Technol.* **2008**, *12*, 73–80.

63. Li, M.; Wu, Y.-J.; Yu, Z.-L.; Sheng, G.-P.; Yu, H.-Q. Nitrogen removal from eutrophic water by floating-bed-grown water spinach (*Ipomoea aquatica* Forsk.) with ion implantation. *Water Res.* **2007**, *41*, 3152–3158. [CrossRef]

64. Wu, H.; Zhang, J.; Li, P.; Zhang, J.; Xie, H.; Zhang, B. Nutrient removal in constructed microcosm wetlands for treating polluted river water in northern China. *Ecol. Eng.* **2011**, *37*, 560–568. [CrossRef]

65. Zhao, F.; Yang, W.; Zeng, Z.; Li, H.; Yang, X.; He, Z.; Gu, B.; Rafiq, M.T.; Peng, H. Nutrient removal efficiency and biomass production of different bioenergy plants in hypereutrophic water. *Biomass Bioenergy* **2012**, *42*, 212–218. [CrossRef]

66. Wu, Q.-T.; Gao, T.; Zeng, S.; Chua, H. Plant-biofilm oxidation ditch for in situ treatment of polluted waters. *Ecol. Eng.* **2006**, *28*, 124–130. [CrossRef]

67. Wang, C.-Y.; Sample, D.J. Assessment of the nutrient removal effectiveness of floating treatment wetlands applied to urban retention ponds. *J. Environ. Manag.* **2014**, *137*, 23–35. [CrossRef]

68. Kadlec, R.H.; Wallace, S. *Treatment Wetlands*; CRC press: Boca Raton, FL, USA, 2008.

69. Li, H.; Zhao, H.-P.; Hao, H.-L.; Liang, J.; Zhao, F.-L.; Xiang, L.-C.; Yang, X.-E.; He, Z.-L.; Stoffella, P.J. Enhancement of nutrient removal from eutrophic water by a plant–microorganisms combined system. *Environ. Eng. Sci.* **2011**, *28*, 543–554. [CrossRef]

70. Lamers, L.P.; Govers, L.L.; Janssen, I.C.; Geurts, J.J.; Van der Welle, M.E.; Van Katwijk, M.M.; Van der Heide, T.; Roelofs, J.G.; Smolders, A.J. Sulfide as a soil phytotoxin—A review. *Front. Plant Sci.* **2013**, *4*, 268. [CrossRef] [PubMed]

71. Pavan, F.; Breschigliaro, S.; Borin, M. Screening of 18 species for digestate phytodepuration. *Environ. Sci. Pollut. Res.* **2015**, *22*, 2455–2466. [CrossRef]

72. Battin, T.J.; Kaplan, L.A.; Newbold, J.D.; Cheng, X.; Hansen, C. Effects of current velocity on the nascent architecture of stream microbial biofilms. *Appl. Environ. Microbiol.* **2003**, *69*, 5443–5452. [CrossRef]

73. Thorén, A.-K. Urea transformation of wetland microbial communities. *Microb. Ecol.* **2007**, *53*, 221–232. [CrossRef]

74. Pang, S.; Zhang, S.; Lv, X.; Han, B.; Liu, K.; Qiu, C.; Wang, C.; Wang, P.; Toland, H.; He, Z. Characterization of bacterial community in biofilm and sediments of wetlands dominated by aquatic macrophytes. *Ecol. Eng.* **2016**, *97*, 242–250. [CrossRef]

75. Branda, S.S.; Vik, Å.; Friedman, L.; Kolter, R. Biofilms: The matrix revisited. *Trends Microbiol.* **2005**, *13*, 20–26. [CrossRef]

76. Latasa, C.; Solano, C.; Penadés, J.R.; Lasa, I. Biofilm-associated proteins. *C. R. Biol.* **2006**, *329*, 849–857. [CrossRef]

77. Lasa, I.; Penadés, J.R. Bap: A family of surface proteins involved in biofilm formation. *Res. Microbiol.* **2006**, *157*, 99–107. [CrossRef] [PubMed]

78. López, D.; Vlamakis, H.; Kolter, R. Biofilms. *Cold Spring Harbor Perspect. Biol.* **2010**, *2*, a000398. [CrossRef] [PubMed]

79. Rice, K.C.; Mann, E.E.; Endres, J.L.; Weiss, E.C.; Cassat, J.E.; Smeltzer, M.S.; Bayles, K.W. The cidA murein hydrolase regulator contributes to DNA release and biofilm development in Staphylococcus aureus. *Proc. Natl. Acad. Sci. USA* **2007**, *104*, 8113–8118. [CrossRef] [PubMed]

80. Branda, S.S.; Chu, F.; Kearns, D.B.; Losick, R.; Kolter, R. A major protein component of the Bacillus subtilis biofilm matrix. *Mol. Microbiol.* **2006**, *59*, 1229–1238. [CrossRef] [PubMed]

81. Herrmann, M.; Saunders, A.M.; Schramm, A. Archaea dominate the ammonia-oxidizing community in the rhizosphere of the freshwater macrophyte Littorella uniflora. *Appl. Environ. Microbiol.* **2008**, *74*, 3279–3283. [CrossRef]

82. Crump, B.C.; Koch, E.W. Attached bacterial populations shared by four species of aquatic angiosperms. *Appl. Environ. Microbiol.* **2008**, *74*, 5948–5957. [CrossRef]

83. Zhou, X.; Wang, G. Nutrient concentration variations during *Oenanthe javanica* growth and decay in the ecological floating bed system. *J. Environ. Sci.* **2010**, *22*, 1710–1717. [CrossRef]

84. Stewart, F.M.; Muholland, T.; Cunningham, A.B.; Kania, B.G.; Osterlund, M.T. Floating islands as an alternative to constructed wetlands for treatment of excess nutrients from agricultural and municipal wastes–results of laboratory-scale tests. *Land Contam. Reclam.* **2008**, *16*, 25–33. [CrossRef]

85. Nihorimbere, V.; Ongena, M.; Smargiassi, M.; Thonart, P. Beneficial effect of the rhizosphere microbial community for plant growth and health. *Biotechnologie Agronomie Société et Environ.* **2011**, *15*, 327–337.

86. Acosta-Martínez, V.; Dowd, S.; Sun, Y.; Allen, V. Tag-encoded pyrosequencing analysis of bacterial diversity in a single soil type as affected by management and land use. *Soil Biol. Biochem.* **2008**, *40*, 2762–2770. [CrossRef]

87. Berg, G.; Smalla, K. Plant species and soil type cooperatively shape the structure and function of microbial communities in the rhizosphere. *FEMS Microbiol. Ecol.* **2009**, *68*, 1–13. [CrossRef] [PubMed]

88. Berendsen, R.L.; Pieterse, C.M.; Bakker, P.A. The rhizosphere microbiome and plant health. *Trends Plant Sci.* **2012**, *17*, 478–486. [CrossRef] [PubMed]

89. el Zahar Haichar, F.; Marol, C.; Berge, O.; Rangel-Castro, J.I.; Prosser, J.I.; Balesdent, J.; Heulin, T.; Achouak, W. Plant host habitat and root exudates shape soil bacterial community structure. *ISME J.* **2008**, *2*, 1221. [CrossRef] [PubMed]

90. Achá, D.; Iniguez, V.; Roulet, M.; Guimaraes, J.R.D.; Luna, R.; Alanoca, L.; Sanchez, S. Sulfate-reducing bacteria in floating macrophyte rhizospheres from an Amazonian floodplain lake in Bolivia and their association with Hg methylation. *Appl. Environ. Microbiol.* **2005**, *71*, 7531–7535. [CrossRef]

91. Zhang, W.; Wu, X.; Liu, G.; Chen, T.; Zhang, G.; Dong, Z.; Yang, X.; Hu, P. Pyrosequencing reveals bacterial diversity in the rhizosphere of three *Phragmites australis* ecotypes. *Geomicrobiol. J.* **2013**, *30*, 593–599. [CrossRef]

92. Van Bodegom, P.; Stams, F.; Mollema, L.; Boeke, S.; Leffelaar, P. Methane oxidation and the competition for oxygen in the rice rhizosphere. *Appl. Environ. Microbiol.* **2001**, *67*, 3586–3597. [CrossRef]

93. Tanaka, Y.; Tamaki, H.; Matsuzawa, H.; Nigaya, M.; Mori, K.; Kamagata, Y. Microbial community analysis in the roots of aquatic plants and isolation of novel microbes including an organism of the candidate phylum OP10. *Microbes Environ.* **2012**, *27*, 149–157. [CrossRef]

94. Unno, Y.; Shinano, T. Metagenomic analysis of the rhizosphere soil microbiome with respect to phytic acid utilization. *Microbes Environ.* **2013**, *28*, 120–127. [CrossRef]

95. Zhang, L.; Shao, H. Heavy metal pollution in sediments from aquatic ecosystems in China. *Clean–Soil Air Water* **2013**, *41*, 878–882. [CrossRef]

96. Uroz, S.; Buée, M.; Murat, C.; Frey-Klett, P.; Martin, F. Pyrosequencing reveals a contrasted bacterial diversity between oak rhizosphere and surrounding soil. *Environ. Microbiol. Rep.* **2010**, *2*, 281–288. [CrossRef] [PubMed]

97. Tribedi, P.; Sil, A.K. Low-density polyethylene degradation by Pseudomonas sp. AKS2 biofilm. *Environ. Sci. Pollut. Res.* **2013**, *20*, 4146–4153. [CrossRef] [PubMed]

98. Wei, B.; Yu, X.; Zhang, S.; Gu, L. Comparison of the community structures of ammonia-oxidizing bacteria and archaea in rhizoplanes of floating aquatic macrophytes. *Microbiol. Res.* **2011**, *166*, 468–474. [CrossRef] [PubMed]

99. Avrahami, S.; Conrad, R.; Braker, G. Effect of ammonium concentration on N_2O release and on the community structure of ammonia oxidizers and denitrifiers. *Appl. Environ. Microbiol.* **2003**, *69*, 3027. [CrossRef]

100. Hempel, M.; Botté, S.E.; Negrin, V.L.; Chiarello, M.N.; Marcovecchio, J.E. The role of the smooth cordgrass *Spartina alterniflora* and associated sediments in the heavy metal biogeochemical cycle within Bahía Blanca estuary salt marshes. *J. Soils Sed.* **2008**, *8*, 289. [CrossRef]

101. Rastogi, G.; Sbodio, A.; Tech, J.J.; Suslow, T.V.; Coaker, G.L.; Leveau, J.H. Leaf microbiota in an agroecosystem: Spatiotemporal variation in bacterial community composition on field-grown lettuce. *ISME J.* **2012**, *6*, 1812. [CrossRef] [PubMed]

102. Jensen, S.I.; Kühl, M.; Priemé, A. Different bacterial communities associated with the roots and bulk sediment of the seagrass *Zostera marina*. *FEMS Microbiol. Ecol.* **2007**, *62*, 108–117. [CrossRef]

103. Faulwetter, J.L.; Burr, M.D.; Parker, A.E.; Stein, O.R.; Camper, A.K. Influence of season and plant species on the abundance and diversity of sulfate reducing bacteria and ammonia oxidizing bacteria in constructed wetland microcosms. *Microb. Ecol.* **2013**, *65*, 111–127. [CrossRef]

104. Chen, N.; Yang, J.-S.; Qu, J.-H.; Li, H.-F.; Liu, W.-J.; Li, B.-Z.; Wang, E.T.; Yuan, H.-L. Sediment prokaryote communities in different sites of eutrophic Lake Taihu and their interactions with environmental factors. *World J. Microbiol. Biotechnol.* **2015**, *31*, 883–896. [CrossRef]

105. Glick, B.R. Using soil bacteria to facilitate phytoremediation. *Biotechnol. Adv.* **2010**, *28*, 367–374. [CrossRef]

106. Ma, Y.; Prasad, M.; Rajkumar, M.; Freitas, H. Plant growth promoting rhizobacteria and endophytes accelerate phytoremediation of metalliferous soils. *Biotechnol. Adv.* **2011**, *29*, 248–258. [CrossRef] [PubMed]

107. Wei, Y.; Hou, H.; ShangGuan, Y.; Li, J.; Li, F. Genetic diversity of endophytic bacteria of the manganese-hyperaccumulating plant Phytolacca americana growing at a manganese mine. *Eur. J. Soil Biol.* **2014**, *62*, 15–21. [CrossRef]

108. Erdei, L.; Mezôsi, G.; Mécs, I.; Vass, I.; Fôglein, F.; Bulik, L. Phytoremediation as a program for decontamination of heavy-metal polluted environment. *Acta Biol. Szeged.* **2005**, *49*, 75–76.

109. Dharni, S.; Srivastava, A.K.; Samad, A.; Patra, D.D. Impact of plant growth promoting *Pseudomonas monteilii* PsF84 and *Pseudomonas plecoglossicida* PsF610 on metal uptake and production of secondary metabolite (monoterpenes) by rose-scented geranium (*Pelargonium graveolens* cv. bourbon) grown on tannery sludge amended soil. *Chemosphere* **2014**, *117*, 433–439.

110. Ma, Y.; Rajkumar, M.; Rocha, I.; Oliveira, R.S.; Freitas, H. Serpentine bacteria influence metal translocation and bioconcentration of *Brassica juncea* and *Ricinus communis* grown in multi-metal polluted soils. *Front. Plant Sci.* **2015**, *5*, 757. [CrossRef] [PubMed]

111. Chadha, N.; Mishra, M.; Rajpal, K.; Bajaj, R.; Choudhary, D.K.; Varma, A. An ecological role of fungal endophytes to ameliorate plants under biotic stress. *Arch. Microbiol.* **2015**, *197*, 869–881. [CrossRef]

112. Marella, S. Bacterial endophytes in sustainable crop production: Applications, recent developments and challenges ahead. *Int. J. Life Sci. Res.* **2014**, *2*, 46–56.

113. Ryan, R.P.; Germaine, K.; Franks, A.; Ryan, D.J.; Dowling, D.N. Bacterial endophytes: Recent developments and applications. *FEMS Microbiol. Lett.* **2008**, *278*, 1–9. [CrossRef]

114. Bacon, C.W.; White, J.F. Functions, mechanisms and regulation of endophytic and epiphytic microbial communities of plants. *Symbiosis* **2016**, *68*, 87–98. [CrossRef]

115. Berg, G.; Hallmann, J. Control of Plant Pathogenic Fungi with Bacterial Endophytes. In *Microbial Root Endophytes*; Springer: Berlin/Heidelberg, Germany, 2006; pp. 53–69.

116. White, J.F.; Kingsley, K.I.; Kowalski, K.P.; Irizarry, I.; Micci, A.; Soares, M.A.; Bergen, M.S. Disease protection and allelopathic interactions of seed-transmitted endophytic pseudomonads of invasive reed grass (*Phragmites australis*). *Plant Soil* **2018**, *422*, 195–208. [CrossRef]

117. Glick, B.R. Plant growth-promoting bacteria: Mechanisms and applications. *Scientifica* **2012**, *2012*, 1–15. [CrossRef] [PubMed]

118. Pinheiro, E.A.; Carvalho, J.M.; dos Santos, D.C.; Feitosa, A.O.; Marinho, P.S.; Guilhon, G.M.S.; Santos, L.S.; de Souza, A.L.; Marinho, A.M. Chemical constituents of Aspergillus sp EJC08 isolated as endophyte from *Bauhinia guianensis* and their antimicrobial activity. *An. Acad. Brasi. Ciênc.* **2013**, *85*, 1247–1253. [CrossRef] [PubMed]

119. Matsuoka, H.; Akiyama, M.; Kobayashi, K.; Yamaji, K. Fe and P solubilization under limiting conditions by bacteria isolated from *Carex kobomugi* roots at the Hasaki coast. *Curr. Microbiol.* **2013**, *66*, 314–321. [CrossRef] [PubMed]

120. Knoth, J.L.; Kim, S.H.; Ettl, G.J.; Doty, S.L. Effects of cross host species inoculation of nitrogen-fixing endophytes on growth and leaf physiology of maize. *Gcb Bioenergy* **2013**, *5*, 408–418. [CrossRef]

121. Jasim, B.; Jimtha John, C.; Shimil, V.; Jyothis, M.; Radhakrishnan, E. Studies on the factors modulating indole-3-acetic acid production in endophytic bacterial isolates from P iper nigrum and molecular analysis of ipdc gene. *J. Appl. Microbiol.* **2014**, *117*, 786–799. [CrossRef]

122. Khan, A.L.; Waqas, M.; Kang, S.-M.; Al-Harrasi, A.; Hussain, J.; Al-Rawahi, A.; Al-Khiziri, S.; Ullah, I.; Ali, L.; Jung, H.-Y. Bacterial endophyte Sphingomonas sp. LK11 produces gibberellins and IAA and promotes tomato plant growth. *J. Microbiol.* **2014**, *52*, 689–695. [CrossRef]

123. Chaturvedi, H.; Singh, V.; Gupta, G. Potential of bacterial endophytes as plant growth promoting factors. *J. Plant Pathol. Microbiol.* **2016**, *7*, 2. [CrossRef]

124. Van Loon, L.; Bakker, P.; Pieterse, C. Systemic resistance induced by rhizosphere bacteria. *Annu. Rev. Phytopathol.* **1998**, *36*, 453–483. [CrossRef]

125. Rajkumar, M.; Prasad, M.N.V.; Swaminathan, S.; Freitas, H. Climate change driven plant–metal–microbe interactions. *Environ. Int.* **2013**, *53*, 74–86. [CrossRef]

126. Yihui, B.; Zhouying, X.; Yurong, Y.; ZHANG, H.; Hui, C.; Ming, T. Effect of dark septate endophytic fungus *Gaeumannomyces cylindrosporus* on plant growth, photosynthesis and Pb tolerance of maize (*Zea mays* L.). *Pedosphere* **2017**, *27*, 283–292.

127. Shi, P.; Zhu, K.; Zhang, Y.; Chai, T. Growth and cadmium accumulation of *Solanum nigrum* L. seedling were enhanced by heavy metal-tolerant strains of *Pseudomonas aeruginosa*. *Water Air Soil Pollut.* **2016**, *227*, 459. [CrossRef]

128. Babu, A.G.; Shea, P.J.; Sudhakar, D.; Jung, I.-B.; Oh, B.-T. Potential use of *Pseudomonas koreensis* AGB-1 in association with *Miscanthus sinensis* to remediate heavy metal (loid)-contaminated mining site soil. *J. Environ. Manag.* **2015**, *151*, 160–166. [CrossRef] [PubMed]

129. Visioli, G.; Vamerali, T.; Mattarozzi, M.; Dramis, L.; Sanangelantoni, A.M. Combined endophytic inoculants enhance nickel phytoextraction from serpentine soil in the hyperaccumulator Noccaea caerulescens. *Front. Plant Sci.* **2015**, *6*, 638. [CrossRef] [PubMed]

130. Afzal, M.; Shabir, G.; Hussain, I.; Khalid, Z.M. Paper and board mill effluent treatment with the combined biological–coagulation–filtration pilot scale reactor. *Bioresour. Technol.* **2008**, *99*, 7383–7387. [CrossRef] [PubMed]

131. Shahid, M.J.; Tahseen, R.; Siddique, M.; Ali, S.; Iqbal, S.; Afzal, M. Remediation of polluted river water by floating treatment wetlands. *Water Supply* **2019**, *19*, 967–977. [CrossRef]

132. Newman, L.A.; Reynolds, C.M. Bacteria and phytoremediation: New uses for endophytic bacteria in plants. *Trends Biotechnol.* **2005**, *23*, 6–8. [CrossRef]

133. Shehzadi, M.; Fatima, K.; Imran, A.; Mirza, M.S.; Khan, Q.M.; Afzal, M. Ecology of bacterial endophytes associated with wetland plants growing in textile effluent for pollutant-degradation and plant growth-promotion potentials. *Plant Biosyst.* **2016**, *150*, 1261–1270. [CrossRef]

134. Faußer, A.C.; Hoppert, M.; Walther, P.; Kazda, M. Roots of the wetland plants *Typha latifolia* and *Phragmites australis* are inhabited by methanotrophic bacteria in biofilms. *Flora-Morphol. Distrib. Funct. Ecol. Plants* **2012**, *207*, 775–782. [CrossRef]

135. Görres, C.-M.; Conrad, R.; Petersen, S.O. Effect of soil properties and hydrology on Archaeal community composition in three temperate grasslands on peat. *FEMS Microbiol. Ecol.* **2013**, *85*, 227–240. [CrossRef]

136. Shpigel, M.; Ben-Ezra, D.; Shauli, L.; Sagi, M.; Ventura, Y.; Samocha, T.; Lee, J. Constructed wetland with Salicornia as a biofilter for mariculture effluents. *Aquaculture* **2013**, *412*, 52–63. [CrossRef]

137. Vymazal, J. Removal of nutrients in various types of constructed wetlands. *Sci. Total Environ.* **2007**, *380*, 48–65. [CrossRef] [PubMed]

138. García-Martínez, M.; López-López, A.; Calleja, M.L.; Marbà, N.; Duarte, C.M. Bacterial community dynamics in a seagrass (*Posidonia oceanica*) meadow sediment. *Estuar. Coasts.* **2009**, *32*, 276–286. [CrossRef]

139. Vymazal, J. Plants used in constructed wetlands with horizontal subsurface flow: A review. *Hydrobiologia* **2011**, *674*, 133–156. [CrossRef]

140. Xu, X.; Thornton, P.E.; Post, W.M. A global analysis of soil microbial biomass carbon, nitrogen and phosphorus in terrestrial ecosystems. *Global Ecol. Biogeogr.* **2013**, *22*, 737–749. [CrossRef]

141. Nelson, E.B.; Karp, M.A. Soil pathogen communities associated with native and non-native P hragmites australis populations in freshwater wetlands. *Ecol. Evol.* **2013**, *3*, 5254–5267. [CrossRef]

142. Ehrenfeld, J.G.; Ravit, B.; Elgersma, K. Feedback in the plant-soil system. *Annu. Rev. Environ. Resour.* **2005**, *30*, 75–115. [CrossRef]

143. Wang, L.; Wu, J.; Ma, F.; Yang, J.; Li, S.; Li, Z.; Zhang, X. Response of arbuscular mycorrhizal fungi to hydrologic gradients in the rhizosphere of *Phragmites australis* (Cav.) Trin ex. Steudel growing in the Sun Island Wetland. *BioMed Res. Int.* **2015**, *2015*, 810124.

144. Curl, E.; Harper, J. Rhizosphere. In *Fauna-microflora Interactions*; John Wiley and Sons Ltd.: Chichester, UK, 1990; pp. 369–388.

145. Petersen, N.R.; Jensen, K. Nitrification and denitrification in the rhizosphere of the aquatic macrophyte *Lobelia dortmanna* L. *Limnol. Oceanogr.* **1997**, *42*, 529–537. [CrossRef]

146. Bodelier, P.; Dedysh, S.N. Microbiology of wetlands. *Front. Microbiol.* **2013**, *4*, 79. [CrossRef]

147. Han, G.; Luo, Y.; Li, D.; Xia, J.; Xing, Q.; Yu, J. Ecosystem photosynthesis regulates soil respiration on a diurnal scale with a short-term time lag in a coastal wetland. *Soil Biol. Biochem.* **2014**, *68*, 85–94. [CrossRef]

148. Koelbener, A.; Ström, L.; Edwards, P.J.; Venterink, H.O. Plant species from mesotrophic wetlands cause relatively high methane emissions from peat soil. *Plant Soil* **2010**, *326*, 147–158. [CrossRef]

149. Philippot, L.; Raaijmakers, J.M.; Lemanceau, P.; Van Der Putten, W.H. Going back to the roots: The microbial ecology of the rhizosphere. *Nat. Rev. Microbiol.* **2013**, *11*, 789. [CrossRef] [PubMed]

150. Weston, D.J.; Timm, C.M.; Walker, A.P.; Gu, L.; Muchero, W.; Schmutz, J.; Shaw, A.J.; Tuskan, G.A.; Warren, J.M.; Wullschleger, S.D. S phagnum physiology in the context of changing climate: Emergent influences of genomics, modelling and host–microbiome interactions on understanding ecosystem function. *Plant, Cell Environ.* **2015**, *38*, 1737–1751. [CrossRef] [PubMed]

151. Husson, O. Redox potential (Eh) and pH as drivers of soil/plant/microorganism systems: A transdisciplinary overview pointing to integrative opportunities for agronomy. *Plant Soil* **2013**, *362*, 389–417. [CrossRef]

152. Vitousek, P.M.; Cassman, K.; Cleveland, C.; Crews, T.; Field, C.B.; Grimm, N.B.; Howarth, R.W.; Marino, R.; Martinelli, L.; Rastetter, E.B. Towards an Ecological Understanding of Biological Nitrogen Fixation. In *The Nitrogen Cycle at Regional to Global Scales*; Springer: Berlin/Heidelberg, Germany, 2002; pp. 1–45.

153. Lamers, L.P.; Van Diggelen, J.M.; Op Den Camp, H.J.; Visser, E.J.; Lucassen, E.C.; Vile, M.A.; Jetten, M.S.; Smolders, A.J.; Roelofs, J.G. Microbial transformations of nitrogen, sulfur, and iron dictate vegetation composition in wetlands: A review. *Front Microbiol.* **2012**, *3*, 156. [CrossRef]

154. Bañeras, L.; Ruiz-Rueda, O.; López-Flores, R.; Quintana, X.D.; Hallin, S. The role of plant type and salinity in the selection for the denitrifying community structure in the rhizosphere of wetland vegetation. *Int. Microbiol.* **2012**, *15*, 89–99.

155. Trias, R.; Ruiz-Rueda, O.; García-Lledó, A.; Vilar-Sanz, A.; López-Flores, R.; Quintana, X.D.; Hallin, S.; Bañeras, L. Emergent macrophytes act selectively on ammonia-oxidizing bacteria and archaea. *Appl. Environ. Microbiol.* **2012**, *78*, 6352–6356. [CrossRef]

156. He, J.-Z.; Shen, J.-P.; Zhang, L.-M.; Di, H.J. A review of ammonia-oxidizing bacteria and archaea in Chinese soils. *Front. Microbiol.* **2012**, *3*, 296.

157. Reddy, K.; Patrick, W.; Lindau, C. Nitrification-denitrification at the plant root-sediment interface in wetlands. *Limnol. Oceanogr.* **1989**, *34*, 1004–1013. [CrossRef]

158. Galloway, J.N.; Schlesinger, W.H.; Levy, H.; Michaels, A.; Schnoor, J.L. Nitrogen fixation: Anthropogenic enhancement-environmental response. *Global Biogeochem. Cycles* **1995**, *9*, 235–252. [CrossRef]

159. López-Guerrero, M.G.; Ormeño-Orrillo, E.; Acosta, J.L.; Mendoza-Vargas, A.; Rogel, M.A.; Ramírez, M.A.; Rosenblueth, M.; Martínez-Romero, J.; Martínez-Romero, E. Rhizobial extrachromosomal replicon variability, stability and expression in natural niches. *Plasmid* **2012**, *68*, 149–158. [CrossRef] [PubMed]

160. Ormeño-Orrillo, E.; Hungria, M.; Martinez-Romero, E. Dinitrogen-fixing prokaryotes. *Prokaryotes Prokaryotic Physiol. Biochem.* **2013**, 427–451.

161. Oyewole, O. Microbial communities and their activities in paddy fields: A review. *J. Vet. Adv.* **2012**, *2*, 74–80.

162. Arima, Y.; Yoshida, T. Nitrogen fixation and denitrification in the roots of flooded crops. *Soil Sci. Plant Nutr.* **1982**, *28*, 483–489. [CrossRef]

163. Knapp, A. The sensitivity of marine N_2 fixation to dissolved inorganic nitrogen. *Front. Microbiol.* **2012**, *3*, 374. [CrossRef] [PubMed]

164. Waisel, Y.; Agami, M. Ecophysiology of Roots of Submerged Aquatic Plants. In *Plant Roots: The Hidden Half*, 2nd ed.; Marcel Dekker, Inc.: New York, NY, USA, 1996; pp. 895–909.

165. Neori, A.; Agami, M. The functioning of rhizosphere biota in wetlands–a review. *Wetlands* **2017**, *37*, 615–633. [CrossRef]

166. Hiraishi, A. Biodiversity of dehalorespiring bacteria with special emphasis on polychlorinated biphenyl/dioxin dechlorinators. *Microbes Environ.* **2008**, *23*, 1–12. [CrossRef] [PubMed]

167. Fennell, D.; Du, S.; Liu, F.; Liu, H.; Haggblom, M. Dehalogenation of Polychlorinated Dibenzo-p-Dioxins and Dibenzofurans, Polychlorinated Biphenyls, and Brominated Flame Retardants, and Potential as a Bioremediation Strategy. In *Environmental Biotechnology and Safety*; Elsevier Inc.: Amsterdam, The Netherlands, 2011; pp. 135–149.

168. Fenchel, T.; Blackburn, H.; King, G.M.; Blackburn, T.H. *Bacterial Biogeochemistry: The Ecophysiology of Mineral Cycling*; Academic press: Cambridge, MA, USA, 2012.

169. Stottmeister, U.; Wießner, A.; Kuschk, P.; Kappelmeyer, U.; Kästner, M.; Bederski, O.; Müller, R.; Moormann, H. Effects of plants and microorganisms in constructed wetlands for wastewater treatment. *Biotechnol. Adv.* **2003**, *22*, 93–117. [CrossRef]

170. Paris, D.F.; Steen, W.C.; Baughman, G.L.; Barnett, J.T. Second-order model to predict microbial degradation of organic compounds in natural waters. *Appl. Environ. Microbiol.* **1981**, *41*, 603–609. [CrossRef]

171. Calheiros, C.S.; Duque, A.F.; Moura, A.; Henriques, I.S.; Correia, A.; Rangel, A.O.; Castro, P.M. Changes in the bacterial community structure in two-stage constructed wetlands with different plants for industrial wastewater treatment. *Bioresour. Technol.* **2009**, *100*, 3228–3235. [CrossRef]

172. Mori, K.; Toyama, T.; Sei, K. Surfactants degrading activities in the rhizosphere of giant duckweed (*Spirodela polyrrhiza*). *Jpn. J. Water Treat. Biol.* **2005**, *41*, 129–140. [CrossRef]

173. Mordukhova, E.A.; Sokolov, S.L.; Kochetkov, V.V.; Kosheleva, I.A.; Zelenkova, N.F.; Boronin, A.M. Involvement of naphthalene dioxygenase in indole-3-acetic acid biosynthesis by Pseudomonas putida. *FEMS Microbiol. Lett.* **2000**, *190*, 279–285. [CrossRef]

174. Jouanneau, Y.; Willison, J.C.; Meyer, C.; Krivobok, S.; Chevron, N.; Besombes, J.-L.; Blake, G. Stimulation of pyrene mineralization in freshwater sediments by bacterial and plant bioaugmentation. *Environ. Sci. Technol.* **2005**, *39*, 5729–5735. [CrossRef] [PubMed]

175. Golubev, S.N.; Schelud'ko, A.V.; Muratova, A.Y.; Makarov, O.E.; Turkovskaya, O.V. Assessing the potential of rhizobacteria to survive under phenanthrene pollution. *Water Air Soil Pollut.* **2009**, *198*, 5–16. [CrossRef]

176. Huang, X.-D.; El-Alawi, Y.; Penrose, D.M.; Glick, B.R.; Greenberg, B.M. A multi-process phytoremediation system for removal of polycyclic aromatic hydrocarbons from contaminated soils. *Environ. Pollut.* **2004**, *130*, 465–476. [CrossRef] [PubMed]

177. Escalante-Espinosa, E.; Gallegos-Martínez, M.; Favela-Torres, E.; Gutiérrez-Rojas, M. Improvement of the hydrocarbon phytoremediation rate by *Cyperus laxus* Lam. inoculated with a microbial consortium in a model system. *Chemosphere* **2005**, *59*, 405–413. [CrossRef]

178. Simpson, D.R. Biofilm processes in biologically active carbon water purification. *Water Res.* **2008**, *42*, 2839–2848. [CrossRef]

179. Ghosh, U.; Weber, A.S.; Jensen, J.N.; Smith, J.R. Granular activated carbon and biological activated carbon treatment of dissolved and sorbed polychlorinated biphenyls. *Water Environ. Res* **1999**, *71*, 232–240. [CrossRef]

180. Guasch, H.; Lehmann, V.; Van Beusekom, B.; Sabater, S.; Admiraal, W. Influence of phosphate on the response of periphyton to atrazine exposure. *Arch. Environ. Contam. Toxicol.* **2007**, *52*, 32–37. [CrossRef]

181. Semrau, J.D.; DiSpirito, A.A.; Yoon, S. Methanotrophs and copper. *FEMS Microbiol. Rev.* **2010**, *34*, 496–531. [CrossRef]

182. Yoon, S. Towards Practical Application of Methanotrophic Metabolism in Chlorinated Hydrocarbon Degradation, Greenhouse Gas Removal, and Immobilization of Heavy Metals. Ph.D. Thesis, University of Michigan, Ann Arbor, MI, USA, 2010.

183. Pandey, V.C.; Singh, J.; Singh, D.; Singh, R.P. Methanotrophs: Promising bacteria for environmental remediation. *Int. J. Environ Sci. Technol.* **2014**, *11*, 241–250. [CrossRef]

184. Han, B.; Zhang, S.; Zhang, L.; Liu, K.; Yan, L.; Wang, P.; Wang, C.; Pang, S. Characterization of microbes and denitrifiers attached to two species of floating plants in the wetlands of Lake Taihu. *PLoS ONE* **2018**, *13*, e0207443. [CrossRef] [PubMed]

185. Gao, Y.; Yi, N.; Wang, Y.; Ma, T.; Zhou, Q.; Zhang, Z.; Yan, S. Effect of *Eichhornia crassipes* on production of N_2 by denitrification in eutrophic water. *Ecol. Eng.* **2014**, *68*, 14–24. [CrossRef]

186. Mullen, M.; Wolf, D.; Ferris, F.; Beveridge, T.; Flemming, C.; Bailey, G. Bacterial sorption of heavy metals. *Appl. Environ. Microbiol.* **1989**, *55*, 3143–3149. [CrossRef] [PubMed]

187. Sessitsch, A.; Kuffner, M.; Kidd, P.; Vangronsveld, J.; Wenzel, W.W.; Fallmann, K.; Puschenreiter, M. The role of plant-associated bacteria in the mobilization and phytoextraction of trace elements in contaminated soils. *Soil Biol. Biochem.* **2013**, *60*, 182–194. [CrossRef] [PubMed]

188. Khan, M.U.; Sessitsch, A.; Harris, M.; Fatima, K.; Imran, A.; Arslan, M.; Shabir, G.; Khan, Q.M.; Afzal, M. Cr-resistant rhizo-and endophytic bacteria associated with *Prosopis juliflora* and their potential as phytoremediation enhancing agents in metal-degraded soils. *Front. Plant Sci.* **2015**, *5*, 755. [CrossRef] [PubMed]

189. Cobbett, C.; Goldsbrough, P. Phytochelatins and metallothioneins: Roles in heavy metal detoxification and homeostasis. *Annu. Rev. Plant Biol.* **2002**, *53*, 159–182. [CrossRef]

190. Kaegi, J.H.; Schaeffer, A. Biochemistry of metallothionein. *Biochemistry* **1988**, *27*, 8509–8515. [CrossRef]

191. Huckle, J.W.; Morby, A.P.; Turner, J.S.; Robinson, N.J. Isolation of a prokaryotic metallothionein locus and analysis of transcriptional control by trace metal ions. *Mol. Microbiol.* **1993**, *7*, 177–187. [CrossRef]

192. Valls, M.; Atrian, S.; de Lorenzo, V.; Fernández, L.A. Engineering a mouse metallothionein on the cell surface of *Ralstonia eutropha* CH34 for immobilization of heavy metals in soil. *Nat. Biotechnol.* **2000**, *18*, 661. [CrossRef]

193. Mejáre, M.; Bülow, L. Metal-binding proteins and peptides in bioremediation and phytoremediation of heavy metals. *Trends Biotechnol.* **2001**, *19*, 67–73. [CrossRef]

194. Shin, M.-N.; Shim, J.; You, Y.; Myung, H.; Bang, K.-S.; Cho, M.; Kamala-Kannan, S.; Oh, B.-T. Characterization of lead resistant endophytic Bacillus sp. MN3-4 and its potential for promoting lead accumulation in metal hyperaccumulator *Alnus firma*. *J. Hazard. Mater.* **2012**, *199*, 314–320. [CrossRef] [PubMed]

195. Mindlin, S.Z.; Bass, I.A.; Bogdanova, E.S.; Gorlenko, Z.M.; Kalyaeva, E.S.; Petrova, M.A.; Nikiforov, V.G. Horizontal transfer of mercury resistance genes in environmental bacterial populations. *Mol. Biol.* **2002**, *36*, 160–170. [CrossRef]

196. Li, T.; Liu, M.; Zhang, X.; Zhang, H.; Sha, T.; Zhao, Z. Improved tolerance of maize (*Zea mays* L.) to heavy metals by colonization of a dark septate endophyte (DSE) *Exophiala pisciphila*. *Sci. Total Environ.* **2011**, *409*, 1069–1074. [CrossRef]

197. Zhang, X.; Li, C.; Nan, Z. Effects of cadmium stress on growth and anti-oxidative systems in *Achnatherum inebrians* symbiotic with *Neotyphodium gansuense*. *J. Hazard. Mater.* **2010**, *175*, 703–709. [CrossRef]

198. Wan, Y.; Luo, S.; Chen, J.; Xiao, X.; Chen, L.; Zeng, G.; Liu, C.; He, Y. Effect of endophyte-infection on growth parameters and Cd-induced phytotoxicity of Cd-hyperaccumulator *Solanum nigrum* L. *Chemosphere* **2012**, *89*, 743–750. [CrossRef]

199. Brown, N.L.; Stoyanov, J.V.; Kidd, S.P.; Hobman, J.L. The MerR family of transcriptional regulators. *FEMS Microbiol. Rev.* **2003**, *27*, 145–163. [CrossRef]

200. Sofu, A.; Sayilgan, E.; Guney, G. Experimental design for removal of Fe (II) and Zn (II) ions by different lactic acid bacteria biomasses. *Int. J. Environ. Res.* **2015**, *9*, 93–100.

201. Li, C.; Wang, S.; Du, X.; Cheng, X.; Fu, M.; Hou, N.; Li, D. Immobilization of iron-and manganese-oxidizing bacteria with a biofilm-forming bacterium for the effective removal of iron and manganese from groundwater. *Bioresour. Technol.* **2016**, *220*, 76–84. [CrossRef]

202. Franzblau, R.E.; Daughney, C.J.; Swedlund, P.J.; Weisener, C.G.; Moreau, M.; Johannessen, B.; Harmer, S.L. Cu (II) removal by *Anoxybacillus flavithermus*–iron oxide composites during the addition of Fe (II) aq. *Geochim. Cosmochim. Acta* **2016**, *172*, 139–158. [CrossRef]

203. Guo, Y.; Huang, T.; Wen, G.; Cao, X. The simultaneous removal of ammonium and manganese from groundwater by iron-manganese co-oxide filter film: The role of chemical catalytic oxidation for ammonium removal. *Chem. Eng. J.* **2017**, *308*, 322–329. [CrossRef]

204. Kim, S.-Y.; Kim, J.-H.; Kim, C.-J.; Oh, D.-K. Metal adsorption of the polysaccharide produced from Methylobacterium organophilum. *Biotechnol. Lett* **1996**, *18*, 1161–1164. [CrossRef]

205. Marchal, M.; Briandet, R.; Koechler, S.; Kammerer, B.; Bertin, P. Effect of arsenite on swimming motility delays surface colonization in Herminiimonas arsenicoxydans. *Microbiology* **2010**, *156*, 2336–2342. [CrossRef]

206. Iyer, A.; Mody, K.; Jha, B. Biosorption of heavy metals by a marine bacterium. *Mar. Pollut. Bull.* **2005**, *50*, 340–343. [CrossRef] [PubMed]

207. Oshima, T.; Kondo, K.; Ohto, K.; Inoue, K.; Baba, Y. Preparation of phosphorylated bacterial cellulose as an adsorbent for metal ions. *React. Funct. Polym.* **2008**, *68*, 376–383. [CrossRef]

208. Ozdemir, G.; Ceyhan, N.; Manav, E. Utilization of an exopolysaccharide produced by Chryseomonas luteola TEM05 in alginate beads for adsorption of cadmium and cobalt ions. *Bioresour. Technol.* **2005**, *96*, 1677–1682. [CrossRef]

209. Ozdemir, G.; Ozturk, T.; Ceyhan, N.; Isler, R.; Cosar, T. Heavy metal biosorption by biomass of *Ochrobactrum anthropi* producing exopolysaccharide in activated sludge. *Bioresour. Technol.* **2003**, *90*, 71–74. [CrossRef]

210. Freire-Nordi, C.S.; Vieira, A.A.H.; Nascimento, O.R. The metal binding capacity of Anabaena spiroides extracellular polysaccharide: An EPR study. *Process Biochem.* **2005**, *40*, 2215–2224. [CrossRef]

211. Pugazhendhi, A.; Boovaragamoorthy, G.M.; Ranganathan, K.; Naushad, M.; Kaliannan, T. New insight into effective biosorption of lead from aqueous solution using *Ralstonia solanacearum*: Characterization and mechanism studies. *J. Clean. Prod.* **2018**, *174*, 1234–1239. [CrossRef]

212. Wu, Y.; Zhao, X.; Jin, M.; Li, Y.; Li, S.; Kong, F.; Nan, J.; Wang, A. Copper removal and microbial community analysis in single-chamber microbial fuel cell. *Bioresour. Technol.* **2018**, *253*, 372–377. [CrossRef]

213. Pugazhendhi, A.; Ranganathan, K.; Kaliannan, T. Biosorptive removal of copper (II) by *Bacillus cereus* isolated from contaminated soil of electroplating industry in India. *Water Air Soil Pollut.* **2018**, *229*, 76. [CrossRef]

214. Wen, X.; Du, C.; Zeng, G.; Huang, D.; Zhang, J.; Yin, L.; Tan, S.; Huang, L.; Chen, H.; Yu, G. A novel biosorbent prepared by immobilized *Bacillus licheniformis* for lead removal from wastewater. *Chemosphere* **2018**, *200*, 173–179. [CrossRef] [PubMed]

215. Pramanik, K.; Mitra, S.; Sarkar, A.; Maiti, T.K. Alleviation of phytotoxic effects of cadmium on rice seedlings by cadmium resistant PGPR strain *Enterobacter aerogenes* MCC 3092. *J. Hazard. Mater.* **2018**, *351*, 317–329. [CrossRef] [PubMed]

216. Choińska-Pulit, A.; Sobolczyk-Bednarek, J.; Łaba, W. Optimization of copper, lead and cadmium biosorption onto newly isolated bacterium using a Box-Behnken design. *Ecotoxicol. Environ. Saf.* **2018**, *149*, 275–283. [CrossRef] [PubMed]

217. Bai, H.-J.; Zhang, Z.-M.; Yang, G.-E.; Li, B.-Z. Bioremediation of cadmium by growing Rhodobacter sphaeroides: Kinetic characteristic and mechanism studies. *Bioresour. Technol.* **2008**, *99*, 7716–7722. [CrossRef]

218. Vijayaraghavan, K.; Yun, Y.-S. Bacterial biosorbents and biosorption. *Biotechnol. Adv.* **2008**, *26*, 266–291. [CrossRef]

219. Volesky, B. Advances in biosorption of metals: Selection of biomass types. *FEMS Microbiol. Rev.* **1994**, *14*, 291–302. [CrossRef]

220. Ma, Y.; Rajkumar, M.; Luo, Y.; Freitas, H. Inoculation of endophytic bacteria on host and non-host plants—effects on plant growth and Ni uptake. *J. Hazard. Mater.* **2011**, *195*, 230–237. [CrossRef]

221. Guo, H.; Luo, S.; Chen, L.; Xiao, X.; Xi, Q.; Wei, W.; Zeng, G.; Liu, C.; Wan, Y.; Chen, J. Bioremediation of heavy metals by growing hyperaccumulaor endophytic bacterium Bacillus sp. L14. *Bioresour. Technol.* **2010**, *101*, 8599–8605. [CrossRef] [PubMed]

222. Chen, L.; Luo, S.; Li, X.; Wan, Y.; Chen, J.; Liu, C. Interaction of Cd-hyperaccumulator *Solanum nigrum* L. and functional endophyte Pseudomonas sp. Lk9 on soil heavy metals uptake. *Soil Biol. Biochem.* **2014**, *68*, 300–308. [CrossRef]

223. Sheng, X.-F.; Xia, J.-J.; Jiang, C.-Y.; He, L.-Y.; Qian, M. Characterization of heavy metal-resistant endophytic bacteria from rape (*Brassica napus*) roots and their potential in promoting the growth and lead accumulation of rape. *Environ. Pollut.* **2008**, *156*, 1164–1170. [CrossRef] [PubMed]

224. Chen, C.; Wang, J.-L. Characteristics of Zn^{2+} biosorption by Saccharomyces cerevisiae. *Biomed. Environ. Sci. BES* **2007**, *20*, 478–482. [PubMed]

225. Talos, K.; Pager, C.; Tonk, S.; Majdik, C.; Kocsis, B.; Kilar, F.; Pernyeszi, T. Cadmium biosorption on native *Saccharomyces cerevisiae* cells in aqueous suspension. *Acta Univ. Sapientiae Agric. Environ* **2009**, *1*, 20–30.

226. Tigini, V.; Prigione, V.; Giansanti, P.; Mangiavillano, A.; Pannocchia, A.; Varese, G.C. Fungal biosorption, an innovative treatment for the decolourisation and detoxification of textile effluents. *Water* **2010**, *2*, 550–565. [CrossRef]

227. Rajkumar, M.; Ae, N.; Freitas, H. Endophytic bacteria and their potential to enhance heavy metal phytoextraction. *Chemosphere* **2009**, *77*, 153–160. [CrossRef]

228. Joshi, P.M.; Juwarkar, A.A. In vivo studies to elucidate the role of extracellular polymeric substances from azotobacter in immobilization of heavy metals. *Environ. Sci. Technol.* **2009**, *43*, 5884–5889. [CrossRef]

229. Visioli, G.; D'Egidio, S.; Vamerali, T.; Mattarozzi, M.; Sanangelantoni, A.M. Culturable endophytic bacteria enhance Ni translocation in the hyperaccumulator *Noccaea caerulescens*. *Chemosphere* **2014**, *117*, 538–544. [CrossRef]

230. Luo, Y.; Guo, W.; Ngo, H.H.; Nghiem, L.D.; Hai, F.I.; Zhang, J.; Liang, S.; Wang, X.C. A review on the occurrence of micropollutants in the aquatic environment and their fate and removal during wastewater treatment. *Sci. Total Environ.* **2014**, *473*, 619–641. [CrossRef]

231. Hempel, M.; Blume, M.; Blindow, I.; Gross, E.M. Epiphytic bacterial community composition on two common submerged macrophytes in brackish water and freshwater. *BMC Microbiol.* **2008**, *8*, 58. [CrossRef]

232. Cai, X.; Gao, G.; Tang, X.; Dong, B.; Dai, J.; Chen, D.; Song, Y. The response of epiphytic microbes to habitat and growth status of *Potamogeton malaianus* Miq. in Lake Taihu. *J. Basic Microbiol.* **2013**, *53*, 828–837.

233. Aung, K.; Jiang, Y.; He, S.Y. The role of water in plant–microbe interactions. *Plant J.* **2018**, *93*, 771–780. [CrossRef] [PubMed]

234. D'Annibale, A.; Leonardi, V.; Federici, E.; Baldi, F.; Zecchini, F.; Petruccioli, M. Leaching and microbial treatment of a soil contaminated by sulphide ore ashes and aromatic hydrocarbons. *Appl. Microbiol. Biotechnol.* **2007**, *74*, 1135–1144. [CrossRef] [PubMed]

235. Tunali, S.; Akar, T.; Özcan, A.S.; Kiran, I.; Özcan, A. Equilibrium and kinetics of biosorption of lead (II) from aqueous solutions by *Cephalosporium aphidicola*. *Sep. Purif. Technol.* **2006**, *47*, 105–112. [CrossRef]

236. Akar, T.; Tunali, S.; Çabuk, A. Study on the characterization of lead (II) biosorption by fungus *Aspergillus parasiticus*. *Appl. Biochem. Biotechnol.* **2007**, *136*, 389–405. [CrossRef] [PubMed]

237. Kelly, D.J.; Budd, K.; Lefebvre, D.D. The biotransformation of mercury in pH-stat cultures of microfungi. *Botany* **2006**, *84*, 254–260. [CrossRef]

238. Ahalya, N.; Ramachandra, T.; Kanamadi, R. Biosorption of heavy metals. *Res. J. Chem. Environ* **2003**, *7*, 71–79.

239. Chihpin, H. Application of Aspergillus oryzae and rhizopus oryzae for Cu (II) removal. *Water Res.* **1996**, *9*, 1985–1990.

240. Shah, M.A. Mycorrhizas in Aquatic Plants. In *Mycorrhizas: Novel Dimensions in the Changing World*; Springer: Berlin/Heidelberg, Germany, 2014; pp. 63–68.

241. Twanabasu, B.R.; Smith, C.M.; Stevens, K.J.; Venables, B.J.; Sears, W.C. Triclosan inhibits arbuscular mycorrhizal colonization in three wetland plants. *Sci. Total Environ.* **2013**, *447*, 450–457. [CrossRef]

242. Wang, S.; Ye, J.; Perez, P.G.; Huang, D.-F. Abundance and diversity of ammonia-oxidizing bacteria in rhizosphere and bulk paddy soil under different duration of organic management. *Afr. J. Microbiol. Res.* **2011**, *5*, 5560–5568.

243. Burke, D.J.; Smemo, K.A.; López-Gutiérrez, J.C.; DeForest, J.L. Soil fungi influence the distribution of microbial functional groups that mediate forest greenhouse gas emissions. *Soil Biol. Biochem.* **2012**, *53*, 112–119. [CrossRef]

244. Mohamed, D.J.; Martiny, J.B. Patterns of fungal diversity and composition along a salinity gradient. *ISME J.* **2011**, *5*, 379. [CrossRef] [PubMed]

245. Liu, W.-L.; Guan, M.; Liu, S.-Y.; Wang, J.; Chang, J.; Ge, Y.; Zhang, C.-B. Fungal denitrification potential in vertical flow microcosm wetlands as impacted by depth stratification and plant species. *Ecol. Eng.* **2015**, *77*, 163–171. [CrossRef]

246. Saha, M.; Sarkar, S.; Sarkar, B.; Sharma, B.K.; Bhattacharjee, S.; Tribedi, P. Microbial siderophores and their potential applications: A review. *Environ. Sci. Pollut. Res.* **2016**, *23*, 3984–3999. [CrossRef] [PubMed]

247. Persson, B. On the mechanism of adhesion in biological systems. *J. Chem. Phy.* **2003**, *118*, 7614–7621. [CrossRef]

248. Nakai, S.; Inoue, Y.; Hosomi, M.; Murakami, A. *Myriophyllum spicatum*-released allelopathic polyphenols inhibiting growth of blue-green algae *Microcystis aeruginosa*. *Water Res.* **2000**, *34*, 3026–3032. [CrossRef]

249. De Stefani, G.; Tocchetto, D.; Salvato, M.; Borin, M. Performance of a floating treatment wetland for in-stream water amelioration in NE Italy. *Hydrobiologia* **2011**, *674*, 157–167. [CrossRef]

250. Afzal, M.; Khan, Q.M.; Sessitsch, A. Endophytic bacteria: Prospects and applications for the phytoremediation of organic pollutants. *Chemosphere* **2014**, *117*, 232–242. [CrossRef] [PubMed]

251. De Meyer, S.E.; De Beuf, K.; Vekeman, B.; Willems, A. A large diversity of non-rhizobial endophytes found in legume root nodules in Flanders (Belgium). *Soil Biol. Biochem.* **2015**, *83*, 1–11. [CrossRef]

252. Yu, X.; Zhang, W.; Lang, D.; Zhang, X.; Cui, G.; Zhang, X. Interactions between endophytes and plants: Beneficial effect of endophytes to ameliorate biotic and abiotic stresses in plants. *J. Plant Biol.* **2019**, *62*, 1–13.

253. Momose, A.; Hiyama, T.; Nishimura, K.; Ishizaki, N.; Ishikawa, S.; Yamamoto, M.; Hung, N.V.P.; Ohtake, N.; Sueyoshi, K.; Ohyama, T. Characteristics of nitrogen fixation and nitrogen release from diazotrophic endophytes isolated from sugarcane (*Saccharum officinarum* L.) stems. *Bull. Facul. Agric. Niigata Univ.* **2013**, *66*, 66.

254. Afzal, M.; Yousaf, S.; Reichenauer, T.G.; Sessitsch, A. The inoculation method affects colonization and performance of bacterial inoculant strains in the phytoremediation of soil contaminated with diesel oil. *Int. J. Phytorem.* **2012**, *14*, 35–47. [CrossRef]

255. Zhao, S.; Zhou, N.; Zhao, Z.-Y.; Zhang, K.; Wu, G.-H.; Tian, C.-Y. Isolation of endophytic plant growth-promoting bacteria associated with the halophyte Salicornia europaea and evaluation of their promoting activity under salt stress. *Curr. Microbiol.* **2016**, *73*, 574–581. [CrossRef] [PubMed]

256. Mohanty, S.R.; Dubey, G.; Kollah, B. Endophytes of *Jatropha curcas* promote growth of maize. *Rhizosphere* **2017**, *3*, 20–28. [CrossRef]

257. Ma, Y.; Rajkumar, M.; Freitas, H. Inoculation of plant growth promoting bacterium Achromobacter xylosoxidans strain Ax10 for the improvement of copper phytoextraction by Brassica juncea. *J. Environ. Manag.* **2009**, *90*, 831–837. [CrossRef] [PubMed]

258. Zhu, L.-J.; Guan, D.-X.; Luo, J.; Rathinasabapathi, B.; Ma, L.Q. Characterization of arsenic-resistant endophytic bacteria from hyperaccumulators *Pteris vittata* and *Pteris multifida*. *Chemosphere* **2014**, *113*, 9–16. [CrossRef]

259. Ijaz, A.; Shabir, G.; Khan, Q.M.; Afzal, M. Enhanced remediation of sewage effluent by endophyte-assisted floating treatment wetlands. *Ecol. Eng.* **2015**, *84*, 58–66. [CrossRef]

260. Rehman, K.; Imran, A.; Amin, I.; Afzal, M. Inoculation with bacteria in floating treatment wetlands positively modulates the phytoremediation of oil field wastewater. *J. Hazard. Mater.* **2018**, *349*, 242–251. [CrossRef]

261. Li, X.-N.; Song, H.-L.; Li, W.; Lu, X.-W.; Nishimura, O. An integrated ecological floating-bed employing plant, freshwater clam and biofilm carrier for purification of eutrophic water. *Ecol. Eng.* **2010**, *36*, 382–390. [CrossRef]

262. Wu, Q.; Hu, Y.; Li, S.; Peng, S.; Zhao, H. Microbial mechanisms of using enhanced ecological floating beds for eutrophic water improvement. *Bioresour. Technol.* **2016**, *211*, 451–456. [CrossRef]

263. Zhao, T.; Fan, P.; Yao, L.; Yan, G.; Li, D.; Zhang, W. Ammonifying bacteria in plant floating island of constructed wetland for strengthening decomposition of organic nitrogen. *Trans. Chin. Soc. Agric. Eng.* **2011**, *27*, 223–226.

264. Feng, Y.; Zhang, H. Experimental study on nitrogen and phosphorus removal efficiency of eutrophical lake by floating combination biological technology. *J. Kunming Univ. Sci. Technol. (Nat. Sci. Ed.)* **2012**, *1*.

265. Cao, W.; Zhang, H.; Wang, Y.; Pan, J. Bioremediation of polluted surface water by using biofilms on filamentous bamboo. *Ecol. Eng.* **2012**, *42*, 146–149. [CrossRef]

266. Bhaskaran, K.; Nadaraja, A.V.; Tumbath, S.; Shah, L.B.; Veetil, P.G.P. Phytoremediation of perchlorate by free floating macrophytes. *J. Hazard. Mater.* **2013**, *260*, 901–906. [CrossRef] [PubMed]

267. Zhao, F.; Zhang, S.; Ding, Z.; Aziz, R.; Rafiq, M.T.; Li, H.; He, Z.; Stoffella, P.J.; Yang, X. Enhanced purification of eutrophic water by microbe-inoculated stereo floating beds. *Pol. J. Environ. Stud.* **2013**, *22*, 957–964.

268. Liu, J.; Wang, F.; Liu, W.; Tang, C.; Wu, C.; Wu, Y. Nutrient removal by up-scaling a hybrid floating treatment bed (HFTB) using plant and periphyton: From laboratory tank to polluted river. *Bioresour. Technol.* **2016**, *207*, 142–149. [CrossRef]
269. Saleem, H.; Rehman, K.; Arslan, M.; Afzal, M. Enhanced degradation of phenol in floating treatment wetlands by plant-bacterial synergism. *Int. J. Phytorem.* **2018**, *20*, 692–698. [CrossRef]
270. Saleem, H.; Arslan, M.; Rehman, K.; Tahseen, R.; Afzal, M. *Phragmites australis*—A helophytic grass—Can establish successful partnership with phenol-degrading bacteria in a floating treatment wetland. *Saudi J. Biol. Sci.* **2018**, *26*, 1179–1186. [CrossRef]
271. Sun, Z.; Xie, D.; Jiang, X.; Fu, G.; Xiao, D.; Zheng, L. Effect of eco-remediation and microbial community using multilayer solar planted floating island (MS-PFI) in the drainage channel. *bioRxiv* **2018**, *1*, 327965.
272. Rehman, K.; Imran, A.; Amin, I.; Afzal, M. Enhancement of oil field-produced wastewater remediation by bacterially-augmented floating treatment wetlands. *Chemosphere* **2019**, *217*, 576–583. [CrossRef]
273. Tara, N.; Iqbal, M.; Mahmood Khan, Q.; Afzal, M. Bioaugmentation of floating treatment wetlands for the remediation of textile effluent. *Water Environ. J.* **2019**, *33*, 124–134. [CrossRef]
274. Tara, N.; Arslan, M.; Hussain, Z.; Iqbal, M.; Khan, Q.M.; Afzal, M. On-site performance of floating treatment wetland macrocosms augmented with dye-degrading bacteria for the remediation of textile industry wastewater. *J. Clean. Prod.* **2019**, *217*, 541–548. [CrossRef]

Review

Application of Floating Aquatic Plants in Phytoremediation of Heavy Metals Polluted Water: A Review

Shafaqat Ali [1,2,*], Zohaib Abbas [1], Muhammad Rizwan [1], Ihsan Elahi Zaheer [1], İlkay Yavaş [3], Aydın Ünay [4], Mohamed M. Abdel-DAIM [5,6], May Bin-Jumah [7], Mirza Hasanuzzaman [8] and Dimitris Kalderis [9]

[1] Department of Environmental Sciences and Engineering, Government College University, Allama Iqbal Road, Faisalabad 38000, Pakistan; zohaib.abbas83@gmail.com (Z.A.); mrazi1532@yahoo.com (M.R.); ihsankhanlashari@gmail.com (I.E.Z.)

[2] Department of Biological Sciences and Technology, China Medical University, Taichung 40402, Taiwan

[3] Department of Plant and Animal Production, Kocarli Vocational High School, Aydın Adnan Menderes University, 09100 Aydın, Turkey; iyavas@adu.edu.tr

[4] Department of Field Crops, Faculty of Agriculture, Aydın Adnan Menderes University, 09100 Aydın, Turkey; aunay@adu.edu.tr

[5] Department of Zoology, College of Science, King Saud University, P.O. Box 2455, Riyadh 11451, Saudi Arabia; abdeldaim.m@vet.suez.edu.eg

[6] Pharmacology Department, Faculty of Veterinary Medicine, Suez Canal University, Ismailia 41522, Egypt

[7] Department of Biology, College of Science, Princess Nourah bint Abdulrahman University, Riyadh 11474, Saudi Arabia; may_binjumah@outlook.com

[8] Department of Agronomy, Sher-e-Bangla Agricultural University, Dhaka 1207, Bangladesh; mhzsauag@yahoo.com

[9] Department of Electronics Engineering, Hellenic Mediterranean University, Chania, 73100 Crete, Greece; dkalderis@chania.teicrete.gr

* Correspondence: shafaqataligill@yahoo.com or shafaqataligill@gcuf.edu.pk

Received: 11 February 2020; Accepted: 24 February 2020; Published: 3 March 2020

Abstract: Heavy-metal (HM) pollution is considered a leading source of environmental contamination. Heavy-metal pollution in ground water poses a serious threat to human health and the aquatic ecosystem. Conventional treatment technologies to remove the pollutants from wastewater are usually costly, time-consuming, environmentally destructive, and mostly inefficient. Phytoremediation is a cost-effective green emerging technology with long-lasting applicability. The selection of plant species is the most significant aspect for successful phytoremediation. Aquatic plants hold steep efficiency for the removal of organic and inorganic pollutants. Water hyacinth (*Eichhornia crassipes*), water lettuce (*Pistia stratiotes*) and Duck weed (*Lemna minor*) along with some other aquatic plants are prominent metal accumulator plants for the remediation of heavy-metal polluted water. The phytoremediation potential of the aquatic plant can be further enhanced by the application of innovative approaches in phytoremediation. A summarizing review regarding the use of aquatic plants in phytoremediation is gathered in order to present the broad applicability of phytoremediation.

Keywords: phytoremediation; heavy metal; aquatic plants; floating aquatic plants; wastewater treatment

---

1. Introduction

Water contaminations, along with limited availability of water, have put a severe burden on the environment. Around 40% population of the world is facing the problem of water scarcity due to climate change, rapid urbanization, food requirement and unchecked consumption of natural resources [1,2].

During the past few decades rapid urbanization, industrialization, agricultural activities, discharge of geothermal waters and olive wastewater especially in olive-cultivating areas enhanced the discharge of polluted wastewater into the environment [3–6]. Wastewater carrying soaring concentrations of pollutants is immensely noxious for aquatic ecosystem and human health [7–9]. Reclamation of wastewater has been the only option left to meet the increasing demand of water in growing industrial and agricultural sectors [10].

Industrial and domestic untreated wastewater contains pesticides, oils, dyes, phenol, cyanides, toxic organics, phosphorous, suspended solids, and heavy metals (HMs) [11]. Heavy metals among these toxic substances can easily be accumulated in the surrounding environment [12]. Commercial activities such as metal processing, mining, geothermal energy plants, automotive, paper, pesticide manufacturing, tanning, dying and plating are held responsible for global contamination of heavy metals [13,14]. Removal of heavy metals from the wastewater is difficult because they exist in different chemical forms. Most metals are not biodegradable, and they can easily pass through different trophic levels to persistently accumulate in the biota [15,16].

Removal of toxic pollutants is extremely important to minize the threat to human health and the surrounding environment. Removal of heavy metals achieved through various techniques such as reverse osmosis [17], ion exchange [18], chemical precipitation [19], adsorption and solvent extraction [20] include enormous operational and maintenance costs and are usually not environmentally friendly [19–22]. These conventional techniques for the remediation of heavy metals are generally costly and time-consuming. These treatment technologies require high capital investment and in the end, generate the problem of sludge disposal [23]. For the remediation of wastewater polluted with heavy metals contaminants, an environmentally friendly and economical treatment technology is needed [24,25]. The current study illustrates an environment-friendly technique phytoremediation for removal of contaminants on a long term basis. Furthermore, this review article summarizes the potential application of aquatic plants in phytoremediation for the treatment of wastewater.

2. Heavy Metals in the Environment

Anthropogenic and geological activities are the main source of heavy-metal pollution. Activities such as mining, military activities, municipal waste, application of fertilizer, discharge of urban effluent, vehicle exhausts, wastewater, waste incineration, fuel production, and smelting cause the production of metal contaminants [26,27]. Natural sources of heavy-metal pollution include erosion, weathering of rocks and volcanic eruption. Parent material during weathering is the primary and initial natural source of heavy metals [28].

Agricultural pesticides and utilization of fertilizers on agricultural soil have raised the concentration of Cd, Zn, Cu and As in soil [29]. A constantly increasing need for agricultural produce has increased the application of pesticides, fertilizers, and herbicides. This excessive use of these agrochemicals may result in the accumulation of these pollutants in plants and the soil as well [30]. Usage of phosphate fertilizer and inorganic fertilizers to control the diseases of crops, grain and vegetable sometime hold an uneven level of Ni, Pb, Zn, Cd, and Cr [31,32]. An enormous quantity of fertilizers is applied to deliver the K, P and N in order to improve the growth of crops, which in turn increase the incidence of cadmium, lead, iron and mercury in substantial high concentrations. Inputs of heavy metal to agricultural land through the excessive use of fertilizers is increasing apprehension about their probable hazard to the environment [33,34].

Wastewater irrigation leads to the buildup of various heavy metals like cadmium, lead, nickel, zinc, etc. Some of these metals like Zn, Cu, Ni, Cd and Pb are frequently present in the subsurface of the soil irrigated with untreated wastewater. Wastewater irrigation for long periods of time increase the concentration of heavy metal in the soil at toxic levels [35]. The unregulated dumping of municipal solid waste is also another main source of raised soil contamination load. Open dumps and land filling are the common practices using worldwide to dispose of municipal solid waste. Despite being a useful source of nutrients, these wastes are also a source of some harmful toxic metals as well. Precarious and overload applications of fertilizers, pesticides and fungicides are very important sources of metal pollution [36]. Metal contamination can also be caused by transportation. Maintenance and deicing operations on roads also generate groundwater/surface contaminants. Corrosion, tread wear, and brake abrasion are well-recorded sources of heavy metals generation linked to highway traffic [37].

3. Phytoremediation

Phytoremediation is considered an effective, aesthetically pleasing, cost effective and environmental friendly technology for the remediation of potentially toxic metals from the environment. Plants in phytoremediation accumulate contaminants through their roots and then translocate these contaminant in the aboveground part of their body [38,39]. The notion of using metal accumulator plants for the removal of heavy metals and several other contaminants in phytoremediation was first introduced in 1983, but this idea has already been implanted for the last 300 years [40]. Phytoremediation is known by different names such as agro-remediation, green remediation, vegetative remediation, green technology and botano remediation [4,41,42].

Use of vegetation, soil and micro biota along with other agrochemical practices makes the vegetative remediation an appealing green technology for the accumulation of different heavy metals [43,44]. The application of in situ and ex-situ remediation is applicable in a phytoremediation process. In situ application is used more commonly because it reduces the multiplication of contaminant in water and airborne waste, which ultimately minimize the risk to the adjacent environment [45]. More than one type of pollutant can be treated on site by the phytoremediation without the need for a disposal site. It also reduces the spread of contamination by preventing soil erosion and leaching [46]. The clean up cost of phytoremediation is far less than other conventional techniques of remediation, which is the utmost advantage of this technique [47]. Phytoremediation is a relatively straightforward technique as it does not require any highly specific personnel and exclusive equipment. This is applicable for the remediation of large scale area where other conventional techniques prove to be extremely inefficient and costly as well [48].

An enormous number of contaminants can be remediated by phytoremediation technology such as insecticides, chlorinated solvents, Polycyclic aromatic hydrocarbons (PAHs), Polychlorinated biphenyl (PCBs), petroleum hydrocarbons, radio nucleosides, surfactants, explosive elements and heavy metals [48,49]. There are a number of plant species that have the ability to accumulate significantly higher concentrations of heavy metals in different parts of the body, such as a leaf, stems and root, without showing any sign of toxicity [50,51].

3.1. Characteristics of Phytoremediation Plants

Plants should have the following characteristics in order to make the phytoremediation an eco-sustainable technology: native and quick growth rate, high biomass yield, the uptake of a large amount of heavy metals, the ability to transport metals in aboveground parts of plant, and a mechanism to tolerate metal toxicity [52–55]. Other factors like pH, solar radiation, nutrient availability and salinity greatly influence the phytoremediation potential and growth of the plant [51,56].

3.1.1. Mechanism of Phytoremediation

Phytoremediation follows different mechanisms such as phytoextraction, phytostabilization, phytovolatilization and rhizofiltration during the uptake or accumulation of heavy metals in the plant [4,41]. The different mechanisms involved in the phytoremediation process are briefly describe below.

Phytoextraction

Phytoextraction is also called phytoaccumulation, and it involves the uptake of heavy metal in the plant roots and then their translocation into an above ground-level portion of the plant like shoots, etc. Once the phytoextraction is done the plant can be harvested and burned for gaining energy and recovering/recycling metal if required from the ash [57,58]. Sometimes phytoremediation and phytoextraction are used synonymously, which is a misconception; phytoextraction is a cleanup technology while phytoremediation is the name of a concept [59]. Phytoextraction is an suitable phytoremediation technique for the remediation of heavy metals from wastewater, sediments and soil [52,60].

Phytostabilization

Phytostabilization involves the use of the plant to restrict the movement of contaminants in the soil. The term phytostabilization is also known as in place deactivation. Remediation of soil, sludge, and sediment can be effectively done by using this technology. It does not interfere with the natural environment and is a much safer alternative option [61,62]. In phytostabilization, plants inhibit or act as a barrier for the percolation of water within the soil. When we need to persevere in our surface water, ground water and restoration of soil quality, this technology is best suited for this purpose because it cuts short the movement of the contaminants [63,64]. Phytostabilization is very effective for a large site, which is heavily affected by the contaminants [65]. Phytostabilization is only a managing approach for inactivating/immobilizing the potentially harmful contaminants. It is not a permanent resolution, because only the movement of metals is restricted, but they continue to stay in the soil [66].

Rhizofiltration

Rhizofiltration involves the use of the plant to ab/adsorb the contaminants, resulting in restricted movement of these contaminants in underground water [67,68]. Roots play a very significant part in rhizofiltration. Factor such as changing pH in the rhizosphere and root exudates helps the precipitation of heavy metal on the surface of the roots. Once the plant has soaked up all the contaminants, they can easily be harvested and disposed [69]. Plants for rhizofiltration should have the ability; to produce a widespread root system, accumulate high concentrations of heavy metals, be easy to handle and have low maintenance cost [42,70]. Both aquatic and terrestrial plants with long fibrous root systems can be used in rhizofiltration [70]. Rhizofiltration is productively used for handling and treatment of the agricultural runoff, industrial discharge, radioactive contaminant, and metals [71]. Heavy metals which are mostly retained in the soil such as cadmium, lead, chromium, nickel, zinc, and copper can be adequately remediated through rhizofiltration [72].

Phytovolatilization

Phytovolatilization is the process in which a plant converts pollutants into a different volatile nature and then their successive release into the surrounding environment with the help of the plant's stomata [48,73]. Plant species like canola and Indian mustard are useful for the phytovolatilization of selenium. Mercury and selenium are the most favorable contaminants that can be remediated in phytovolatilization [74]. One of the greatest advantages of phytovolatilization is that it does not require any additional management once the plantation is done. Other benefits are minimizing soil erosion, no disturbance to the soil, unrequited harvesting, and the disposal of plant biomass [75]. Bacteria present in the rhizosphere also help in the biotransformation of the contaminant, which eventually boosts the rate of phytovolatilization.

3.2. Advances in Phytoremediation

3.2.1. Chemical Assisted Phytoremediation

The phytoremediation potential depends upon the phyto-availability of different heavy metals present in the soil [76]. The application of specific chemicals has proved to be a successful technique to boost the bioavailability of heavy metals to plants [41]. Organic fertilizers and chelating reagents are commonly used to decrease the pH of soils, which ultimately enhance the bioavailability and bioaccumulation in plants. In tobacco, decreased pH by application of a chelating reagent showed increased accumulation of Cd. The application of ethylenediaminetetraacetic acid (EDTA) boosted the phytoextraction and bioaccumulation of Cd, Zn, and Pb in various studies [77,78]. Some other chelating agents, diethylene triamine penta-acetic acid (DTPA) and ethylene glycol tetra-acetic acid (AGTA), also have been proved efficient chelators to enhance the phytoavailability and phytoextraction of heavy metals [79]. Organic acids such as malic acid, acetic acid, citric acid and oxalic acid have been proved effective chelating agents. The phytoremediation potential of plants may also be enhanced by strengthening plants to tolerate heavy-metal stress and toxicity. Application of salicylic acid (SA), has been found effective to alleviate metal stress in the plant, resulting in enhanced phytoremediation potential of plants [80,81].

Application of different chemicals also has some drawbacks. The applied chemical may cause the toxicity in plants, may leach to groundwater, and may disturb the translocation of heavy metals in plants. The applied chemicals often may form complexes with heavy metals, which have non-biodegrade abilities, leading to a source of secondary pollution [82]. The application of chelators may disturb the plant growth and development. It may result in decreased growth of roots, shoot, and biomass due to the toxic effects of chelators [83]. The negative impacts of chelators can be minimized by the application of a proper amount of the chelators, cautious application, and proper understanding of the water seepage mechanism [84]. The organic acids have advantages over synthetic chelators being economical and easily biodegradable and environment-friendly [85,86].

3.2.2. Microbial Assisted Phytoremediation

Plant-associated microorganisms have a key role in the remediation of heavy metals from soils [87]. These microorganisms influence the availability and accumulation of heavy metals in soil and plants. Recently, bio-augmentation of plants with particular and adapted microbes has been extensively studied in phytoremediation [38,53,88]. Plant growth-promoting rhizobacteria (PGPR) proved to increase biomass production, disease resistance, and reduce metal induced toxicity in bio-augmented plants [89]. Similarly, endophytic bacteria also play a very prominent part in phytoremediation [90,91]. The plant–endophyte interaction, fortify the plants to tolerate both biotic and abiotic stress [92]. Endophytes have developed several mechanisms to alleviate metal toxicity in plants. These methods include efflux of toxic metal ions from the cell, the transformation of metal ions into less-hazardous forms, sequestration, precipitation, adsorption, and biomethylation [93]. Application of rhizospheric and endophytic bacteria in soils/plants improves plant growth and boosts the phytoremediation potential of plants by enhancing metals availability, metals uptake, accumulation, reduced metal stress in plants. Furthermore, the rhizospheric and endophytic bacteria also enhance the phytoremediation potential of plants by enhancing soil fertility by the production of growth regulators and the provision of essential nutrients [94–96]. The mycorrhizal fungi in the root zone form an association with the roots of plants, and have a beneficial role in phytoremediation [97]. This plant-fungi association enhance the availability of essential plant nutrients through their hyphal network, modify the root exudates, alter soil pH and stimulate the bioavailability of various heavy metals to associated plants [98,99].

3.2.3. Transgenic Plants

The application of transgenic plants in phytoremediation is a novel approach to enhance the effectiveness of phytoremediation. Specific genes in transgenic plants increase the metabolism, accumulation and uptake of definite pollutants [94,100]. The ideal plant to engineer for phytoremediation should possess characteristics; high biomass yield adopted to local and target environment and well-established transformation protocol. Transgenic plants also enhance the detoxification process of organic pollutants and the addition of toxic compounds in the food chain [100,101]. Firstly, transgenic plants were introduced for the remediation of inorganic pollutants; now they are effectively used to remove organic pollutants from contaminated media [102]. *Nicotiana tabaccum* and *Arabidopsis thaliana* are an example of transgenic plants firstly practiced for effective removal of heavy metals, cadmium, and mercury, respectively [103,104]. Transgenic plants have been proved efficient for the treatment of phenolic, chlorinated, and explosives contaminants [105,106].

Plants can be engineered to degrade the organic pollutants in the rhizosphere. In this, transgenic plants do not uptake and accumulate the pollutants; rather, incorporated genes secrete enzymes which degrade organic pollutants in the rhizospheric zone [107]. This approach also solves the problem of plant harvesting and handling loaded with toxic metals, as all the metal detoxification and removal process occurs in the rhizosphere by roots [108]. The transgenic *Arabidopsis* plants enhanced the degradation of 2,3-dihydroxybiphenyl (2,3-DHB). Similarly, transgenic tobacco plants speed up the detoxification of 1-chlorobuatne in the rhizospheric zone [109]. This ability of transgenic plants is attributed to the increased diversity of the microbial community, increased metabolic activity, the release of root exudates and enzymes and increased contact between roots and contaminants [110,111].

3.2.4. Non-Living Plant Biomass

Non-living plant biomass can be profitably used for metal uptake and metal recovery. Successive use of dried and dead biomass of plants (as simple biosorbent substance) to remove the metals from water has gained popularity over the past few years because it is easy to handle and is a cost-effective natural approach [112,113]. Water hyacinth's (*Eichornia crassipes*) dried roots showed the potential to remove cadmium and lead effectively from wastewater [114,115]. Biomass of different aquatic plant species such as *Eichhornia crassipes*, *Potamogetonlucens*, and *Salvinia herzegoi* was reported to be successfully used as an exceptional biosorbent material for the removal of Cr, Ni, Cd, Zn, Cu, and Pb effectively in various studies [116,117].

4. Aquatic Plants and Phytoremediation

The aquatic ecosystem is a cost-effective and resourceful clean up technique for phytoremediation of a large contaminated area. Aquatic plants act as a natural absorber for contaminants and heavy metals [118]. Removal of different heavy metals along with other contaminants through the application of aquatic plants is the most proficient and profitable method [52,119]. Constructed wetlands along with aquatic plants were extensively applied throughout the world for the treatment of wastewater [120,121]. The selection of aquatic plant species for the accumulation of heavy metal is a very important matter to enhance the phytoremediation [71,122].

Over the years, aquatic plants have gained an overwhelming reputation because of their capacity to clean up contaminated sites throughout the world [120,123]. Aquatic plants always develop an extensive system of roots which helps them and makes them the best option for the accumulation of contaminants in their roots and shoots [124,125]. The growth and cultivation of aquatic plants are time-consuming, which may restrict the growing demand of phytoremediation [126]. Nevertheless, this shortcoming is substituted by the number of advantages that this technology possesses for the treatment of wastewater [100,127].

4.1. Types of Aquatic Plants

4.1.1. Free-Floating Aquatic Plants

These are the plants with floating leaves and submerged roots. Some of the free-floating aquatic plants are well recognized for their capability to eliminate the metals from the contaminated environment: water hyacinth (*Eichhornia crassipes*) [128], water ferns (*Salvinia minima*) [129], duckweeds (*Lemna minor*, *Spirodelaintermedia*), [130,131], water lettuce (*Pistiastratoites*), [132], water cress (*Nasturtium officinale*) [133]. The potential of these free floating aquatic plant for the elimination of heavy metals is comprehensively studied in different studies [99,134,135]. Active transport of heavy metals in free-floating aquatic plants occurs from the roots, from where metals are transferred to other parts of the plant body. Passive transport is associated with the direct contact of the plant body with the pollution medium. In passive transport, heavy metals mainly accumulate in upper parts of the plant body [136]. Water hyacinth, duckweed and water lettuce are the most frequently used free-floating plants for the remediation of heavy metals from wastewater [137–140]. The aptitude of different aquatic plants to mitigate different heavy metals is mentioned in Table 1.

Table 1. Accumulation potential of various aquatic plants.

Aquatic Plant	Common Name	Metals/Metalloids	Reference
Eichhornia crassipes	Water hyacinth	Pb, Hg, Cu, Cr, Ni, Zn.	Molisani et al. [141]; Hu et al. [142]
Pistia stratiotes	Water lettuce	Cr, Zn, Fe, Mn, Cu	Maine et al. [136]; Miretzky et al. [143]
Salvinia minima	water spangles	As, Ni, Cr, Cd	Olguin et al. [135]; Sooknah, [144]
Salvinia herzogii	Water fern	Cd, Cr	Maine et al. [136]; Sunñe et al. [145]
Lemna minor	Duckweed	Cr, As, Ni, Cu, Pb	Kara [146]; Ater et al. [147]; Basile et al. [148]
Spirodela intermedia	Duckweed	Fe, Zn, Mn, Cu, Cr, Pb	Miretzky et al. [143]; Cardwell et al. [149]
Nasturtium officinale	Water cress	Cr, Ni, Zn, Cu,	Kara [146]; Zurayk et al. [150]
Myriophyllum spicatum	Parrot feathers	Pb, Cd, Fe, Cu	Sivaci et al. [151]; Branković et al. [152]
Ceratophyllum demersum	Hornwort	As, Cd, Cr, Pb	Bunluesin et al. [153]; El-Khatib et al. [154]
Potamogeton crispus	Pondweed	Cu, Fe, Ni, Zn, and Mn	Borisova et al. [155]
Potamogeton pectinatus	American pondweed	Cd, Pb, Cu, Zn	Singh et al. [156]; Penga et al. [157]
Typha latifolia	common cattail	Zn, Mn, Ni, Fe, Pb, Cu	Hejna et al. [158]; Qian et al. [159]; Sasmaz et al. [160]
Mentha aquatica	Water mint	Pb, Cd, Fe, Cu	Branković et al. [152]; Kamal et al. [161]
Vallisneria spiralis	Tape grass	Ar	Giri [162]
spartina alterniflora	Cordgrass	Cu. Cr, Zn, Ni, Mn, Cd, Pb, As.	Aksorn and Visoottiviseth [163]; Hempel et al. [164]
Phragmites australis	Common reed	Fe, Cu, Cd, Pb, Zn	Ganjalia et al. [165]; Ha and Anh [166]
Scirpus	Bulrush	Cd, Fe, Al.	Kutty and Al-Mahaqeri [167]
Polygonum hydropiperoides	Smartweed	Cu, Pb, Zn	Rudin et al., [168]

4.1.2. Water Hyacinth (*Eichhornia Crassipes*)

Water hyacinth (*Eichhornia crassipes*) is a free-floating aquatic plant which belongs to the family of Pontedericeae that is closely correlated with the lily family. Water hyacinth is the most widespread invasive vascular plant of the world. It has an extensive dark blue root system along with curved, straight leaves. The roots contain a stolon from which new plants are produced [169]. Water hyacinth possesses the unique ability to grow in heavily polluted environments and successively extract pollutants [134]. It has the advanced tendency of remediating different pollutants like organic material, heavy metals, total suspended solids, total dissolved solids, and nutrients [170–172]. Removal of nutrients and heavy metals are vastly reliant on the optimal growth rate of water hyacinth [169,173].

Water hyacinth (*Eichhornia crassipes*) is recommended to treat industrial wastewater, domestic wastewater, sewage effluents, and sludge ponds because it has (1) high absorption rate of different organic and inorganic contaminants (2) can tolerate an extremely polluted environment and (3) has a gigantic production rate of biomass [174]. *Eichhornia crassipes* has greater ability to remediate contaminants like arsenic, zinc, mercury, nickel, copper and lead from industrial and domestic wastewater streams [175–177].

Water hyacinth's derived ash and activated carbon showed good accumulation capacity of different HMs like cooper, nickel, zinc and chromium. It also holds the benefit of having the least biological sludge production and creation of bio-sorbent, which facilitate metal recovery [178]. Major industries like paper, food processing, textile, leather, cosmetics, and dyes manufacturing results in the release of dye contaminants into the environments. Dyes are most stable and stand firm against oxidizing agents, which in the end enhance water pollution. The widespread root system and tolerance against these dyes help water hyacinth to effectively accumulate the reactive dyes [114,179]. Water hyacinth shows significance removal efficiency for Cd, Pb, Cu, Zn, Fe, As, Mn, Cr, As, Al and Hg as reported in different recent studies [180–183]. Shoot powder of water hyacinth removed Cr and Cu by 99.98% and 99.96% when exposed to tannery effluents [184]. Recent research studies conducted to check the removal efficiency of water hyacinth for heavy metals are given in Table 2.

Table 2. Recent studies on uptake of heavy metals by water hyacinth.

Metals/Metalloids	Results	Conditions	Reference
Ni	Concentrations of Ni Areal parts-(0.29 ± 0.02 mg/kg) Roots-(3.34 ± 0.26 mg/kg)	1, 2, 3 and 4 mg L^{-1} concentration of nickel.	González et al. [24]
Cd	Initial concentration of cadmium was 0.3 while Cd in leaves of the plant was 31 ± 3.	Cadmium exposure at 1000 and 130 ug/L.	Shuvaeva et al. [180]
Al, Pb, AS, Cd, Cu	Removal rate: Al-(73%) Pb-(73%,) As-(74%) Cd-(82.8%) Cu-(78.6%)	Wastewater from steel effluents	Aurangzeb et al. [181]
Cd, Hg, Pb, Ni	Removal rate: Cd-(97.5%) Hg-(99.9%) Pb-(83.4%) Ni-(95.1%)	Initial concentrations of Cd: 0.24, Hg: 4.971, Pb: 1.199, Ni: 3.34 in industrial wastewater	Fazal et al. [182]
Cr, Cu	Tannery effluents	Removal rate. Cr-(99.98%) Cu-(99.96%)	Sarkar et al. [184]
Cd, Zn, Cu, Pb	Removal rate: Cd-(98%) Zn-(84%) Cu-(99%) Pb-(98%)	Anaerobic packed bed reactors system	Sekomo et al. [185].
Cr, Zn	Removal efficiency of Cr. (63%) on 3rd day, (80%) on 9th day Removal efficiency of Zn. (67%) on 9th day, (96%) on 12th day, (100%) on 15th day.	Stock solutions	Swarnalatha and Radhakrishnan [186]
Pb, Cu, Mn, Cd	Uptake in leaves Pb-(3.40–5.06 mg/kg) Cu-(6.41–13.5 mg/kg) Mn-(62.9–67.9 mg/kg) Cd-(0.037–0.13 mg/kg)	Wastewater from mining.	Prasad and Maiti [187]

Table 2. *Cont.*

Metals/Metalloids	Results	Conditions	Reference
Mo, Ag, Ba, Pb, Cd	TF Mo-(0.85 ± 0.14) Ag-(0.18 ± 0.04) Ba-(0.12 ± 0.03) Pb-(0.06 ± 0.01) Cd-(0.65 ± 0.09)	Gold mine waste water.	Romanova et al. [188]
Zn, Cd, Cu, Pb	Removal rate: Zn-(93.5%) Cd-(95.16%) Cu-(58.23%) Pb-(98.33%)	Stock solutions.	Li et al. [189]

4.1.3. Water Lettuce

Water lettuce (*Pistia stratiotes* L.) fit in the aracae/arum family. Water cabbage, Nile cabbage, water lettuce, jalkhumbhi, and shellflower are some of the other common names of these plants. It is mostly found in lakes, stream, and ponds [190]. *Pistia stratiotes* have 20 cm-wide and 10–20 cm-long pale green leaves. Whitish hair covers the lower surface of the plant. It has underwater hanging structure underneath the floating leaves [191]. Water lettuce possesses extraordinary tolerance over an extensive range of pH and temperature [192]. Extension and proliferation of water lettuce occur with the production of daughter plants. *P. Stratiotes* also produces seeds which remain present in water; their germination occurs during the wet seasons [193].

Water lettuce (*P. Stratiotes*) is an excellent contender for the phytoremediation of contaminants as it is more prone than other aquatic vegetation [194,195]. The plant has the capacity of reducing/removing nutrients such as biological oxygen demand (BOD), chemical oxygen demand (COD), dissolved oxygen (DO), pH, total Kjeldahl nitrogen (TKN), ammonia (NH_3), nitrite (NO_2^-), nitrate (NO_3^-) and phosphate (PO_4^{3-}), from drinking and surface water, storm water, sewage water and industrial wastewater [196–198]. The small size of *P. Stratiotes* by contrast with water hyacinth showed better removing capacity for the Zinc and mercury from industrial wastewater stream [199].

Uptake of Cu, Zn, Fe, Cr and Cd does not have any harmful effect on the plant which makes *P. stratiotes* eligible to be used as a hyperaccumulators plant for the mitigation of organic contaminants and heavy metals from wastewater on a broad scale [200,201]. The biomass of *P. stratiotes* reduces more than 70% of zinc and cadmium from the contaminated solution during the experiment [202]. Water lettuce is an excellent accumulator of Pb, Zn, Cu, Cd, Mg, Fe, and Mn as reported by different recent studies given in Table 3.

Table 3. Recent studies on uptake of heavy metals by water lettuce.

Metals/Metalloids	Condition	Results	Reference
Fe, Mn, Cr, Pb, Cu Zn, Ni, Co	Three sites of Al-Sero drain Giza, selected for the collection of plant and water sample.	High value of BCF and RP observed positive correlation exist between Fe and Cu with root and shoots of plant.	Galal et al. [71]
Cd	20 to 50 g of plant applied in container having 10 L river water and cadmium exposure of 1000 and 130 (g/L) concentrations.	BCF-(1270 ± 250). Initial concentration of cadmium was 0, 3 while Cd concentration in leaves of the plants was 32 ± 3.	Shuvaeva et al. [180]
Pb, Cu	120 g of plant applied in 10 litre of steel industry effluents.	Removal rate: Pb-(70.7%) Cu-(66.5%)	Aurangzeb et al. [181]

Table 3. *Cont.*

Metals/Metalloids	Condition	Results	Reference
Fe, Mn, Na, Ni, Pb, Cr, Cu, Zn, Al, Ca, Cd, Co, K, Mg	Plants covered two storm water detention ponds	50% accumulation of Ca, Co, Cd, Mn, Zn and Mg in roots. More than 50% absorbance by roots for Pb, Ni, Cu, Cr and Al.	Lu et al. [197]
Cd, Zn	Initial concentration of Zinc, 1.8, 18, 50, 79, 105 mg/L, initial concentration of Cadmium 0.01, 0.1, 1, 10 mg/L.	Removal rate. 70% reduction for both Zinc and Cadmium.	Rodrigues et al. [202]
As	Initial concentration of As applied 0, 5, 10, 15 and 20 µM.	High absorbance of arsenic observed in the roots of the plants.	Farnese et al. [203]
Pb (II)	Stock solution (2000 mg/L)	Removal of Pb (II) 96%.	Volf et al. [204]

4.1.4. Duck Weed

Duckweed is a free-floating aquatic plant which floats on the surface of slow-moving and still water. This plant belongs to family Araceae but is frequently classified in subfamily *Lemnoideae*. This family of free-floating plant species consists of five genera such as (1) *Wolffia*, (2) *Wolffiella*, (3) *Spirodela*, (4) *Lemna*, and (5) *Landoltia*, having no less than 40 species recognized [205]. Duckweed is also known as a water lens. These are richly found in ditches, canals, and ponds; these are smaller and faster-growing plants on the earth. They can survive in high pH (3.5 to 10.5) and temperature 7 to 35 °C) [206].

The capability of duckweed plant to develop in polluted site with tremendous variation in pH, temperature, nutrient level makes them effective for use in phytoremediation [207]. Duckweed can eliminate a vast variety of different heavy metals, inorganic and organic contaminants, pesticides, nutrients arise from agricultural runoff, sewage, industrial and domestic wastewater [131,138,208]. Duckweed can easily inhibit the growth of algae and fungi in different ponds because it has the ability to cover the ponds due to its widespread high growth rate. They also diminish nitrogen from these ponds by taking up ammonia and denitrification [209]. Removal of the nutrient with the application of duckweed biomass will help to upgrade the quality of water and degradation of water ecology. Duckweed shows higher removal aptitude for chemical oxygen demand (COD), biological oxygen demand (BOD), total nitrogen (TN), total suspended solid (TSS) and NH_3^-N from wastewater under favorable environmental circumstances [131,207].

Much higher elimination of different HMs such as As, Cr, Cu, Zn, Ag, Hg, Pb and Cd has been done through different species of duck weed including *L. minor*, *L. Trisulca* and *L. gibba* from wastewater [4,210,211] *S. polyrhiza*, *L. gibba* and *L. minor* examined for their remediation efficiency of boron, arsenic, and uranium. *Spirodelapolyrhiza* was investigated and found to be a good phyto-remediator of arsenic [212], *L. gibba* was found to be appropriate for the remedy of boron with a lower concentration of 2 mgL^{-1} without any harmful effect on biomass [213]. It can also accumulate uranium (120%), boron (40%), and arsenic (133%) [214]. *L. minor* found to be an excellent contender for the remediation of arsenic [215]. In comparison with other macrophytes, duckweed is the most suitable plant for phytoremediation.

Use of duckweed for the remediation of nutrient pollutants and HMs from industrial and agricultural wastewater was reported in previous reports [216,217]. Several researchers have reported that duckweed (*L. minor*) could take up a huge concentration of heavy metals such as nickel, manganese, zinc, uranium, arsenic, and copper [218]. *Lemna minor L* shows an increase in chromium uptake percentage of 6.1%, 26.5%, 20.5%, 20.2% at a different exposure concentration of chromium stress [219].

Duckweed has the ability to conserve nature by acting as a hyperaccumulator plant in phytoremediation technology. Table 4 shows recent studies of heavy-metal uptake from wastewater by duckweed.

Table 4. Recent studies on uptake of heavy metals by duckweed species.

Metals/Metalloids	Condition	Results	References
Cr	0, 10, 100, 200 μM Cr concentration	increase in chromium uptake percentage by *L. Minor* 6.10%, 26.5%, 20.5%, 20.2%	Sallah-ud-Din et al. [219]
Cr, Pb	2, 4, 10 and 15 mg/L concentrations with using lab water.	Removalrate Cr-(86.2–94.8%) Pb-(91.0–96.4%).	Abdallah [220]
Cd	0.5, 1.0, 1.5, 2.0, 2.5, and 3.0 mg/L concentrations applied.	Removal rate Cd-(42–78%)	Chaudhuri et al. [221]
Ar	Initial artificial concentrations of 0.5, 1.0, and 2.0 mg/L.	Removal of arsenic more than 70% at 0.5 mg/L on 15th day of experiment.	Goswami et al. [222]
Fe	Concentrations of 100% and 50%, for (7, 14, and 21 days)	Maximum recommend = 5 mg/L Fe at 7 days.	Teixeira et al. [223]
Co, Cu, Fe, Cd, Ni, Mo, Mn, Zn, Cr, Se	Mining wastewater in rich with selenium	Removal rate: Co-(87%) Se-(55%) 35–60% removal rate for rest of the heavy metals.	Flores-Miranda et al. [224]
Pb, Cd	Artificial by concentration of (2, 5 and 10 mg/L)	Removal rate (1) Pb-(98.1%) in 10 mg/L at 7 pH. (60.1%) in 2 mg/L at 9 pH. (2) Cd-(84.8 %) in 2 mg/L at pH 7. (41.6%) in (10 mg/L at pH 9.	Verma and Suthar [225]
Cu, Zn, Cd	Initial concentrations Cu-(4.10 mg/L) Zn-(4.30 mg/L) Cd-(7.30 mg/L)	Cu-(0.381 ± 0.021 mg/g) Zn-(0.557 ± 0.009 mg/g) Cd-(1.251 ± 0.041 mg/g)	Török et al. [226]
Cd, Cu, Pb, Ni	Municipal and industrial wastewater.	Removal rate Ni (99%) 80% removal percentage for rest of metals.	Bokhari et al. [227]

4.1.5. *Salvinia* (Water Fern)

Water Fern (*Salvinia auriculatais*), a small free-floating macrophyte, is extensively scattered in aquatic ecosystems. It has the ability to reproduce quickly and have the ability to settle widespread colonies in areas in no time. *Salvinia* can double its population within around 3 to 5 days under suitable conditions [228]. A substantial growth rate, ease to handle, wide distribution and sensitivity to various noxious entities support the application of *Salvinia* for used as a bio-indicator of pollution index and for phytoremediation as well [229].

Water Fern (*Salvinia*) species, especially *S. natans*, are potentially used in phytoremediation as it has an enormous capacity for removal of HMs due to the rapid growth rate and tolerance to toxic pollutants [230,231]. It can effectively be used for the treatment of different kinds of wastewater and waste produced in the constructed wetland [232]. Roots of *Salvinia* have a higher rate of metal accumulation. Accumulation of metal in *S. natans* and *S. minima* reduces As, with increasing concentration of phosphate while heavy-metal uptake increase with the addition of sulfur [233].

The presence of favorable environmental conditions along with the existence of certain nutrients and chelators will determine the fate of *Salvinia* in the hyperaccumulation of heavy metals [234]. Among different species of *Salvinia*, *S. minima* shows high bioaccumulation factor (BCF) for the accumulation of cadmium and lead [235]. *S. minima* has successfully been used for the remediation of high-strength synthetic organic wastewater [236]. Different species of the *Salvinia* (water fern) are an excellent accumulator of Fe, Cd, Ni, Mn, Zn and Pb as reported by several studies, and their details are given in Table 5.

Table 5. Recent studies on uptake of heavy metals by *Salvinia* species.

Metals/Metalloids	Condition	Results	Reference
Cd, Ni, Pb, Zn	Initial concentrations Cd-(0.03 mg/L) Ni-(0.40 mg/L) Pb-(1.00 mg/L) Zn-(1.00 mg/L)	Removal rate: Zn-(0.4046 mg/m^{-2}) Ni-(0.0595 mg/m^{-2}) Cd-(0.0045 mg/m^{-2}) Pb-(0.1423 mg/m^{-2})	Iha and Bianchini [129]
Zn, Cu, Ni and Cr	15 mg/L initial concentration. 10 g biomass of five plants.	Removal rate: Zn-(84.8%) Cu-(73.8%) Ni-(56.8%) Cr-(41.4%)	Dhir et al. [231]
Ni	0, 20, 40, 80, 160 M concentrations of NiCl2	Accumulation of Ni 16.3 mg/g	Fuentes et al. [237]
Cu, Cr, Pb, Cd	Initial concentration Cu-(1.092 ± 0.026) Cr-(2.201 ± 0.0024) Pb-(2.974 ± 0.018) Cd-(0.251 ± 0.017)	After treatment Cu-(2.035 ± 0.014) Cr-(1.052 ± 0.022) Pb-(1.924 ± 0.012) Cd-(0.018 ± 0.018)	Ranjitha et al. [238]
Pb, Ni, Cu, Zn, Mn, Fe, Cr, Cd	coal mine effluents	Removal rate Pb-(96.96%) Ni-(97.01%) Cu-(96.77%) Zn-(96.38%) Mn-(96.22%) Fe-(94.12%) Cr-(92.85%) Cd-(80.99%)	Lakra et al. [239]

4.2. Submerged Aquatic Plants

In submerged aquatic plants, leaves are the main part for metal uptake. Passive movement of the cuticle results in the absorption of heavy metals. Polygllalacturonic acid of the cell wall and negatively charged cutin and pectin polymers of cuticle results in the sucking inward of minerals. Movement of Positive metal ions takes place due to this inward enhanced charged density [59]. They have the ability to remove heavy metals from water and sediments [240–242]. Some of the famous submerged plants such as parrot feather (*Myriophyllum spicatum*), coontail or hornwort (*Ceratophyllumdemersum*), pondweed (*Potamogeton Crispus*), American pondweed (*Potamogetonpectinatus*), *Mentha Aquatica*, *Vallisneria spiralis* and water mint are well known for their ability to accumulate Zn, Cr, Fe, Cu, Cd, Ni, Hg and Pb [152,154,155,157,243].

4.3. Emergent Aquatic Plants

These plants are usually found on submerged soil where the water table is 0.5 m below the soil. Accumulation of HMs in emergent plants varies from plant to plant, they have the skill to bio-concentrate most of the metals in below ground-level roots from water and sediments, while some of the emergent plants, distribute the burden of metals in aerial parts as well. For example, smooth cordgrass (*spartinaalterniflora*) take up heavy metals in leaves [164], common reed (*Phragmites australis*) bears most of the heavy metal burden in the roots of the plant [166]. Sequestration and detoxification of heavy metals occur at the cellular level in these plants [244]. Cattail (*Typha latifolia*), bulrush (*Scirpus* spp.), common reed (*Phragmites*) and smartweed (*Polygonumhydropiperoides*) are the best emergent aquatic plant that can effectively be used for the phytoremediation of several HMs like Cd, Fe, Pb, Cr, Zn, Ni, Cu [160,167,168].

5. Significance of Aquatic Plants for Phytoremediation of Wastewater

Phytoremediation of heavy metals with aquatic plants has gained significant consideration due to its elegance and cost-effectiveness [39,227]. The earlier worker has demonstrated that aquatic plants have the capability to eliminate HMs from different kinds of wastewater [140,225,245,246]. Aquatic plants remove heavy metals via absorption or through surface adsorption and integrate them into their system, and then accumulation them in certain bounded forms [247,248]. Effluents from wastewater mitigated through the aquatic plants, thus causing less harm to the surrounding environment. A wide array of aquatic plants like water hyacinths, *Salvinia* sp., water lettuce, giant duckweed, and *Azolla* sp. have displayed tremendous ability for the phytoremediation of numerous kinds of wastewater [249,250]. This review briefly describes the effectiveness of these aquatic plants for the remediation of different types of wastewater.

5.1. Phytoremediation of Municipal Wastewater

Municipal wastewater possesses significant risk for the aquatic environment as it is a main cause of heavy metal pollution. Zn, Cu, Ni, Pb and Hg are potentially more noxious metals and they may cause chronic and acute health effects, bioaccumulation and phytotoxicity [251–253]. Application of aquatic plants for the removal of heavy metals from municipal wastewater, sewage water, spillage areas, and other polluted sites has become a common practice and experimental technique [254,255]. Aquatic plants can be used as bio-accumulators as they have the ability to accumulate high concentrations of HMs in their biomass [256,257]. Root and shoot tissues of *Typha domingensis* showed maximum accumulation of Zn, Cd, Ni, Fe and Mn during the first 48 h of study, planted in pots filled with municipal wastewater. [258]. *L. gibba* was studied for the accumulation of arsenic, boron, and uranium from the municipal wastewater as an alternative removal method. Results revealed that U, As and B were rapidly absorbed by the plant during the first 2 days of a 7-day experimental study [214]. Two rooted macrophytes *Typha angustifolia* and *Phragmites australis* removed 14–85% of heavy metals such as zinc, lead, arsenic, nickel, iron, copper, aluminum and magnesium from municipal wastewater in a hydroponic study [259]. Similarly, aquatic plants *Typha latifolia* and *Phragmites australis* showed excellent removal efficiency of heavy metals from the municipal wastewater. Both these aquatic plants showed higher removal rate for aluminum (96%), copper (91%), lead (88%) and zinc (85%) and slightly less removal rate for iron (44%), boron (40%) and cobalt (31%) [260].

5.2. Phytoremediation of Industrial Wastewater

Discharge of industrial waste into soil and water signifies a more critical threat to human health, living organisms, and other resources [261]. Phytoremediation, along with newly developed engineering and biological strategies, has facilitated the successful removal of HMs from industrial wastewater through both phytostabilization and phytoextraction [262]. Twelve aquatic plants were tested for their phytoremediation capability for different HMs originating from the industrial wastewater in Swabi district, Pakistan. Results demonstrated that these aquatic plants significantly removed heavy metals from industrial wastewater with excellent removal efficiencies: Cd (90%), Cr (89%), Fe (74.1%), Pb (50%), Cu (48.3%) and Ni (40.9%), respectively [263].

Southern cattail (*Typha domingensi*) showed maximum accumulation of zinc, aluminum, iron, and lead, especially in roots rather than leaves from the industrial wastewater pond. Rhizofiltration was found as dominant mechanisms, which explained the phytoremediation potential of *Typha domingensis* [264]. Promising aquatic plants showed better strength for the phytoremediation of the industrial effluents than other plants. Aquatic macrophytes *Marsileaquadrifolia*, *Hydrillaverticillata* and *Ipomeaaquati* cashowed much better accumulation potential and translocation factor (TF) value for HMs (Fe, Cr, Zn, Pb, As, Hg, Cd and Cu,) from the industrial effluents as compared to the terrestrial plants *Sesbaniacannabina*, *Eclipta alba* and algal species (*Phormidiumpapyraceum*, *Spirulina platensis*) [265]. A high diversity of aquatic plants has the following advantages for remediation: high removal efficiency, better habitat, and distribution resilience [47,58]. *T. domingensis* is such a dominant aquatic plant species having a high tolerance to a toxic environment and proficient in accumulation of HMs. Maine et al. [266] also reported that *T. domingensis* showed much better survival and removal efficiency for iron, zinc, nickel, and chromium released from industrial wastewater of metallurgy plant over other higher diversified aquatic plants. Such a type of aquatic plant can be used on a large scale to study the long-term removal performance.

5.3. Phytoremediation of Textile Wastewater

Wastewater from the textile industries is considered as most polluted wastewater among other industrial sectors [267]. Printing and dyeing process of textile industry effluents produce both organic and inorganic contaminants. Heavy metals in textile effluents are more toxic as they are more dangerous for public health [268]. Mahmood et al. [269] investigated the feasibility of *E. crassipes* for the eradication of copper, chromium, and zinc from five different textile industries from Lahore district, Pakistan. *E. crassipes* effectively removed 94.78% Cr, 96.88% Zn, and 94.44% Cu from the industrial wastewater sample during the investigation period of 96 h. In another study, *E. crassipes* removed 94.87% of cadmium from the textile wastewater [114]. It is well documented that amongst different aquatic plants, water hyacinth is the superlative contender for the phytoremediation of textile industry effluents [270]. Aquatic plants *Pistia stratiotes*, *Azollapinnata*, and *Salvinia, molesta* were found very competent for the elimination of Fe, Cu and Mn at 25% concentration of the textile effluents [271]. A hairy root system of aquatic plants plays a vital part in the remediation of pollutants from wastewater in phytoremediation [272].

Roy et al. [273] investigated the remediation capability of three free-floating (*Eichhornia crassipes*, *Pistia stratiotes*, *Spirodelapolyrhiza*) aquatic plants for Cu, Pb, As and Cr from textile wastewater effluents. These macrophytes expressed an extensive uptake tendency for heavy metals, and *Eichhornia crassipes* was detected as the most competent contender in the remediation of HMs due to its fibrous widespread root system. Similarly, Ajayi and Ogunbayo, [274] also reported the effectiveness of water hyacinth in remediation of Fe, Cu, and Cd from textile effluents. High removal percentage (70–90%) for various heavy metals such as copper, chromium, zinc, iron, and lead from textile wastewater was observed with the water hyacinth as reported by different researchers [269,275,276]. Duckweed (*Lemna minor*) also showed great potential for the removal of Cr, Zn, Pb, and Cd from the textile wastewater [185]. It has been reported previously that the application of an aquatic macrophytes treatment system (AMTS) is beneficial for the remediation of textile wastewater [277].

5.4. Phytoremediation of Mining Effluents

Mining activities harmfully affect the whole environment and put an incredible burden on local fauna and flora. The process of mining operations includes the discharge of an enormous amount of toxic effluents into the aquatic environment [278]. Effluents of mining activities hold a much higher concentration of different pollutants like calcium carbonate, TDS, TSS and heavy metals [279]. Heavy metals originating from the mining effluents are very persistent in nature and can easily accumulate in the soil, water, sediment and also have the ability to enter the food chain via bioaccumulation and assimilation thus affecting the health of human and animals [280]. Various methods have been developed around the world to remove the HMs. Phytoremediation is such a method that showed promising results in the successful remediation of heavy metals originated from the mining effluents by employing aquatic macrophytes [281].

As indicated by Sasmaz et al. [246], aquatic plants were very effective in the removal of remove HMs from mining effluents. Mishra et al. [282] explore the potential of three aquatic plants *Spirodela Polyrhiza, Eichhornia crassipes* and *Lemna minor* for the effective elimination of heavy metals. *Eichhornia crassipes* removed much a higher percentage of heavy metals than the other two macrophytes. *Eichhornia crassipes* eliminated Fe, Cr, Cu by 70.5%, 69.1%, 76.9% from the mining effluents. Similarly, *Eichhornia crassipes* effectively removed Cr (VI) by 99.5% from industrial mine effluents in 15 days of the experimental period [283]. Aquatic plant *Limnocharis flava* significantly remove the Hg from the mining effluents in a pilot scale experimental study of 30 days [284]. Most widely used aquatic plants used in the phytoremediation of mine effluents are floating, submerged and emergent. Emergent plants usually promote the elimination of HMs from the mine effluents via collective processes like retention and uptake of heavy metals over their respective tissues [285]. The emergent plant used in the remediation of heavy metals from mine effluents include *Phragmites australis, P. australis, P. karka, P. australis and T. dominguensis* [286,287]. Floating aquatic plants cannot improve adsorption via the substrate, however, they promote adsorption process to the plant biomass. Successfully used floating aquatic plants in the treatment of mine effluents include *Salvinia natans, Pistia stratiotes, Eichornia crassipes* [239,288]. Submerged aquatic macrophytes like *Ceratophylum demersum, Cabomba piauhyensis, Egeria densa, Myriophylum spicatum* and *Hydrilla verticillata* are recommended to be used for the phytoremediation of mine effluents as they have shown outstanding ability to accumulate HMs in their whole body biomass [289,290].

5.5. Phytoremediation of Landfill Leachate

Landfilling and open dumping are the most common way of treating municipal solid waste (MSW) worldwide [291]. Leachate forms as a result of interaction among waste in landfill, water from the soil, and different types of other liquid contaminants disposed of in the landfill. Intermittent and non-uniform percolation of moisture content occurrs via solid waste in landfill, which eventually leads towards the generation of landfill leachate [292]. Generated landfill leachate if not properly managed, can easily lead towards numerous adverse health and environmental impacts [293]. One of the major constraint in the management of landfill leachate is the lack of effective treatment methods for the huge amount of landfill leachate generated worldwide [294]. Different chemical and physicochemical approaches have been used to eradicate pollutants from the leachate. Unfortunately, these methods are generally expensive and complicated as well. Economically viable and environmentally friendly option is a priority in landfill leachate management. Jones et al. [295] reported that plant-based remediation technology i.e., phytoremediation is very successful in the treatment of landfill leachate. Aquatic plants such as *Gynerium sagittatum (Gs), Colocasia esculenta (Ce), Heliconia psittacorum (He)* have shown tremendous phytoremediation potential for the remediation of landfill leachate [296].

Aquatic plants have the ability to withstand the high pollution load of the landfill leachate without showing any sign of a significant cutback in biomass and growth rate [297]. *Eichhornia crassipes* in a floating system has shown the tremendous capacity of removing diverse HMs such as Cu, Ni, Pb, Cd and Cr from the landfill leachate, thus reducing the pollution density of the landfill leachate [298]. Many researchers have utilized *Eichhornia crassipes* for the successful eradication of contaminants from the landfill leachate [299,300]. Ugya and Priatamby [301] also reported high removal efficiencies for various heavy metals form the landfill leachate generated from the landfill site with the assistance of *Pistia stratiotes*. Application of Duckweed (*Lemna minor*) also showed a significant reduction in copper, zinc, lead, nickel, and iron from landfill leachate [131].

Extensive root systems of aquatic plants enhance the capacity of these aquatic macrophytes to extract large concentration of HMs via the root system and then transport them to aboveground parts of the plant body [302]. The depth of root zone in aquatic plants is a vital impediment in successful phytoremediation of landfill leachate. Extensive root systems of the aquatic plants serve as main entry route of heavy metals, and these plants mostly store these HMs in roots then leaves and stem as observed in earlier reports [282,303]. Grisey et al. [304] reported that *P. australis* accumulates large concentration of HMs from landfill leachate in its root zone more resourcefully than other part of the body. Similarly *P. Cyperus Papyrus* also showed maximum accumulation of lead in its root than shoots during the treatment of the Kiteezi landfill site, Uganda [305]. Parallel outcomes were also examined by [306]. Thus it is highly recommended that planting of aquatic plants should be promoted for the treatment of landfill leachate in order to avoid seepage of heavy metals and other contaminants from landfill leachate onto aquifer to contaminate the water bodies during ultimate discharge and runoff of leachate [301].

6. Role of Aquatic Plants in Constructed Wetlands

Remediation of wastewater through constructed wetland has been magnificently executed over the last few decades worldwide as an appropriate management choice for wastewater [307]. Constructed wetlands (CWs) are designed to treat distinct form of wastewater within the controlled environment. A broad range of wastewaters such as agricultural [308], municipal [309], landfill leachate [296], storm water [310] and industrial wastewater [311] can be remediated in constructed wetlands. The constructed wetland provides a comparatively simple and cheap solution for controlling water contamination without disturbing resources of natural wetlands [312]. Aquatic plants are an imperative constituent in CWs for the remediation of wastewater. Aquatic plants in CWs have two significant indirect functions: (1) leaves and stem of the aquatic macrophytes ehance the surface area for significant attachment of microbial communities, (2) aquatic plants have the aptitude to transport gases like oxygen down towards the root zone to allow their roots to subsist in the anaerobic environment [313]. Rhizosphere excessively support the microbial communities that handle the necessary alteration of metallic ions, different compounds, and nutrients [314]. Therefore, the application of aquatic macrophytes in CWs helps in the remediation of wastewater polluted with different contaminants and also acts as a sink for the contaminants [285].

Elimination of heavy metals in CWs depends on the kind of metallic elements, their ionic form, season, substrate condition, and kind of plant species [315]. Dense population of aquatic plants in CWs considerably increased the effectiveness of HM remediation from the wastewater [316,317]. Aquatic plants play a precise and vigorous part in maintaining the biochemistry of wetlands via their active and passive circulation of essential ingredients [318]. Heavy-metal concentration in wetland aquatic macrophytes generally decreased in the following order: root > leaves > stems [319,320]. However, the concentration of heavy metals does not deliver sufficient evidence regarding the uptake of heavy metals in aquatic plants in wetlands. In wetlands, uptake of heavy metals depends heavily upon the biomass of the particular aquatic plant [321]. *E. crassipes* is one such plant which has the capability to double its biomass within a few days under favorable conditions. Most recently, Rai [322] reported the water hyacinth (*E. crassipes*) as the most appropriate wetland plant for the phytoremediation of metals from wastewater. Use of *E. crassipes* in the constructed wetland to remove heavy metals has been recommended as the best choice in order to make use of *E. crassipes* (nuisance weed) effectively throughout the world. Sukumaran et al. [323] reported the utilization of *E. crassipes* for the phytoremediation of Cd, Pb, Cu, Ar from industrial discharge by applying constructed wetland technology. *E. crassipes* showed much higher remediation potential for Cu, Ni, Fe, Cd, Zn, Cr than the other two free-floating aquatic plants *Pistia Stratiotes*, *Spirodela polyrhiza* during a 15-day experiment.

Ladislas et al. [324] indicated the accumulation of Cd, Zn and Ni in aquatic floating macrophytes *J. effusus* and *C. riparia* growing in wetlands receiving storm water. The ratio of HMs concentration was significant in roots then shoots. Dan et al. [325] examined the accumulation of different HMs such as Fe, Cd, Zn, Ni, Pb and Cr by *Juncus effuses* and *Phragmites australis* from landfill leachate through a lab-scale constructed wetland. Both aquatic plants showed much higher removal efficacy for the targeted metals. Similarly, Leung et al. [285] also reported high removal percentages for heavy metals with three aquatic plants (*Phragmites australis*, *Cyperus malaccensis* and *Typha latifolia*,) in CWs receiving wastewater from the mining industries. The phytoremediation potential of three aquatic wetland plants i.e., *Cyperus alternifolius, Cynodon dactylon* and *Typha latifolia,* was examined for the transfer and translocation of HMs from the root zone to upper parts of the body in a constructed wetland receiving refinery wastewater. Results affirmed that the highest concentration of Cr, Zn, Cd, Pb and Fe were accumulated by roots of the plants followed by leaves and stem [326]. Similarly, *Typha latifolia* showed maximum removal efficiency of 96%, 95% and 80% for Cd, Cr and Pb correspondingly in a laboratory-scale constructed wetland unit. [327]. A CW with *Phragmites australis* was assessed for the phytoremediation of municipal wastewater. Most of the metals were significantly removed from the municipal wastewater with reasonable efficiencies. Results demonstrated that *Phragmites australis* accumulated most of the heavy metals in their belowground part, and only a minor fraction of metals translocated two aboveground biomass of the plant [328]. Vymazal et al. [329] observed a maximum amount of HMs usually found in belowground biomass, while the lowest concentration of HMs were detected in aboveground biomass of the wetland plants.

Hadad et al. [330] also stated a similar trend of much higher accumulation of HMs in the root zone than upper parts of the plant. Plants frequently tolerate the high concentration of metals because they restrict the accumulation and absorption to the leaves upholding a constant and comparatively low concentration of HMs in aboveground parts of the plant. Research over the past decade has shown aquatic plants contribute significantly in the elimination of heavy metals through constructed wetland technology [266,318,331,332]. Discrepancies with the remediation of HMs through aquatic wetland plants in CWs might be attributed to various aspects including the type of wetland, inflow load of heavy metals and type of wetland plants. Nevertheless, the type of aquatic plant employed in the CW system is one of the main prominent element in the remediation of metals from wastewater [333]. Further investigation is needed to increase the removal efficiency of these aquatic plants within constructed wetlands.

7. Other Advantages of Aquatic Plants

The study of phytoremediation reveals that the aquatic macrophytes have the advantage over other plants in the remediation of heavy metals [71,131,257]. The widespread availability, rapid growth rate, high biomass, cost-effectiveness and tolerance to toxic pollutants make them the best suited, available phytoremediation plants. Purifications system using these aquatic plants have gained more attention worldwide because of their capacity to accumulate and remove of a persistent organic pollutant from water bodies [48,131].

Involvement of appropriate phytoremediation technology needs intervallic harvesting of the plant biomass in order to assimilate and confiscate heavy metals and nutrients from water bodies. Conversion of biomass into exclusive material is the significant factor in promoting this technique for the treatment of contaminants. An aquatic plant's biomass can latter on be used as animal feed, useful in the production of biogas and compost as reported in many studies [25]. Bio-sorption along with bioaccumulation of *Lemna minor* biomass was inspected which indicates its possible use as animal feed [131,334]. Aquatic plants possess sugar in the shape of starch, cellulose, and hemicelluloses. Carbohydrate hydrolysis of this fermentable sugar results in the production of lactic acid, ethanol, and other important products. Therefore, sugar present in aquatic plants is a new promising feature supporting their role in the eco-sustainable environment. Aquatic plants i.e., *Pistia stratiotes* and *Eichhornia crassipes,* have been reported to produce sugar during their enzymatic hydrolysis process [335]. Free-floating aquatic plants (*Azolla* spp., *Wolffia* spp., *Spirodela* sp. and *Duckweeds*) can be used as a food source for water bird. They also provide shelter for insect larvae and small mollusks. Fishes also use the mats of these plants as cover and use their shade for reproduction [336]. Aquatic plants can be efficiently used to improve the aquaculture for fishponds. Elimination of nitrogenous waste of aquatic plants is also an added benefit for their application in aquaculture, i.e., *Canna generalis* L., *Typha angustifolia. Echinoduruscordifolius,* and *Cyperus involucratus* removed ammonia, nitrate and nitrite efficiently [337].

8. Conclusions and Future Prospects

Heavy metals in our environment as a persistent pollutant needs absolute elimination for a completely remedial objective. Utilization of phytoremediation seems to be a less disruptive, economical and environmentally sound clean-up technology. Choice of appropriate plant is the most significant feature in phytoremediation. Aquatic plants perform very vibrant roles in the remediation of heavy metals from the polluted site with equal ease to other hyperaccumulator plants. Application of aquatic plants both in bioaccumulation (with living plant biomass) and bio-sorption (with dead plant biomass) can be done successfully for the eradication of heavy metals.

Comprehensive interaction, transport, and chelator activities regulate the storage and accumulation of heavy metals by the aquatic macrophytes. Genetic engineering enhances the accumulation and tolerance capacity of plants, which shows its exceptional application in improving the effectiveness of phytoremediation. In plants, at the molecular level, different extensive steps have been evaluated that favor the transgenic methods in order to plead with the changeover metal fraction of plants. Genetically engineered plants show high tolerance and metal uptake capacity and, as a result, gene manipulation has successfully been investigated in terrestrial plants, but, genetic engineering of aquatic plants to enhance their heavy-metal uptake capacity is in its preliminary phases.

Disposal of plant biomass later on, can be used for the production of biogas and also can be used as animal feed. The application of aquatic plants in phytoremediation like other conventional physical and chemical techniques does not require any post-filtration and can be effectively used to treat a large volume of polluted water and soil. Based on the present review, the benefits of using aquatic plants to treat contaminants are huge, because this technology does not only treat the contaminants but is cost-effective and visually pleasing as well as being advantageous for the sustainability of whole ecosystems.

Funding: The authors are thankful for the financial support from the Government College University, Faisalabad. This research work was funded by the Deanship of Scientific Research at Princess Nourah bint Abdulrahman University through the Fast-Track Research Funding Program.

Acknowledgments: The authors are thankful for the financial support from the Government College University, Faisalabad. This research work was funded by the Deanship of Scientific Research at Princess Nourah bint Abdulrahman University through the Fast Track Research Funding Program.

Conflicts of Interest: The authors declare that there is no conflict of interests regarding the publication of this paper.

References

1. Calzadilla, A.; Rehdanz, K.; Tol, R.S. Water scarcity and the impact of improved irrigation management: A computable general equilibrium analysis. *Agric. Econ.* **2011**, *42*, 305–323. [CrossRef]
2. Connell, D.W. *Pollution in Tropical Aquatic Systems*; CRC Press: Boca Raton, FL, USA, 2018.
3. Aguliar, M.J. Olive oil mill wastewater for soil nitrogen and carbon conservation. *J. Environ. Manag.* **2009**, *90*, 2845–2848. [CrossRef] [PubMed]
4. Rahman, M.A.; Hasegawa, H. Aquatic arsenic: Phytoremediation using floating macrophytes. *Chemosphere* **2011**, *83*, 633–646. [CrossRef] [PubMed]
5. Renuka, N.; Sood, A.; Ratha, S.S.K.; Prasanna, R.; Ahluwalia, A.S. Evaluation of microalgal consortia for treatment of primary treated sewage effluent and biomass production. *J. Appl. Phycol.* **2013**, *25*, 1529–1537. [CrossRef]
6. Goncalves, A.L.; Pires, J.C.; Simões, M. A review on the use of microalgal consortia for wastewater treatment. *Algal Res.* **2017**, *24*, 403–415. [CrossRef]
7. Ahmed, M.B.; Zhou, J.L.; Ngo, H.H.; Guo, W.; Thomaidis, N.S.; Xu, J. Progress in the biological and chemical treatment technologies for emerging contaminant removal from wastewater: A critical review. *J. Hazard Mater.* **2017**, *323*, 274–298. [CrossRef]
8. Carstea, E.M.; Bridgeman, J.; Baker, A.; Reynolds, D.M. Fluorescence spectroscopy for wastewater monitoring: A review. *Water Res.* **2016**, *95*, 205–219. [CrossRef]

9. Mendoza, R.E.; García, I.V.; de Cabo, L.; Weigandt, C.F.; de Iorio, A.F. The interaction of heavy metals and nutrients present in soil and native plants with arbuscular mycorrhizae on the riverside in the Matanza-Riachuelo River Basin (Argentina). *Sci. Total Environ.* **2015**, *505*, 555–564. [CrossRef]

10. Tee, P.F.; Abdullah, M.O.; Tan, I.A.W.; Rashid, N.K.A.; Amin, M.A.M.; Nolasco-Hipolito, C.; Bujang, K. Review on hybrid energy systems for wastewater treatment and bio-energy production. *Renew. Sustain. Energy Rev.* **2016**, *54*, 235–246. [CrossRef]

11. Pakdel, P.M.; Peighambardoust, S.J. A review on acrylic based hydrogels and their applications in wastewater treatment. *J. Environ. Manag.* **2018**, *217*, 123–143. [CrossRef]

12. An, B.; Lee, C.-G.; Song, M.-K.; Ryu, J.-C.; Lee, S.; Park, S.-J.; Lee, S.H. Applicability and toxicity evaluation of an adsorbent based on jujube for the removal of toxic heavy metals. *React. Funct. Polym.* **2015**, *93*, 138–147. [CrossRef]

13. Peligro, F.R.; Pavlovic, I.; Rojas, R.; Barriga, C. Removal of heavy metals from simulated wastewater by in situ formation of layered double hydroxides. *Chem. Eng. J.* **2016**, *306*, 1035–1040. [CrossRef]

14. Raval, N.P.; Shah, P.U.; Shah, N.K. Adsorptive removal of nickel (II) ions from aqueous environment: A review. *J. Environ. Manag.* **2016**, *179*, 1–20. [CrossRef] [PubMed]

15. Gall, J.E.; Boyd, R.S.; Rajakaruna, N. Transfer of heavy metals through terrestrial food webs: A review. *Environ. Monit. Assess.* **2015**, *187*, 201. [CrossRef]

16. Zhu, C.; Tian, H.; Cheng, K.; Liu, K.; Wang, K.; Hua, S.; Zhou, J. Potentials of whole process control of heavy metals emissions from coal-fired power plants in China. *J. Clean. Prod.* **2016**, *114*, 343–351. [CrossRef]

17. Al-Alawy, A.F.; Salih, M.H. Comparative Study between Nanofiltration and Reverse Osmosis Membranes for the Removal of Heavy Metals from Electroplating Wastewater. *J. Eng.* **2017**, *23*, 1–21.

18. Levchuk, I.; Màrquez, J.J.R.; Sillanpää, M. Removal of natural organic matter (NOM) from water by ion exchange—A review. *Chemosphere* **2018**, *192*, 90–104. [CrossRef]

19. Huang, H.; Liu, J.; Zhang, P.; Zhang, D.; Gao, F. Investigation on the simultaneous removal of fluoride, ammonia nitrogen and phosphate from semiconductor wastewater using chemical precipitation. *Chem. Eng. J.* **2017**, *307*, 696–706. [CrossRef]

20. Burakov, A.E.; Galunin, E.V.; Burakova, I.V.; Kucherova, A.E.; Agarwal, S.; Tkachev, A.G.; Gupta, V.K. Adsorption of heavy metals on conventional and nanostructured materials for wastewater treatment purposes: A review. *Ecotoxicol. Environ. Saf.* **2018**, *148*, 702–712. [CrossRef]

21. Kulkarni, P.; Olson, N.D.; Paulson, J.N.; Pop, M.; Maddox, C.; Claye, E.; Mongodin, E.F. Conventional wastewater treatment and reuse site practices modify bacterial community structure but do not eliminate some opportunistic pathogens in reclaimed water. *Sci. Total Environ.* **2018**, *639*, 1126–1137. [CrossRef]

22. Olguín, E.J.; Sánchez-Galván, G. Heavy metal removal in phytofiltration and phycoremediation: The need to differentiate between bioadsorption and bioaccumulation. *New Biotechnol.* **2012**, *30*, 3–8. [CrossRef] [PubMed]

23. Grandclément, C.; Seyssiecq, I.; Piram, A.; Wong-Wah-Chung, P.; Vanot, G.; Tiliacos, N.; Doumenq, P. From the conventional biological wastewater treatment to hybrid processes, the evaluation of organic micropollutant removal: A review. *Water Res.* **2017**, *111*, 297–317. [CrossRef] [PubMed]

24. González-González, A.; Cuadros, F.; Ruiz-Celma, A.; López-Rodríguez, F. Influence of heavy metals in the biomethanation of slaughterhouse waste. *J. Clean. Prod.* **2014**, *65*, 473–478. [CrossRef]

25. Shahid, M.J.; Arslan, M.; Ali, S.; Siddique, M.; Afzal, M. Floating Wetlands: A Sustainable Tool for Wastewater Treatment. *Clean-Soil Air Water* **2018**, *46*, 1800120. [CrossRef]

26. Hargreaves, A.J.; Constantino, C.; Dotro, G.; Cartmell, E. Campo. Fate and removal of metals in municipal wastewater treatment: A review. *Environ. Technol. Rev.* **2018**, *7*, 1–18. [CrossRef]

27. Zhong, L.; Liu, L.; Yang, J. Assessment of heavy metals contamination of paddy soil in Xiangyin county, China. In Proceedings of the 19th World Congress of Soil Science: Soil Solutions for a Changing World, Brisbane, Australia, 1–6 August 2010; Symposium 4.1.2 Management and protection of receiving environments.

28. El-Gammal, M.; Ali, R.; Samra, R.A. Assessing heavy metal pollution in soils of Damietta Governorate, Egypt. In Proceedings of the International Conference on Advances in Agricultural, Biological and Environmental Sciences, Dubai, UAE, 14–15 October 2014; Volume 1, pp. 116–124.

29. Zarcinas, B.A.; Ishak, C.F.; McLaughlin, M.J.; Cozens, G. Heavy metals in soils and crops in Southeast Asia. *Environ. Geochem. Health* **2004**, *26*, 343–357. [CrossRef]

30. Sahibin, A.; Zulfahmi, A.; Lai, K.; Errol, P.; Talib, M. Heavy metals content of soil under vegetables cultivation in Cameron Highland. In Proceedings of the Regional Symposium on Environment and Natural Resources, Kuala Lumpur, Malaysia, 10–11 April 2002.

31. Nagajyoti, P.C.; Lee, K.D.; Sreekanth, T. Heavy metals, occurrence and toxicity for plants: A review. *Environ. Chem. Lett.* **2010**, *8*, 199–216. [CrossRef]

32. Uddin, M.K. A review on the adsorption of heavy metals by clay minerals, with special focus on the past decade. *Chem. Eng. J.* **2017**, *308*, 438–462. [CrossRef]

33. Czarnecki, S.; Düring, R.A. Influence of long-term mineral fertilization on metal contents and properties of soil samples taken from different locations in Hesse, Germany. *Soil* **2015**, *1*, 23–33. [CrossRef]

34. Sardar, K.; Ali, S.; Hameed, S.; Afzal, S.; Fatima, S.; Shakoor, M.B.; Tauqeer, H.M. Heavy metals contamination and what are the impacts on living organisms. *Greener J. Environ. Manag. Public Saf.* **2013**, *2*, 172–179.

35. Paul, D. Research on heavy metal pollution of river Ganga: A review. *Ann. Agrar. Sci.* **2017**, *15*, 278–286. [CrossRef]

36. Islam, M.A.; Romić, D.; Akber, M.A.; Romić, M. Trace metals accumulation in soil irrigated with polluted water and assessment of human health risk from vegetable consumption in Bangladesh. *Environ. Geochem. Health* **2018**, *40*, 59–85. [CrossRef] [PubMed]

37. Sanchez-Martin, M.; Sanchez-Camazano, M.; Lorenzo, L. Cadmium and lead contents in suburban and urban soils from two medium-sized cities of Spain: Influence of traffic intensity. *Bull. Environ. Contam. Toxicol.* **2000**, *64*, 250–257. [CrossRef] [PubMed]

38. Ashraf, S.; Afzal, M.; Naveed, M.; Shahid, M.; Zahir, Z.A. Endophytic bacteria enhance remediation of tannery effluent in constructed wetlands vegetated with Leptochloa fusca. *Int. J. Phytoremediat.* **2018**, *20*, 121–128. [CrossRef]

39. Sharma, S.; Singh, B.; Manchanda, V. Phytoremediation: Role of terrestrial plants and aquatic macrophytes in the remediation of radionuclides and heavy metal contaminated soil and water. *Environ. Sci. Pollut. Res.* **2015**, *22*, 946–962. [CrossRef]

40. Blaylock, M. *Phytoremediation of Contaminated Soil and Water: Field Demonstration of Phytoremediation of Lead Contaminated Soils*; Lewis Publishers: Boca Raton, FL, USA, 2008.

41. Sarwar, N.; Imran, M.; Shaheen, M.R.; Ishaque, W.; Kamran, M.A.; Matloob, A.; Hussain, S. Phytoremediation strategies for soils contaminated with heavy metals: Modifications and future perspectives. *Chemosphere* **2017**, *171*, 710–721. [CrossRef]

42. Kushwaha, A.; Hans, N.; Kumar, S.; Rani, R. A critical review on speciation, mobilization and toxicity of lead in soil-microbe-plant system and bioremediation strategies. *Ecotoxicol. Environ. Saf.* **2018**, *147*, 1035–1045. [CrossRef]

43. Helmisaari, H.-S.; Salemaa, M.; Derome, J.; Kiikkilä, O.; Uhlig, C.; Nieminen, T. Remediation of heavy metal–contaminated forest soil using recycled organic matter and native woody plants. *J. Environ. Qual.* **2007**, *36*, 1145–1153. [CrossRef]

44. Mahar, A.; Wang, P.; Ali, A.; Awasthi, M.K.; Lahori, A.H.; Wang, Q.; Zhang, Z. Challenges and opportunities in the phytoremediation of heavy metals contaminated soils: A review. *Ecotoxicol. Environ. Saf.* **2016**, *126*, 111–121. [CrossRef]

45. Ensley, B.D. *Phytoremediation of Toxic Metals: Using Plants to Clean up the Environment*; John Wiley & s Sons, Incorporated: Hoboken, NJ, USA, 2000.

46. Sová, A.H.; Jakabová, S.; Simon, L.; Pernyeszi, T.; Majdik, C. Induced phytoextraction of lead from contaminated soil. *Agric. Environ.* **2009**, *1*, 116–122.

47. Gerhardt, K.E.; Gerwing, P.D.; Greenberg, B.M. Opinion: Taking phytoremediation from proven technology to accepted practice. *Plant Sci.* **2017**, *256*, 170–185. [CrossRef] [PubMed]

48. Leguizamo, M.A.O.; Gómez, W.D.F.; Sarmiento, M.C.G. Native herbaceous plant species with potential use in phytoremediation of heavy metals, spotlight on wetlands—A review. *Chemosphere* **2017**, *168*, 1230–1247. [CrossRef] [PubMed]

49. Hussain, F.; Hussain, I.; Khan, A.H.A.; Muhammad, Y.S.; Iqbal, M.; Soja, G.; Yousaf, S. Combined application of biochar, compost, and bacterial consortia with Italian ryegrass enhanced phytoremediation of petroleum hydrocarbon contaminated soil. *Environ. Exp. Bot.* **2018**, *153*, 80–88. [CrossRef]

50. Bennett, L.E.; Burkhead, J.L.; Hale, K.L.; Terry, N.; Pilon, M.; Pilon-Smits, E.A. Analysis of transgenic Indian mustard plants for phytoremediation of metal-contaminated mine tailings. *J. Environ. Qual.* **2003**, *32*, 432–440. [CrossRef] [PubMed]

51. Reeves, R.D.; Baker, A.J.; Jaffré, T.; Erskine, P.D.; Echevarria, G.; van der Ent, A. A global database for plants that hyperaccumulate metal and metalloid trace elements. *New Phytol.* **2018**, *218*, 407–411. [CrossRef]

52. Ali, H.; Khan, E.; Sajad, M.A. Phytoremediation of heavy metals—Concepts and applications. *Chemosphere* **2013**, *91*, 869–881. [CrossRef]

53. Arslan, M.; Imran, A.; Khan, Q.M.; Afzal, M. Plant–bacteria partnerships for the remediation of persistent organic pollutants. *Environ. Sci. Pollut. Res.* **2017**, *24*, 4322–4336. [CrossRef]

54. Burges, A.; Alkorta, I.; Epelde, L.; Garbisu, C. From phytoremediation of soil contaminants to phytomanagement of ecosystem services in metal contaminated sites. *Int. J. Phytoremediat.* **2018**, *20*, 384–397. [CrossRef]

55. Cunningham, S.D.; Ow, D.W. Promises and prospects of phytoremediation. *Plant Physiol.* **1996**, *110*, 715. [CrossRef]

56. Tewes, L.J.; Stolpe, C.; Kerim, A.; Krämer, U.; Müller, C. Metal hyperaccumulation in the Brassicaceae species Arabidopsis halleri reduces camalexin induction after fungal pathogen attack. *Environ. Exp. Bot.* **2018**, *153*, 120–126. [CrossRef]

57. Erakhrumen, A.A. Phytoremediation: An environmentally sound technology for pollution prevention, control and remediation in developing countries. *Educ. Res. Rev.* **2017**, *2*, 151–156.

58. Chandra, R.; Kumar, V.; Tripathi, S.; Sharma, P. Heavy metal phytoextraction potential of native weeds and grasses from endocrine-disrupting chemicals rich complex distillery sludge and their histological observations during in-situ phytoremediation. *Ecol. Eng.* **2018**, *111*, 143–156. [CrossRef]

59. Prasad, M.; Greger, M.; Aravind, P. Biogeochemical cycling of trace elements by aquatic and wetland plants: Relevance to phytoremediation. *Trace Elem. Environ. Biogeochem. Biotechnol Bioremediat.* **2005**, *1*, 451–474.

60. Kocoń, A.; Jurga, B. The evaluation of growth and phytoextraction potential of Miscanthus x giganteus and Sida hermaphrodita on soil contaminated simultaneously with Cd, Cu, Ni, Pb, and Zn. *Environ. Sci. Pollut. Res.* **2017**, *24*, 4990–5000. [CrossRef] [PubMed]

61. Cundy, A.B.; Bardos, R.; Church, A.; Puschenreiter, M.; Friesl-Hanl, W.; Müller, I.; Vangronsveld, J. Developing principles of sustainability and stakeholder engagement for "gentle" remediation approaches: The European context. *J. Environ. Manag.* **2013**, *129*, 283–291. [CrossRef]

62. Najeeb, U.; Ahmad, W.; Zia, M.H.; Zaffar, M.; Zhou, W. Enhancing the lead phytostabilization in wetland plant *Juncus effusus* L. through somaclonal manipulation and EDTA enrichment. *Arab. J. Chem.* **2017**, *10*, 3310–3317. [CrossRef]

63. Jadia, C.D.; Fulekar, M. Phytoremediation of heavy metals: Recent techniques. *Afr. J. Biotechnol.* **2009**, *8*, 921–928.

64. Labidi, S.; Firmin, S.; Verdin, A.; Bidar, G.; Laruelle, F.; Douay, F.; Sahraoui, A.L.H. Nature of fly ash amendments differently influences oxidative stress alleviation in four forest tree species and metal trace element phytostabilization in aged contaminated soil: A long-term field experiment. *Ecotoxicol. Environ. Saf.* **2017**, *138*, 190–198. [CrossRef]

65. Mahdavian, K.; Ghaderian, S.M.; Torkzadeh-Mahani, M. Accumulation and phytoremediation of Pb, Zn, and Ag by plants growing on Koshk lead–zinc mining area, Iran. *J. Soils Sediments* **2017**, *17*, 1310–1320. [CrossRef]

66. Zeng, P.; Guo, Z.; Cao, X.; Xiao, X.; Liu, Y.; Shi, L. Phytostabilization potential of ornamental plants grown in soil contaminated with cadmium. *Int. J. Phytoremediat.* **2018**, *20*, 311–320. [CrossRef]

67. Abhilash, P.; Jamil, S.; Singh, N. Transgenic plants for enhanced biodegradation and phytoremediation of organic xenobiotics. *Biotechnol. Adv.* **2009**, *27*, 474–488. [CrossRef] [PubMed]

68. Benavides, L.C.L.; Pinilla, L.A.C.; Serrezuela, R.R.; Serrezuela, W.F.R. Extraction in Laboratory of Heavy Metals Through Rhizofiltration using the Plant Zea Mays (maize). *Int. J. Appl. Environ. Sci.* **2018**, *13*, 9–26.

69. Zhu, Y.; Zayed, A.; Qian, J.; De Souza, M.; Terry, N. Phytoaccumulation of trace elements by wetland plants: II. Water hyacinth. *J. Environ. Qual.* **1999**, *28*, 339–344. [CrossRef]

70. Raskin, I.; Ensley, B.D. *Phytoremediation of Toxic Metals*; John Wiley and Sons: Hoboken, NJ, USA, 2000.

71. Galal, T.M.; Eid, E.M.; Dakhil, M.A.; Hassan, L.M. Bioaccumulation and rhizofiltration potential of *Pistia stratiotes* L. for mitigating water pollution in the Egyptian wetlands. *Int. J. Phytoremediat.* **2018**, *20*, 440–447. [CrossRef] [PubMed]

72. Sreelal, G.; Jayanthi, R. Review on phytoremediation technology for removal of soil contaminant. *Indian J. Sci. Res.* **2017**, *14*, 127–130.

73. Ghosh, M.; Singh, S. A review on phytoremediation of heavy metals and utilization of it's by products. *Asian J. Energy Environ.* **2005**, *6*, 18.

74. Karami, A.; Shamsuddin, Z.H. Phytoremediation of heavy metals with several efficiency enhancer methods. *Afr. J. Biotechnol.* **2010**, *9*, 3689–3698.

75. Cristaldi, A.; Conti, G.O.; Jho, E.H.; Zuccarello, P.; Grasso, A.; Copat, C.; Ferrante, M. Phytoremediation of contaminated soils by heavy metals and PAHs. A brief review. *Environ. Technol. Innov.* **2017**, *8*, 309–326. [CrossRef]

76. Lombi, E.; Zhao, F.; Dunham, S.; McGrath, S. Phytoremediation of heavy metal—Contaminated soils. *J. Environ. Qual.* **2001**, *30*, 1919–1926. [CrossRef]

77. Farid, M.; Ali, S.; Shakoor, M.B.; Bharwana, S.A.; Rizvi, H.; Ehsan, S.; Tauqeer, H.M.; Iftikhar, U.; Hannan, F. EDTA assisted phytoremediation of cadmium, lead and zinc. *Int. J. Agron. Plant Prod.* **2013**, *4*, 2833–2846.

78. Hadi., F.; Ali., N.; Ahmad, A. Enhanced phytoremediation of cadmium-contaminated soil by *Parthenium hysterophorus* plant: Effect of gibberellic acid (GA3) and synthetic chelator, alone and in combinations. *Biorem. J.* **2014**, *18*, 46–55. [CrossRef]

79. Pereira, B.F.F.; Abreu, C.A.D.; Herpin, U.; Abreu, M.F.D.; Berton, R.S. Phytoremediation of lead by jack beans on a Rhodic Hapludox amended with EDTA. *Sci. Agric.* **2010**, *67*, 308–318. [CrossRef]

80. Popova, L.P.; Maslenkova, L.T.; Ivanova, A.; Stoinova, Z. *Role of Salicylic Acid in Alleviating Heavy Metal Stress Environmental Adaptations and Stress Tolerance of Plants in the Era of Climate Change*; Springer: Berlin/Heidelberg, Germany, 2012; pp. 447–466.

81. Shaheen, M.R.; Ayyub, C.M.; Amjad, M.; Waraich, E.A. Morpho-physiological evaluation of tomato genotypes under high temperature stress conditions. *J. Sci. Food Agric.* **2016**, *96*, 2698–2704. [CrossRef] [PubMed]

82. Nowack, B.; Schulin, R.; Robinson, B.H. Critical assessment of chelant-enhanced metal phytoextraction. *Environ. Sci. Technol.* **2006**, *40*, 5225–5232. [CrossRef] [PubMed]

83. Nedelkoska, T.; Doran, P. Characteristics of heavy metal uptake by plant species with potential for phytoremediation and phytomining. *Miner. Eng.* **2000**, *13*, 549–561. [CrossRef]

84. Navari-Izzo, F.; Quartacci, M.F. Phytoremediation of metals—Tolerance mechanisms against oxidative stress. *Minerva Biotecnol.* **2001**, *13*, 73–83.

85. Doty, S.L.; Freeman, J.L.; Cohu, C.M.; Burken, J.G.; Firrincieli, A.; Simon, A.; Blaylock, M.J. Enhanced degradation of TCE on a superfund site using endophyte-assisted poplar tree phytoremediation. *Environ. Sci. Technol.* **2017**, *51*, 10050–10058. [CrossRef]

86. Souza, L.A.; Piotto, F.A.; Nogueirol, R.C.; Azevedo, R.A. Use of non-hyperaccumulator plant species for the phytoextraction of heavy metals using chelating agents. *Sci. Agric.* **2013**, *70*, 290–295. [CrossRef]

87. Ma, Y.; Oliveira, R.S.; Nai, F.; Rajkumar, M.; Luo, Y.; Rocha, I.; Freitas, H. The hyperaccumulator Sedum plumbizincicola harbors metal-resistant endophytic bacteria that improve its phytoextraction capacity in multi-metal contaminated soil. *J. Environ. Manag.* **2015**, *156*, 62–69. [CrossRef]

88. Fang, T.; Bao, S.; Sima, X.; Jiang, H.; Zhu, W.; Tang, W. Study on the application of integrated eco-engineering in purifying eutrophic river waters. *Ecol. Eng.* **2016**, *94*, 320–328. [CrossRef]

89. Afzal, M.; Khan, Q.M.; Sessitsch, A. Endophytic bacteria: Prospects and applications for the phytoremediation of organic pollutants. *Chemosphere* **2014**, *117*, 232–242. [CrossRef] [PubMed]

90. Hussain, Z.; Arslan, M.; Malik, M.H.; Mohsin, M.; Iqbal, S.; Afzal, M. Integrated perspectives on the use of bacterial endophytes in horizontal flow constructed wetlands for the treatment of liquid textile effluent: Phytoremediation advances in the field. *J. Environ. Manag.* **2018**, *224*, 387–395. [CrossRef] [PubMed]

91. Khan, M.U.; Sessitsch, A.; Harris, M.; Fatima, K.; Imran, A.; Arslan, M.; Afzal, M. Cr-resistant rhizo-and endophytic bacteria associated with Prosopis juliflora and their potential as phytoremediation enhancing agents in metal-degraded soils. *Front. Plant Sci.* **2015**, *5*, 755. [CrossRef] [PubMed]

92. Ashraf, S.; Afzal, M.; Rehman, K.; Naveed, M.; Zahir, Z.A. Plant-endophyte synergism in constructed wetlands enhances the remediation of tannery effluent. *Water Sci. Technol.* **2018**, *77*, 1262–1270. [CrossRef] [PubMed]

93. Ma, Y.; Rajkumar, M.; Zhang, C.; Freitas, H. Beneficial role of bacterial endophytes in heavy metal phytoremediation. *J. Environ. Manag.* **2016**, *174*, 14–25. [CrossRef] [PubMed]

94. Doty, S.L. Enhancing phytoremediation through the use of transgenics and endophytes. *New Phytol.* **2008**, *179*, 318–333. [CrossRef]
95. Phetcharat, P.; Duangpaeng, A. Screening of endophytic bacteria from organic rice tissue for indole acetic acid production. *Procedia Eng.* **2012**, *32*, 177–183. [CrossRef]
96. Zhang, X.; Wang, H.; He, L.; Lu, K.; Sarmah, A.; Li, J.; Huang, H. Using biochar for remediation of soils contaminated with heavy metals and organic pollutants. *Environ. Sci. Pollut. Res.* **2013**, *20*, 8472–8483. [CrossRef]
97. Coninx, L.; Martinova, V.; Rineau, F. Mycorrhiza-assisted phytoremediation. *Adv. Bot. Res.* **2017**, *83*, 127–188.
98. Chen, B.; Li, X.; Tao, H.; Christie, P.; Wong, M. The role of arbuscular mycorrhiza in zinc uptake by red clover growing in a calcareous soil spiked with various quantities of zinc. *Chemosphere* **2003**, *50*, 839–846. [CrossRef]
99. Muthusaravanan, S.; Sivarajasekar, N.; Vivek, J.; Paramasivan, T.; Naushad, M.; Prakashmaran, J.; Al-Duaij, O.K. Phytoremediation of heavy metals: Mechanisms, methods and enhancements. *Environ. Chem. Lett.* **2018**, *16*, 1339–1359. [CrossRef]
100. Koźmińska, A.; Wiszniewska, A.; Hanus-Fajerska, E.; Muszyńska, E. Recent strategies of increasing metal tolerance and phytoremediation potential using genetic transformation of plants. *Plant Biotechnol. Rep.* **2018**, *12*, 1–14. [CrossRef] [PubMed]
101. Van Aken, B. Transgenic plants for phytoremediation: Helping nature to clean up environmental pollution. *Trends Biotechnol.* **2008**, *26*, 225–227. [CrossRef] [PubMed]
102. Pilon-Smits, E. Phytoremediation. *Annu. Rev. Plant Biol.* **2005**, *56*, 15–39. [CrossRef] [PubMed]
103. Misra, S.; Gedamu, L. Heavy metal tolerant transgenic *Brassica napus* L. and *Nicotiana tabacum* L. plants. *Theor. Appl. Genet.* **1989**, *78*, 161–168. [CrossRef]
104. Rugh, C.L.; Wilde, H.D.; Stack, N.M.; Thompson, D.M.; Summers, A.O.; Meagher, R.B. Mercuric ion reduction and resistance in transgenic Arabidopsis thaliana plants expressing a modified bacterial merA gene. *Proc. Natl. Acad. Sci. USA* **1996**, *93*, 3182–3187. [CrossRef]
105. Eapen, S.; Singh, S.; D'souza, S. Advances in development of transgenic plants for remediation of xenobiotic pollutants. *Biotechnol. Adv.* **2007**, *25*, 442–451. [CrossRef]
106. Macek, T.; Kotrba, P.; Svatos, A.; Novakova, M.; Demnerova, K.; Mackova, M. Novel roles for genetically modified plants in environmental protection. *Trends Biotechnol.* **2008**, *26*, 146–152. [CrossRef]
107. Kawahigashi, H. Transgenic plants for phytoremediation of herbicides. *Curr. Opin. Biotechnol.* **2009**, *20*, 225–230. [CrossRef]
108. Darrah, P.; Jones, D.; Kirk, G.; Roose, T. Modelling the rhizosphere: A review of methods for 'upscaling'to the whole-plant scale. *Eur. J. Soil Sci.* **2006**, *57*, 13–25. [CrossRef]
109. Uchida, E.; Ouchi, T.; Suzuki, Y.; Yoshida, T.; Habe, H.; Yamaguchi, I.; Nojiri, H. Secretion of bacterial xenobiotic-degrading enzymes from transgenic plants by an apoplastic expressional system: An applicability for phytoremediation. *Environ. Sci. Technol.* **2005**, *39*, 7671–7677. [CrossRef] [PubMed]
110. Kidd, P.; Prieto-Fernández, A.; Monterroso, C.; Acea, M. Rhizosphere microbial community and hexachlorocyclohexane degradative potential in contrasting plant species. *Plant Soil* **2008**, *302*, 233–247. [CrossRef]
111. Yadav, K.K.; Gupta, N.; Kumar, A.; Reece, L.M.; Singh, N.; Rezania, S.; Khan, S.A. Mechanistic understanding and holistic approach of phytoremediation: A review on application and future prospects. *Ecol. Eng.* **2018**, *120*, 274–298. [CrossRef]
112. Han, Z.; Guo, Z.; Zhang, Y.; Xiao, X.; Xu, Z.; Sun, Y. Adsorption-pyrolysis technology for recovering heavy metals in solution using contaminated biomass phytoremediation. *Resour. Conserv. Recy.* **2018**, *129*, 20–26. [CrossRef]
113. Kaewsarn, P. Biosorption of copper (II) from aqueous solutions by pre-treated biomass of marine algae *Padina* sp. *Chemosphere* **2002**, *47*, 1081–1085. [CrossRef]
114. Priya, E.S.; Selvan, P.S. Water hyacinth (*Eichhornia crassipes*)–An efficient and economic adsorbent for textile effluent treatment—A review. *Arab. J. Chem.* **2017**, *10*, 3548–3558. [CrossRef]
115. Wang, G.; Fuerstenau, M.; Smith, R. Sorption of heavy metals onto nonliving water hyacinth roots. *Min. Proc. Ext. Met. Rev.* **1998**, *19*, 309–322. [CrossRef]
116. Wang, T.; Weissman, J.; Ramesh, G.; Varadarajan, R. Parameters for removal of toxic heavy metals by water milfoil (*Myriophyllum spicatum*). *Bull. Environ. Contam. Toxicol.* **1996**, *57*, 779–786. [CrossRef]

117. Wang, Y.; Meng, D.; Fei, L.; Dong, Q.; Wang, Z. A novel phytoextraction strategy based on harvesting the dead leaves: Cadmium distribution and chelator regulations among leaves of tall fescue. *Sci. Total Environ.* **2019**, *650*, 3041–3047. [CrossRef]

118. Pratas, J.; Paulo, C.; Favas, P.J.; Venkatachalam, P. Potential of aquatic plants for phytofiltration of uranium-contaminated waters in laboratory conditions. *Ecol. Eng.* **2014**, *69*, 170–176. [CrossRef]

119. Guittonny-Philippe, A.; Petit, M.-E.; Masotti, V.; Monnier, Y.; Malleret, L.; Coulomb, B.; Laffont-Schwob, I. Selection of wild macrophytes for use in constructed wetlands for phytoremediation of contaminant mixtures. *J. Environ. Manag.* **2015**, *147*, 108–123. [CrossRef] [PubMed]

120. Gorito, A.M.; Ribeiro, A.R.; Almeida, C.M.R.; Silva, A.M. A review on the application of constructed wetlands for the removal of priority substances and contaminants of emerging concern listed in recently launched EU legislation. *Environ. Pollut.* **2017**, *227*, 428–443. [CrossRef] [PubMed]

121. Mesa, J.; Mateos-Naranjo, E.; Caviedes, M.; Redondo-Gómez, S.; Pajuelo, E.; Rodríguez-Llorente, I. Scouting contaminated estuaries: Heavy metal resistant and plant growth promoting rhizobacteria in the native metal rhizoaccumulator *Spartina maritima*. *Mar. Pollut. Bull.* **2015**, *90*, 150–159. [CrossRef] [PubMed]

122. Fritioff, Å.; Greger, M. Aquatic and terrestrial plant species with potential to remove heavy metals from stormwater. *Int. J. Phytoremediat.* **2003**, *5*, 211–224. [CrossRef] [PubMed]

123. Gopal, B. Perspectives on wetland science, application and policy. *Hydrobiologia* **2003**, *490*, 1–10. [CrossRef]

124. Mays, P.; Edwards, G. Comparison of heavy metal accumulation in a natural wetland and constructed wetlands receiving acid mine drainage. *Ecol. Eng.* **2001**, *16*, 487–500. [CrossRef]

125. Stoltz, E.; Greger, M. Accumulation properties of As, Cd, Cu, Pb and Zn by four wetland plant species growing on submerged mine tailings. *Environ. Exp. Bot.* **2002**, *47*, 271–280. [CrossRef]

126. Said, M.; Cassayre, L.; Dirion, J.-L.; Nzihou, A.; Joulia, X. Behavior of heavy metals during gasification of phytoextraction plants: Thermochemical modelling Computer aided. *Chem. Eng.* **2015**, *37*, 341–346.

127. Syukor, A.A.; Zularisam, A.; Ideris, Z.; Ismid, M.M.; Nakmal, H.; Sulaiman, S.; Nasrullah, M. Performance of Phytogreen Zone for BOD5 and SS Removal for Refurbishment Conventional Oxidation Pond in an Integrated Phytogreen System. *World Acad. Sci. Eng. Technol.* **2014**, *8*, 159.

128. Gunathilakae, N.; Yapa, N.; Hettiarachchi, R. Effect of arbuscular mycorrhizal fungi on the cadmium phytoremediation potential of *Eichhornia crassipes* (Mart.) solms. *Groundw. Sustain. Dev.* **2018**. [CrossRef]

129. Iha, D.S.; Bianchini, I., Jr. Phytoremediation of Cd, Ni, Pb and Zn by Salvinia minima. *Int. J. Phytoremediat.* **2015**, *17*, 929–935. [CrossRef] [PubMed]

130. da-Silva, C.J.; Canatto, R.A.; Cardoso, A.A.; Ribeiro, C.; Oliveira, J.A. Arsenic-hyperaccumulation and antioxidant system in the aquatic macrophyte *Spirodela intermedia* W. Koch (Lemnaceae). *Theor. Exp. Plant Physiol.* **2017**, *29*, 203–213. [CrossRef]

131. Daud, M.; Ali, S.; Abbas, Z.; Zaheer, I.E.; Riaz, M.A.; Malik, A.; Zhu, S.J. Potential of Duckweed (*Lemna minor*) for the Phytoremediation of Landfill Leachate. *J. Chem.* **2018**, 1–9. [CrossRef]

132. Abbas, Z.; Arooj, F.; Ali, S.; Zaheer, I.E.; Rizwan, M.; Riaz, M.A. Phytoremediation of landfill leachate waste contaminants through floating bed technique using water hyacinth and water lettuce. *Int. J. Phytoremediat.* **2019**, *21*, 1356–1367. [CrossRef] [PubMed]

133. Shi, J.; Xiang, Z.; Peng, T.; Li, H.; Huang, K.; Liu, D.; Huang, T. Effects of melatonin-treated Nasturtiumofficinale on the growth and cadmium accumulation of subsequently grown rice seedlings. *Int. J. Environ. Anal. Chem.* **2020**. [CrossRef]

134. Maine, M.A.; Duarte, M.V.; Suñé, N.L. Cadmium uptake by floating macrophytes. *Water Res.* **2001**, *35*, 2629–2634. [CrossRef]

135. Olguín, E.; Hernández, E.; Ramos, I. The effect of both different light conditions and the pH value on the capacity of Salvinia minima Baker for removing cadmium, lead and chromium. *Acta Biotechnol.* **2002**, *22*, 121–131. [CrossRef]

136. Maine, M.A.; Suñé, N.L.; Lagger, S.C. Chromium bioaccumulation: Comparison of the capacity of two floating aquatic macrophytes. *Water Res.* **2004**, *38*, 1494–1501. [CrossRef]

137. Anaokar, G.; Sutar, T.; Mali, A.; Waghchoure, K.; Jadhav, R.; Walunj, S. Low Cost Municipal Wastewater Treatment Using Water Hyacinth. *J. Water Resour. Pollut. Stud.* **2018**, *3*, 2.

138. Chen, G.; Huang, J.; Fang, Y.; Zhao, Y.; Tian, X.; Jin, Y.; Zhao, H. Microbial community succession and pollutants removal of a novel carriers enhanced duckweed treatment system for rural wastewater in Dianchi lake basin. *Bioresour. Technol.* **2018**, *276*, 8–17. [CrossRef]

139. Hua, J.; Zhang, C.; Yin, Y.; Chen, R.; Wang, X. Phytoremediation potential of three aquatic macrophytes in manganese-contaminated water. *Water Environ. J.* **2012**, *26*, 335–342. [CrossRef]

140. Singh, D.; Gupta, R.; Tiwari, A. Potential of duckweed (*Lemna minor*) for removal of lead from wastewater by phytoremediation. *J. Pharm. Res.* **2012**, *5*, 1578–1582.

141. Molisani, M.M.; Rocha, R.; Machado, W.; Barreto, R.C.; Lacerda, L.D. Mercury contents in aquatic macrophytes from two reservoirs in the Paraiba do Sul: Guandu river system, SE Brazil. *Braz. J. Biol.* **2006**, *66*, 101–107. [CrossRef]

142. Hu, C.; Zhang, L.; Hamilton, D.; Zhou, W.; Yang, T.; Zhu, D. Physiological responses induced by copper bioaccumulation in *Eichhornia crassipes* (Mart.). *Hydrobiologia* **2007**, *579*, 211–218. [CrossRef]

143. Miretzky, P.; Saralegui, A.; Cirelli, A.F. Aquatic macrophytes potential for the simultaneous removal of heavy metals (Buenos Aires, Argentina). *Chemosphere* **2004**, *57*, 997–1005. [CrossRef] [PubMed]

144. Sooknah, H. A review of the mechanisms of pollutant removal in water hyacinth systems. *Sci. Technol. Res. J.* **2000**, *6*, 49–57.

145. Suñé, N.; Sánchez, G.; Caffaratti, S.; Maine, M.A. Cadmium and chromium removal kinetics from solution by two aquatic macrophytes. *Environ. Pollut.* **2007**, *145*, 467–473. [CrossRef]

146. Kara, Y. Bioaccumulation of copper from contaminated wastewater by using *Lemna minor* (Aquatic green plant). *Bullet. Environ. Contam. Toxicol.* **2004**, *72*, 467–471. [CrossRef]

147. Ater, M.; Ali, A.N.; Kasmi, H. Tolerance and accumulation of copper and chromium in two duckweed species: *Lemna minor* L. and *Lemna gibba* L. *Rev. Sci.* **2006**, *19*, 57–67.

148. Basile, A.; Sorbo, S.; Conte, B.; Cobianchi, R.C.; Trinchella, F.; Capasso, C.; Carginale, V. Toxicity, accumulation, and removal of heavy metals by three aquatic macrophytes. *Int. J. Phytoremediat.* **2012**, *14*, 374–387. [CrossRef]

149. Cardwell, A.; Hawker, D.; Greenway, M. Metal accumulation in aquatic macrophytes from southeast Queensland, Australia. *Chemosphere* **2002**, *48*, 653–663. [CrossRef]

150. Zurayk, R.; Sukkariyah, B.; Baalbaki, R.; Ghanem, D.A. Chromium phytoaccumulation from solution by selected hydrophytes. *Int. J. Phytoremediat.* **2001**, *3*, 335–350. [CrossRef]

151. Sivaci, E.R.; Sivaci, A.; Sökmen, M. Biosorption of cadmium by *Myriophyllum spicatum* L. and Myriophyllum triphyllum orchard. *Chemosphere* **2004**, *56*, 1043–1048. [CrossRef] [PubMed]

152. Branković, S.; Pavlović-Muratspahić, D.; Topuzović, M.; Glišić, R.; Milivojević, J.; Đekić, V. Metals concentration and accumulation in several aquatic macrophytes. *Biotechnol. Biotechnol. Equip.* **2012**, *26*, 2731–2736. [CrossRef]

153. Bunluesin, S.; Kruatrachue, M.; Pokethitiyook, P.; Lanza, G.; Upatham, E.; Soonthornsarathool, V. Plant screening and comparison of *Ceratophyllum demersum* and *Hydrilla verticillata* for cadmium accumulation. *Bull. Environ. Contam. Toxicol.* **2004**, *73*, 591–598. [CrossRef] [PubMed]

154. El-Khatib, A.; Hegazy, A.; Abo-El-Kassem, A.M. Bioaccumulation potential and physiological responses of aquatic macrophytes to Pb pollution. *Int. J. Phytoremediat.* **2014**, *16*, 29–45. [CrossRef] [PubMed]

155. Borisova, G.; Chukina, N.; Maleva, M.; Prasad, M. *Ceratophyllum demersum* L. and Potamogeton alpinus Balb. From Iset'river, Ural region, Russia differ in adaptive strategies to heavy metals exposure–a comparative study. *Int. J. Phytoremediat.* **2014**, *16*, 621–633. [CrossRef]

156. Singh, N.; Pandey, G.; Rai, U.; Tripathi, R.; Singh, H.; Gupta, D. Metal accumulation and ecophysiological effects of distillery effluent on *Potamogeton pectinatus* L. *Bull. Environ. Contam. Toxicol.* **2005**, *74*, 857–863. [CrossRef]

157. Peng, K.; Luo, C.; Lou, L.; Li, X.; Shen, Z. Bioaccumulation of heavy metals by the aquatic plants *Potamogeton pectinatus* L. and *Potamogeton malaianus* Miq and their potential use for contamination indicators and in wastewater treatment. *Sci. Total Environ.* **2008**, *392*, 22–29. [CrossRef]

158. Hejna, M.; Moscatelli, A.; Stroppa, N.; Onelli, E.; Pilu, S.; Baldi, A.; Rossi, L. Bioaccumulation of heavy metals from wastewater through a *Typha latifolia* and *Thelypteris palustris* phytoremediation system. *Chemosphere* **2020**, *241*, 125018. [CrossRef]

159. Qian, J.-H.; Zayed, A.; Zhu, Y.-L.; Yu, M.; Terry, N. Phytoaccumulation of trace elements by wetland plants: III. Uptake and accumulation of ten trace elements by twelve plant species. *J. Environ. Qual.* **1999**, *28*, 1448–1455. [CrossRef]

160. Sasmaz, A.; Obek, E.; Hasar, H. The accumulation of heavy metals in *Typha latifolia* L. grown in a stream carrying secondary effluent. *Ecol. Eng.* **2008**, *33*, 278–284. [CrossRef]

161. Kamal, M.; Ghaly, A.; Mahmoud, N.; Cote, R. Phytoaccumulation of heavy metals by aquatic plants. *Environ. Int.* **2004**, *29*, 1029–1039. [CrossRef]

162. Giri, A.K. Bioaccumulation Potential and Toxicity of Arsenite Using Rooted-Submerged Vallisneria Spiralis in a Hydroponic Culture and its Characterization Studies. *J. Adv. Sci. Res.* **2019**, *10*, 17–22.

163. Aksorn, E.; Visoottiviseth, P. Selection of Suitable Emergent Plants for Removal of Arsenic from Arsenic Contaminated Water. *Sci. Asia* **2004**, *30*, 105–113. [CrossRef]

164. Hempel, M.; Botté, S.E.; Negrin, V.L.; Chiarello, M.N.; Marcovecchio, J.E. The role of the smooth cordgrass *Spartina alterniflora* and associated sediments in the heavy metal biogeochemical cycle within Bahía Blanca estuary salt marshes. *J. Soils Sediments* **2008**, *8*, 289. [CrossRef]

165. Ganjali, S.; Tayebi, L.; Atabati, H.; Mortazavi, S. Phragmites australis as a heavy metal bioindicator in the Anzali wetland of Iran. *Toxicol. Environ. Chem.* **2014**, *96*, 1428–1434. [CrossRef]

166. Ha, N.T.H.; Anh, B.T.K. The removal of heavy metals by iron mine drainage sludge and Phragmites australis. In Proceedings of the IOP Conference Series: Earth and Environmental Science, Bandung, Indonesia, 20–22 September 2016; IOP Publishing: Bristol, UK, 2017; Volume 71, p. 012022.

167. Kutty, A.A.; Al-Mahaqeri, S.A. An Investigation of the Levels and Distribution of Selected Heavy Metals in Sediments and Plant Species within the Vicinity of Ex-Iron Mine in Bukit Besi. *J. Chem.* **2016**, *2016*, 2096147. [CrossRef]

168. Rudin, S.M.; Murray, D.W.; Whitfeld, T.J. Retrospective analysis of heavy metal contamination in Rhode Island based on old and new herbarium specimens. *Appl. Plant Sci.* **2017**, *5*, 1600108. [CrossRef]

169. Rezania, S.; Ponraj, M.; Din, M.F.M.; Songip, A.R.; Sairan, F.M.; Chelliapan, S. The diverse applications of water hyacinth with main focus on sustainable energy and production for new era: An overview. *Renew Sustain. Energy Rev.* **2015**, *41*, 943–954. [CrossRef]

170. Kumar, H.S.; Bose, A.; Raut, A.; Sahu, S.K.; Raju, M. Evaluation of Anthelmintic Activity of *Pistia stratiotes* Linn. *J. Basic Clin. Pharm.* **2010**, *1*, 103. [PubMed]

171. Muthunarayanan, V.; Santhiya, M.; Swabna, V.; Geetha, A. Phytodegradation of textile dyes by water hyacinth (*Eichhornia crassipes*) from aqueous dye solutions. *Int. J. Environ. Sci.* **2011**, *1*, 1702.

172. Yadav, S.; Jadhav, A.; Chonde, S.; Raut, P. Performance Evaluation of Surface Flow Constructed Wetland System by Using *Eichhornia crassipes* for Wastewater Treatment in an Institutional Complex. *Univers. J. Environ. Res. Technol.* **2011**, *1*, 435–441.

173. Rezania, S.; Din, M.F.M.; Taib, S.M.; Dahalan, F.A.; Songip, A.R.; Singh, L.; Kamyab, H. The efficient role of aquatic plant (water hyacinth) in treating domestic wastewater in continuous system. *Int. J. Phytoremediat.* **2016**, *18*, 679–685. [CrossRef]

174. Jafari, N. Ecological and socio-economic utilization of water hyacinth (*Eichhornia crassipes* Mart Solms). *J. Appl. Sci. Environ. Manag.* **2010**, *14*, 43–49. [CrossRef]

175. Dixit, S.; Tiwari, S. Effective utilization of an aquatic weed in an eco-friendly treatment of polluted water bodies. *J. Appl. Sci. Environ. Manag.* **2007**, *11*, 41–44. [CrossRef]

176. Hussain, S.T.; Mahmood, T.; Malik, S.A. Phytoremediation technologies for Ni^{++} by water hyacinth. *Afr. J. Biotechnol.* **2010**, *9*, 8648–8660.

177. Mahamadi, C. Water hyacinth as a biosorbent: A review. *Afr. J. Environ. Sci. Technol.* **2011**, *5*, 1137–1145. [CrossRef]

178. Mahmood, T.; Malik, S.A.; Hussain, S.T. Biosorption and recovery of heavy metals from aqueous solutions by *Eichhornia crassipes* (water hyacinth) ash. *Biol. Resour.* **2010**, *5*, 1244–1256.

179. Shah, R.A.; Kumawat, D.; Singh, N.; Wani, K.A. Water hyacinth (*Eichhornia crassipes*) as a remediation tool for dye effluent pollution. *Int. J. Sci. Nat.* **2010**, *1*, 172–178.

180. Shuvaeva, O.V.; Belchenko, L.A.; Romanova, T.E. Studies on cadmium accumulation by some selected floating macrophytes. *Int. J. Phytoremediat.* **2013**, *15*, 979–990. [CrossRef] [PubMed]

181. Aurangzeb, N.; Nisa, S.; Bibi, Y.; Javed, F.; Hussain, F. Phytoremediation potential of aquatic herbs from steel foundry effluent. *Braz. J. Chem. Eng.* **2014**, *31*, 881–886. [CrossRef]

182. Fazal, S.; Zhang, B.; Mehmood, Q. Biological treatment of combined industrial wastewater. *Ecol. Eng.* **2015**, *84*, 551–558. [CrossRef]

183. Mahalakshmi, R.; Sivapragasam, C.; Vanitha, S. *Comparison of BOD 5 Removal in Water Hyacinth and Duckweed by Genetic Programming Information and Communication Technology for Intelligent Systems*; Springer: Berlin/Heidelberg, Germany, 2019; pp. 401–408.

184. Sarkar, M.; Rahman, A.; Bhoumik, N. Remediation of chromium and copper on water hyacinth (E. *crassipes*) shoot powder. *Water Resour. Indust.* **2017**, *17*, 1–6. [CrossRef]

185. Sekomo, C.B.; Kagisha, V.; Rousseau, D.; Lens, P. Heavy metal removal by combining anaerobic upflow packed bed reactors with water hyacinth ponds. *Environ. Technol.* **2012**, *33*, 1455–1464. [CrossRef]

186. Swarnalatha, K.; Radhakrishnan, B. Studies on removal of Zinc and Chromium from aqueous solutions using water Hyacinth. *Pollut* **2015**, *1*, 193–202.

187. Prasad, B.; Maiti, D. Comparative study of metal uptake by *Eichhornia crassipes* growing in ponds from mining and nonmining areas—A field study. *Bioremediat. J.* **2016**, *20*, 144–152. [CrossRef]

188. Romanova, T.E.; Shuvaeva, O.V.; Belchenko, L.A. Phytoextraction of trace elements by water hyacinth in contaminated area of gold mine tailing. *Int. J. Phytoremediat.* **2016**, *18*, 190–194. [CrossRef]

189. Li, Q.; Chen, B.; Lin, P.; Zhou, J.; Zhan, J.; Shen, Q.; Pan, X. Adsorption of heavy metal from aqueous solution by dehydrated root powder of long-root *Eichhornia crassipes*. *Int. J Phytoremediat.* **2016**, *18*, 103–109. [CrossRef]

190. Quattrocchi, U. *CRC World Dictionary of Plant Names: Common Names, Scientific Names, Eponyms, Synonyms, and Etymology*; CRC Press: Boca Raton, FL, USA, 2017; p. 728.

191. Dipu, S.; Kumar, A.A.; Thanga, V.S.G. Phytoremediation of dairy effluent by constructed wetland technology. *Environment* **2011**, *31*, 263–278. [CrossRef]

192. Lima, L.; Pelosi, B.; Silva, M.; Vieira, M. Lead and chromium biosorption by *Pistia stratiotes* biomass. *Chem. Eng. Trans.* **2013**, *32*, 1045–1050.

193. Acevedo, R.; Nicolson, D. Araceae. *Contrib. US Natl. Herb.* **2005**, *52*, 44.

194. Forni, C.; Patrizi, C.; Migliore, L. Floating aquatic macrophytes as a decontamination tool for antimicrobial drugs. In *Soil and Water Pollution Monitoring, Protection and Remediation*; Springer: Dordrecht, The Netherlands, 2006; pp. 3–23.

195. Yasar, A.; Zaheer, A.; Tabinda, A.B.; Khan, M.; Mahfooz, Y.; Rani, S.; Siddiqua, A. Comparison of Reed and Water Lettuce in Constructed Wetlands for Wastewater Treatment. *Water Environ. Res.* **2018**, *90*, 129–135. [CrossRef]

196. Akinbile, C.; Yusoff, M.S. Assessing water hyacinth (*Eichhornia crassopes*) and lettuce (*Pistia stratiotes*) effectiveness in aquaculture wastewater treatment. *Int. J. Phytoremed.* **2012**, *14*, 201–211. [CrossRef]

197. Sarwar, T.; Shahid, M.; Khalid, S.; Shah, A.H.; Ahmad, N.; Naeem, M.A.; Bakhat, H.F. Quantification and risk assessment of heavy metal build-up in soil–plant system after irrigation with untreated city wastewater in Vehari, Pakistan. *Environ. Geochem. Health* **2019**, 1–17. [CrossRef]

198. Polomski, R.F.; Taylor, M.D.; Bielenberg, D.G.; Bridges, W.C.; Klaine, S.J.; Whitwell, T. Nitrogen and phosphorus remediation by three floating aquatic macrophytes in greenhouse-based laboratory-scale subsurface constructed wetlands. *Water Air Soil Pollut.* **2009**, *197*, 223–232. [CrossRef]

199. Skinner, K.; Wright, N.; Porter-Goff, E. Mercury uptake and accumulation by four species of aquatic plants. *Environ. Pollut.* **2007**, *145*, 234–237. [CrossRef]

200. Eloy, G.-G.; Marta, R.; Gertjan, M.; Miquel, C.; Rosina, G. Quantitative risk assessment of norovirus and adenovirus for the use of reclaimed water to irrigate lettuce in Catalonia. *Water Res.* **2019**, *153*, 91–99.

201. Mishra, V.K.; Tripathi, B. Concurrent removal and accumulation of heavy metals by the three aquatic macrophytes. *Bioresour. Technol.* **2008**, *99*, 7091–7097. [CrossRef]

202. Rodrigues, A.C.D.; do Amaral Sobrinho, N.M.B.; dos Santos, F.S.; dos Santos, A.M.; Pereira, A.C.C.; Lima, E.S.A. Biosorption of toxic metals by water lettuce (*Pistia stratiotes*) biomass. *Water Air Soil Pollut.* **2017**, *228*, 156. [CrossRef]

203. Farnese, F.; Oliveira, J.; Lima, F.; Leão, G.; Gusman, G.; Silva, L. Evaluation of the potential of *Pistia stratiotes* L. (water lettuce) for bioindication and phytoremediation of aquatic environments contaminated with arsenic. *Braz. J. Biol.* **2014**, *74*, 108–112. [CrossRef]

204. Volf, I.; Rakoto, N.G.; Bulgariu, L. Valorization of *Pistia stratiotes* biomass as biosorbent for Lead (II) ions removal from aqueous media. *Sep. Sci. Technol.* **2015**, *50*, 1577–1586. [CrossRef]

205. Les, D.H.; Crawford, D.J.; Landolt, E.; Gabel, J.D.; Kimball, R.T. Phylogeny and systematics of *Lemnaceae*, the duckweed family. *Syst. Bot.* **2002**, *27*, 221–240.

206. Radić, S.; Stipaničev, D.; Cvjetko, P.; Mikelić, I.L.; Rajčić, M.M.; Širac, S.; Pavlica, M. Ecotoxicological assessment of industrial effluent using duckweed (*Lemna minor* L.) as a test organism. *Ecotoxicology* **2010**, *19*, 216.

207. Krishna, K.B.; Polprasert, C. An integrated kinetic model for organic and nutrient removal by duckweed-based wastewater treatment (DUBWAT) system. *Ecol. Eng.* **2008**, *34*, 243–250. [CrossRef]

208. Mkandawire, M.; Dudel, E.G. Are *Lemna* spp. effective phytoremediation agents. *Bioremediat. Biodivers. Bioavailab.* **2007**, *1*, 56–71.

209. Alaerts, G.; Mahbubar, R.; Kelderman, P. Performance analysis of a full-scale duckweed-covered sewage lagoon. *Water Res.* **1996**, *30*, 843–852. [CrossRef]

210. Obek, E.; Sasmaz, A. Bioaccumulation of aluminum by *Lemna gibba* L. from secondary treated municipal wastewater effluents. *Bull. Environ. Contam. Toxicol.* **2011**, *86*, 217–220. [CrossRef]

211. Singh, D.; Tiwari, A.; Gupta, R. Phytoremediation of lead from wastewater using aquatic plants. *J. Agric. Technol.* **2012**, *8*, 1–11. [CrossRef]

212. Rahman, M.A.; Hasegawa, H.; Ueda, K.; Maki, T.; Okumura, C.; Rahman, M.M. Arsenic accumulation in duckweed (*Spirodela polyrhiza* L.): A good option for phytoremediation. *Chemosphere* **2007**, *69*, 493–499. [CrossRef]

213. Marín, C.M.D.C.; Oron, G. Boron removal by the duckweed *Lemna gibba*: A potential method for the remediation of boron-polluted waters. *Water Res.* **2007**, *41*, 4579–4584. [CrossRef]

214. Sasmaz, A.; Obek, E. The accumulation of arsenic, uranium, and boron in *Lemna gibba* L. exposed to secondary effluents. *Ecol. Eng.* **2009**, *35*, 1564–1567. [CrossRef]

215. Alvarado, S.; Guédez, M.; Lué-Merú, M.P.; Nelson, G.; Alvaro, A.; Jesús, A.C.; Gyula, Z. Arsenic removal from waters by bioremediation with the aquatic plants Water Hyacinth (*Eichhornia crassipes*) and Lesser Duckweed (*Lemna minor*). *Bioresour. Technol.* **2008**, *99*, 8436–8440. [CrossRef]

216. Mohedano, R.A.; Costa, R.H.; Tavares, F.A.; Belli Filho, P. High nutrient removal rate from swine wastes and protein biomass production by full-scale duckweed ponds. *Bioresour. Technol.* **2012**, *112*, 98–104. [CrossRef]

217. Zhang, K.; Chen, Y.P.; Zhang, T.T.; Zhao, Y.; Shen, Y.; Huang, L.; Guo, J.S. The logistic growth of duckweed (*Lemna minor*) and kinetics of ammonium uptake. *Environ. Technol.* **2014**, *35*, 562–567. [CrossRef]

218. Böcük, H.; Yakar, A.; Türker, O.C. Assessment of *Lemna gibba* L. (duckweed) as a potential ecological indicator for contaminated aquatic ecosystem by boron mine effluent. *Ecol. Indic.* **2013**, *29*, 538–548.

219. Sallah-Ud-Din, R.; Farid, M.; Saeed, R.; Ali, S.; Rizwan, M.; Tauqeer, H.M.; Bukhari, S.A.H. Citric acid enhanced the antioxidant defense system and chromium uptake by *Lemna minor* L. grown in hydroponics under Cr stress. *Environ. Sci. Pollut. Res.* **2017**, *24*, 17669–17678. [CrossRef]

220. Abdallah, M.A.M. Phytoremediation of heavy metals from aqueous solutions by two aquatic macrophytes, *Ceratophyllum demersum* and *Lemna gibba* L. *Environ. Technol.* **2012**, *33*, 1609–1614. [CrossRef]

221. Chaudhuri, D.; Majumder, A.; Misra, A.K.; Bandyopadhyay, K. Cadmium removal by *Lemna minor* and *Spirodela polyrhiza*. *Int. J. Phytoremediat.* **2014**, *16*, 1119–1132. [CrossRef]

222. Goswami, C.; Majumder, A.; Misra, A.K.; Bandyopadhyay, K. Arsenic uptake by *Lemna minor* in hydroponic system. *Int. J. Phytoremediat.* **2014**, *16*, 1221–1227. [CrossRef]

223. Teixeira, S.; Vieira, M.; Marques, J.E.; Pereira, R. Bioremediation of an iron-rich mine effluent by *Lemna minor*. *Int. J. Phytoremediat.* **2014**, *16*, 1228–1240. [CrossRef]

224. Del Carmen Flores-Miranda, M.; Luna-Gonz alez, A.; Cort es-Espinosa, D.V.; onica Escamilla-Montes, R. Bacterial fermentation of *Lemna* sp. as a potential substitute of fish meal in shrimp diets. *Afr. J. Microbiol. Res.* **2014**, *8*, 1516–1526.

225. Verma, R.; Suthar, S. Lead and cadmium removal from water using duckweed–*Lemna gibba* L.: Impact of pH and initial metal load. *Alexander Eng. J.* **2015**, *54*, 1297–1304. [CrossRef]

226. Török, A.; Gulyás, Z.; Szalai, G.; Kocsy, G.; Majdik, C. Phytoremediation capacity of aquatic plants is associated with the degree of phytochelatin polymerization. *J. Hazard. Mater.* **2015**, *299*, 371–378. [CrossRef]

227. Bokhari, S.H.; Ahmad, I.; Mahmood-Ul-Hassan, M.; Mohammad, A. Phytoremediation potential of *Lemna minor* L. for heavy metals. *Int. J. Phytoremediat.* **2016**, *18*, 25–32. [CrossRef]

228. Henry-Silva, G.G.; Camargo, A.F.M. Efficiency of aquatic macrophytes to treat Nile tilapia pond effluents. *Sci. Agric.* **2006**, *63*, 433–438. [CrossRef]

229. Gardner, J.L.; Al-Hamdani, S.H. Interactive effects of aluminum and humic substances on Salvinia. *J. Aquat. Plant Manag.* **1997**, *35*, 30–34.

230. Dhir, B. Salvinia: An aquatic fern with potential use in phytoremediation. *Environ. Int. J. Sci. Technol.* **2009**, *4*, 23–27.

231. Dhir, B.; Sharmila, P.; Saradhi, P.P.; Sharma, S.; Kumar, R.; Mehta, D. Heavy metal induced physiological alterations in *Salvinia natans*. *Ecotoxicol. Environ. Saf.* **2011**, *74*, 1678–1684. [CrossRef]

232. Abd-Elnaby, A.; Egorov, M. Efficiency of different particle sizes of dried *Salvinia natans* in the removing of Cu (II) and oil pollutions from water. *J. Water Chem. Technol.* **2012**, *34*, 143–146. [CrossRef]

233. Hoffmann, T.; Kutter, C.; Santamaria, J. Capacity of Salvinia minima Baker to tolerate and accumulate as and Pb. *Eng. Life Sci.* **2004**, *4*, 61–65. [CrossRef]

234. Olguin, E.; Rodriguez, D.; Sanchez, G.; Hernandez, E.; Ramirez, M. Productivity, protein content and nutrient removal from anaerobic effluents of coffee wastewater in *Salvinia minima* ponds, under subtropical conditions. *Acta Biotechnol.* **2003**, *23*, 259–270. [CrossRef]

235. Olguín, E.J.; Sánchez-Galván, G.; Pérez-Pérez, T.; Pérez-Orozco, A. Surface adsorption, intracellular accumulation and compartmentalization of Pb (II) in batch-operated lagoons with *Salvinia minima* as affected by environmental conditions, EDTA and nutrients. *J. Ind. Microbiol. Biotechnol.* **2005**, *32*, 577–586. [CrossRef]

236. Olguín, E.J.; Sánchez-Galván, G.; Pérez-Pérez, T. Assessment of the phytoremediation potential of *Salvinia minima* Baker compared to *Spirodela polyrrhiza* in high-strength organic wastewater. *Water Air Soil Pollut.* **2007**, *181*, 135–147. [CrossRef]

237. Fuentes, I.I.; Espadas-Gil, F.; Talavera-May, C.; Fuentes, G.; Santamaría, J.M. Capacity of the aquatic fern (*Salvinia minima* Baker) to accumulate high concentrations of nickel in its tissues, and its effect on plant physiological processes. *Aquat. Toxicol.* **2014**, *155*, 142–150. [CrossRef]

238. Ranjitha, J.; Raj, A.; Kashyap, R.; Vijayalakshmi, S.; Donatus, M. Removal of heavy metals from industrial effluent using *Salvinia molesta*. *Int. J. Chem. Technol. Res.* **2016**, *9*, 608–613.

239. Lakra, K.C.; Lal, B.; Banerjee, T. Decontamination of coal mine effluent generated at the Rajrappa coal mine using phytoremediation technology. *Int. J. Phytoremediat.* **2017**, *19*, 530–536. [CrossRef]

240. Keskinkan, O.; Goksu, M.; Yuceer, A.; Basibuyuk, M.; Forster, C. Heavy metal adsorption characteristics of a submerged aquatic plant (*Myriophyllum spicatum*). *Process Biochem.* **2003**, *39*, 179–183. [CrossRef]

241. Rai, U.; Tripathi, R.; Vajpayee, P.; Pandey, N.; Ali, M.; Gupta, D. Cadmium accumulation and its phytotoxicity in *Potamogeton pectinatus* L.(*Potamogetonaceae*). *Bull. Environ. Contam. Toxicol.* **2003**, *70*, 0566–0575. [CrossRef]

242. Saygıdeğer, S.; Doğan, M. Lead and Cadmium Accumulation and Toxicity in the Presence of EDTA in *Lemna minor* L. and *Ceratophyllum demersum* L. *Bull. Environ. Contam. Toxicol.* **2004**, *73*, 182–189. [CrossRef]

243. Casagrande, G.C.R.; dos Reis, C.; Arruda, R.; de Andrade, R.L.T.; Battirola, L.D. Bioaccumulation and Biosorption of Mercury by *Salvinia biloba* Raddi (Salviniaceae). *Water Air Soil Pollut.* **2018**, *229*, 166. [CrossRef]

244. Windham, L.; Weis, J.S.; Weis, P. Patterns and processes of mercury release from leaves of two dominant salt marsh macrophytes, *Phragmites australis* and *Spartina alterniflora*. *Estuaries* **2001**, *24*, 787–795. [CrossRef]

245. Tangahu, B.V.; Abdullah, S.R.S.; Basri, H.; Idris, M.; Anuar, N.; Mukhlisin, M. A Review on Heavy Metals (As, Pb, and Hg) Uptake by Plants through Phytoremediation. *Int. J. Chem. Eng.* **2011**, *31*, 939161. [CrossRef]

246. Sasmaz, A.; Mete, I.; Sasmaz, D.M. Removal of Cr, Ni and Co in the water of chromium mining areas by using *Lemna gibba* L. and *Lemna minor* L. *Water Environ. J.* **2016**, *30*, 235–242. [CrossRef]

247. Rai, U.N.; Sinha, S.; Tripathi, R.D.; Chandra, P. Wastewater treatability potential of some aquatic macrophytes: Removal of heavy metals. *Ecol. Eng.* **1995**, *5*, 5–12. [CrossRef]

248. Sas-Nowosielska, A.; Galimska-Stypa, R.; Kucharski, R.; Zielonka, U.; Małkowski, E.; Gray, L. Remediation aspect of microbial changes of plant rhizosphere in mercury contaminated soil. *Environ. Monit. Assess.* **2008**, *137*, 101–109. [CrossRef] [PubMed]

249. Soda, S.; Hamada, T.; Yamaoka, Y.; Ike, M.; Nakazato, H.; Saeki, Y.; Sakurai, Y. Constructed wetlands for advanced treatment ofwastewater with a complex matrix from a metal-processing plant: Bioconcentration and translocation factors ofvarious metals in *Acorus gramineus* and *Cyperus alternifolius*. *Ecol. Eng.* **2012**, *39*, 63–70. [CrossRef]

250. Rodríguez, M.; Brisson, J. Pollutant removal efficiency ofnative versus exotic common reed (*Phragmites australis*) in North American treatment wetlands. *Ecol. Eng.* **2015**, *74*, 364–370. [CrossRef]

251. Alhadrami, H.A.; Mbadugha, L.; Paton, G.I. Hazard and risk assessment of human exposure to toxic metals using in vitro digestion assay. *Chem. Spec. Bioavailab.* **2016**, *28*, 78–87. [CrossRef]

252. Mânzatu, C.; Nagy, B.; Ceccarini, A.; Iannelli, R.; Giannarelli, S.; Majdik, C. Laboratory tests for the phytoextraction of heavy metals from polluted harbor sediments using aquatic plants. *Mar. Pollut. Bull.* **2015**, *101*, 605–611. [CrossRef]

253. Wacławek, S.; Lutze, H.V.; Grübel, K.; Padil, V.V.; Černík, M.; Dionysiou, D.D. Chemistry of persulfates in water and wastewater treatment: A review. *Chem. Eng. J.* **2017**, *330*, 44–62. [CrossRef]

254. Akpor, O.; Muchie, M. Remediation of heavy metals in drinking water and wastewater treatment systems: Processes and applications. *Int. J. Phy. Sci.* **2010**, *5*, 1807–1817.

255. Carolin, C.F.; Kumar, P.S.; Saravanan, A.; Joshiba, G.J.; Naushad, M. Efficient techniques for the removal of toxic heavy metals from aquatic environment: A review. *J. Environ. Chem. Eng.* **2017**, *5*, 2782–2799. [CrossRef]

256. Bonanno, G.; Borg, J.A.; Di Martino, V. Levels of heavy metals in wetland and marine vascular plants and their biomonitoring potential: A comparative assessment. *Sci. Total Environ.* **2017**, *576*, 796–806. [CrossRef] [PubMed]

257. Bravo, S.; Amorós, J.; Pérez-de-los-Reyes, C.; García, F.; Moreno, M.; Sánchez-Ormeño, M.; Higueras, P. Influence of the soil pH in the uptake and bioaccumulation of heavy metals (Fe, Zn, Cu, Pb and Mn) and other elements (Ca, K, Al, Sr and Ba) in vine leaves, Castilla-La Mancha (Spain). *J. Geochem. Explor.* **2017**, *174*, 79–83. [CrossRef]

258. Mojiri, A. Phytoremediation of heavy metals from municipal wastewater by Typhadomingensis. *Afr. J. Microbiol. Res.* **2012**, *6*, 643–647.

259. Pedescoll, A.; Sidrach-Cardona, R.; Hijosa-Valsero, M.; Bécares, E. Design parameters affecting metals removal in horizontal constructed wetlands for domestic wastewater treatment. *Ecol. Eng.* **2015**, *80*, 92–99. [CrossRef]

260. Morari, F.; Dal Ferro, N.; Cocco, E. Municipal wastewater treatment with *Phragmites australis* L. and *Typha latifolia* L. for irrigation reuse. Boron and heavy metals. *Water Air Soil Pollut.* **2015**, *226*, 56. [CrossRef]

261. Francová, A.; Chrastný, V.; Šillerová, H.; Vítková, M.; Kocourková, J.; Komárek, M. Evaluating the suitability of different environmental samples for tracing atmospheric pollution in industrial areas. *Environ. Pollut.* **2017**, *220*, 286–297. [CrossRef]

262. Cheraghi, M.; Lorestani, B.; Khorasani, N.; Yousefi, N.; Karami, M. Findings on the phytoextraction and phytostabilization of soils contaminated with heavy metals. *Biol. Trace Elem. Res.* **2011**, *144*, 1133–1141. [CrossRef]

263. Khan, S.; Ahmad, I.; Shah, M.T.; Rehman, S.; Khaliq, A. Use of constructed wetland for the removal of heavy metals from industrial wastewater. *J. Environ. Manag.* **2009**, *90*, 3451–3457. [CrossRef] [PubMed]

264. Hegazy, A.; Abdel-Ghani, N.; El-Chaghaby, G. Phytoremediation of industrial wastewater potentiality by Typha domingensis. *Int. J. Environ. Sci. Technol.* **2011**, *8*, 639–648. [CrossRef]

265. Ahmad, A.; Ghufran, R.; Zularisam, A. Phytosequestration of metals in selected plants growing on a contaminated Okhla industrial areas, Okhla, New Delhi, India. *Water Air Soil Pollut.* **2011**, *217*, 255–266. [CrossRef]

266. Maine, M.; Hadad, H.; Sánchez, G.; Di Luca, G.; Mufarrege, M.; Caffaratti, S.; Pedro, M. Long-term performance of two free-water surface wetlands for metallurgical effluent treatment. *Ecol. Eng.* **2017**, *98*, 372–377. [CrossRef]

267. Awomeso, J.; Taiwo, A.; Gbadebo, A.; Adenowo, J. Studies on the pollution of water body by textile industry effluents in Lagos, Nigeria. *J. Appl. Sci. Environ. Sanit.* **2010**, *5*, 353–359.

268. Soares, P.A.; Souza, R.; Soler, J.; Silva, T.F.; Souza, S.M.G.U.; Boaventura, R.A.; Vilar, V.J. Remediation of a synthetic textile wastewater from polyester-cotton dyeing combining biological and photochemical oxidation processes. *Sep. Purif. Technol.* **2017**, *172*, 450–462. [CrossRef]

269. Mahmood, Q.; Zheng, P.; Islam, E.; Hayat, Y.; Hassan, M.; Jilani, G.; Jin, R. Lab scale studies on water hyacinth (*Eichhornia crassipes* Marts Solms) for biotreatment of textile wastewater. *Casp. J. Environ. Sci.* **2005**, *3*, 83–88.

270. Sanmuga Priya, E.; Senthamil Selvan, P.; Umayal, A. Biodegradation studies on dye effluent and selective remazol dyes by indigenous bacterial species through spectral characterisation. *Desalin. Water Treat.* **2015**, *55*, 241–251. [CrossRef]

271. Manjunath, S.; Kousar, H. Phytoremediation of Textile Industry Effluent using free floating macrophyte Azolla pinnata. *Int. J. Environ. Sci.* **2016**, *5*, 68–71.

272. Majumder, A.; Ray, S.; Jha, S. Hairy Roots and Phytoremediation. *Biol. Process. Plant Syst.* **2018**, 549–572.

273. Roy, C.; Jahan, M.; Rahman, S. Characterization and treatment of textile wastewater by aquatic plants (macrophytes) and algae. *Eur. J. Sustain. Dev. Res.* **2018**, *2*, 29.

274. Ajayi, T.O.; Ogunbayo, A.O. Achieving environmental sustainability in wastewater treatment by phytoremediation with water hyacinth (*Eichhornia crassipes*). *J. Sustain. Dev.* **2012**, *5*, 80–90. [CrossRef]

275. Kolawole, B.S. Cleaning of Effluent from Textile Industry by Water Hyacinth. Bachelor's Thesis, University of Agriculture, Abeokuta, Nigeria, 2001.

276. Mokhtar, H.; Morad, N.; Ahmad Fizri, F.F. Hyperaccumulation of copper by two species of aquatic plants. In Proceedings of the International Conference on Environment Science and Engineering, Bali Island, Indonesia, 1–3 April 2011; IACSIT Press: Singapore, 2011.

277. Tusief, M.Q.; Malik, M.H.; Asghar, H.N.; Mohsin, M.; Mahmood, N. Bioremediation of Textile Wastewater through Floating Treatment Wetland System. *Int. J. Agric. Biol.* **2019**, *22*, 821–826.

278. Goswami, S. Impact of Coal Mining on Environment. *Eur. Res.* **2015**, *92*, 185–196.

279. Mishra, V.K.; Tripathi, B.D.; Kim, K. Removal and accumulation of mercury by aquatic. *J. Hazard. Mater.* **2009**, *172*, 749–754. [CrossRef] [PubMed]

280. Ebenebe, P.C.; Shale, K.; Sedibe, M. South African Mine Effluents: Heavy Metal Pollution and Impact on the Ecosystem. *Int. J. Chem. Sci.* **2017**, *15*, 198.

281. Mishra, V.K.; Shukla, R. Aquatic Macrophytes for the Removal of Heavy Metals from Coal Mining Effluent. In *Phytoremediation*; Springer: Berlin/Heidelberg, Germany, 2016; Volume 3, pp. 143–156.

282. Mishra, V.K.; Upadhyay, A.R.; Pandey, S.K.; Tripathi, B.D. Concentrations of heavy metals and aquatic macrophytes of Govind Ballabh Pant Sagar an anthropogenic lake affected by coal mining effluent. *Environ. Monit. Assess.* **2008**, *141*, 49–58. [CrossRef]

283. Saha, P.; Shinde, O.; Sarkar, S. Phytoremediation of industrial mines wastewater using water hyacinth. *Int. J. Phytoremediat.* **2016**, *19*, 87–96. [CrossRef]

284. Marrugo-Negrete, J.; Enamorado-Montes, G.; Durango-Hernandez, J.; Pinedo-Hernandez, J.; Díez, S. Removal of mercury from gold mine effluents using *Limnocharis flava* in constructed wetlands. *Chemosphere* **2017**, *167*, 188–192. [CrossRef]

285. Leung, H.M.; Duzgoren-Aydin, N.S.; Au, C.K.; Krupanidhi, S.; Fung, K.Y.; Cheung, K.C.; Wong, Y.K.; Peng, X.L.; Ye, Z.H.; Yung, K.K.L. Monitoring and assessment of heavy metal contamination in a constructed wetland in Shaoguan (Guangdong Province, China): Bioaccumulation of Pb, Zn, Cu and Cd in aquatic and terrestrial components. *Environ. Sci. Pollut. Res. Int.* **2017**, *24*, 9079–9088. [CrossRef] [PubMed]

286. Türker, O.C.; Böcük, H.; Yakar, A. The phytoremediation ability of a polyculture constructed wetland to treat boron from mine effluent. *J. Hazard. Mater.* **2013**, *252*, 132–141. [CrossRef] [PubMed]

287. Younger, P.L.; Henderson, R. Synergistic wetland treatment of sewage and mine water: Pollutant removal performance of the first full-scale system. *Water Res.* **2014**, *15*, 74–82. [CrossRef] [PubMed]

288. Mardalena, M.; Faizal, M.; Napoleon, A. The Absorption of Iron (Fe) and Manganese (Mn) from Coal Mining Wastewater with Phytoremediation Technique Using Floating Fern (*Salvinia natans*), Water Lettuce (*Pistia stratiotes*) and Water Hyacinth (*Eichornia crassipes*). *Biol. Res. J.* **2018**, *4*, 1–7.

289. Abu Bakar, A.F.; Yusoff, I.; Fatt, N.T.; Othman, F.; Ashraf, M.A. Arsenic, zinc, and aluminium removal from gold mine wastewater effluents and accumulation by submerged aquatic plants (*Cabomba piauhyensis*, *Egeria densa*, and *Hydrilla verticillata*). *BioMed Res. Int.* **2013**, *2013*, 890803–890810. [CrossRef]

290. Pat-Espadas, A.M.; Portales, R.L.; Amabilis-Sosa, L.E.; Gómez, G.; Vidal, G. Review of Constructed Wetlands for Acid Mine Drainage Treatment. *Water* **2018**, *10*, 1685. [CrossRef]

291. Nagendran, R.; Selvam, A.; Joseph, K.; Chiemchaisri, C. Phytoremediation and rehabilitation of municipal solid waste landfills and dumpsites: A brief review. *Waste Manag.* **2006**, *26*, 1357–1369. [CrossRef]

292. Hughes, K.; Christy, A.; Heimlich, J. *Landfill Types and Liner Systems; Ohio State University Extension Fact Sheet CDFS-138-05*; The Ohio State University: Columbus, OH, USA, 2008; p. 4.

293. Njoku, P.O.; Edokpayi, J.N.; Odiyo, J.O. Health and environmental risks of residents living close to a landfill: A case study of Thohoyandou Landfill, Limpopo Province, South Africa. *Int. J. Environ. Res. Public Health* **2019**, *16*, 2125. [CrossRef]

294. Aziz, H.A.; Ramli, S.F. Recent development in sanitary landfilling and landfill leachate treatment in Malaysia. *Int. J. Environ. Eng.* **2018**, *9*, 201–229. [CrossRef]

295. Jones, D.L.; Williamson, K.L.; Owen, A.G. Phytoremediation of landfill leachate. *Waste Manag.* **2006**, *26*, 825–837. [CrossRef]

296. Madera-Parra, C.A.; Peña-Salamanca, E.J.; Peña, M.R.; Rousseau, D.P.L.; Lens, P.N.L. Phytoremediation of Landfill Leachate with *Colocasia esculenta*, *Gynerum sagittatum* and *Heliconia psittacorum* in Constructed Wetlands. *Int. J. Phytoremediat.* **2015**, *17*, 16–24. [CrossRef] [PubMed]

297. Jaya, N.; Amir, A.; Mohd-Zaki, Z. Removal of Nitrate by *Eichhornia crassipes* sp. in Landfill Leachate. In Proceedings of the International Civil and Infrastructure Engineering Conference, Kota Kinabalu, Malaysia, 28 September–1 October 2014; Springer: Berlin/Heidelberg, Germany, 2015; pp. 1043–1052.

298. El-Gendy, A.S.; Biswas, N.; Bewtra, J.K. Municipal Landfill Leachate Treatment for Metal Removal Using Water Hyacinth in a Floating Aquatic System. *Water Environ. Res.* **2006**, *78*, 951–964. [CrossRef] [PubMed]

299. Akinbile, C.O.; Yusoff, M.S.; Shian, L.M. Leachate Characterization and Phytoremediation Using Water Hyacinth (*Eichorrnia crassipes*) in Pulau Burung, Malaysia. *Bioremediat. J.* **2012**, *16*, 9–18. [CrossRef]

300. Omondi, E.A.; Ndiba, P.K.; Njuru, P.G. Phytoremediation of Polychlorobiphenyls (PCB's) in Landfill E-Waste Leachate with Water Hyacinth (E. *Crassipes*). *Int. J. Sci. Technol. Res.* **2015**, *4*, 147–156.

301. Ugya, A.Y.; Priatamby, A. Phytoremediation of Landfill Leachates Using *Pistia stratiotes*: A Case Study of Kinkinau U/Ma'azu Kaduna, Nigeria. *J. Agric. Biol. Environ. Stat.* **2016**, *2*, 60–63.

302. Wani, R.A.; Ganai, B.A.; Shah, M.A.; Uqab, B. Heavy Metal Uptake Potential of Aquatic Plants through Phytoremediation Technique—A Review. *J. Bioremediat. Biodegrad.* **2017**, *8*, 4. [CrossRef]

303. Duman, F.; Cicek, M.; Sezen, G. Seasonal changes of metal accumulation and distribution in common club rush (*Schoenoplectus lacustris*) and common reed (*Phragmites australis*). *Ecotoxicology* **2007**, *16*, 457–463. [CrossRef]

304. Grisey, E.; Laffray, X.; Contoz, O.; Cavalli, E.; Mudry, J.; Aley, L. The Bioaccumulation Performance of Reeds and Cattails in a Constructed Treatment Wetland for Removal of Heavy Metals in Landfill Leachate Treatment (Etueffont, France). *Water Air Soil Pollut.* **2012**, *223*, 1723–1741. [CrossRef]

305. Mugisa, D.J.; Banadda, N.; Kiggundu, N.; Asuman, R. Lead uptake of water plants in water stream at Kiteezi landfill site, Kampala (Uganda). *Afr. J. Environ. Sci. Technol.* **2015**, *9*, 502–507.

306. Odong, R.; Kansiime, F.; Omara, J.; Kyambadde, J. The potential of four tropical wetland plants for the treatment of abattoir effluent. *Int. J. Environ. Technol. Manag.* **2013**, *16*, 203–222. [CrossRef]

307. Wang, M.; Zhang, D.Q.; Dong, J.W.; Tan, S.K. Constructed wetlands for wastewater treatment in cold climate—A review. *J. Environ. Sci.* **2017**, *57*, 293–311. [CrossRef] [PubMed]

308. Vymazal, J.; Březinová, T. The use of constructed wetlands for removal of pesticides from agricultural runoff and drainage: A review. *Environ. Int.* **2015**, *75*, 11–20. [CrossRef] [PubMed]

309. Abou-Elela, S.I.; Golinielli, G.; Abou-Taleb, E.M.; Hellal, M.S. Municipal wastewater treatment in horizontal and vertical flows constructed wetlands. *Ecol. Eng.* **2013**, *61*, 460–468. [CrossRef]

310. Griffiths, L.N.; Mitsch, W.J. Removal of nutrients from urban stormwater runoff by storm-pulsed and seasonally pulsed created wetlands in the subtropics. *Ecol. Eng.* **2017**, *108*, 414–424. [CrossRef]

311. Saeed, T.; Muntahaa, S.; Rashid, M.; Sun, G.; Hasna, A. Industrial wastewater treatment in constructed wetlands packed with construction materials and agricultural by-products. *J. Clean. Prod.* **2018**, *189*, 442–453. [CrossRef]

312. Rizzo, A.; Bresciani, R.; Martinuzzi, N.; Masi, F. Online Monitoring of a Long-Term Full-Scale Constructed Wetland for the Treatment of Winery Wastewater in Italy. *Appl. Sci.* **2020**, *10*, 555. [CrossRef]

313. Brix, H. Do macrophytes play a role in constructed treatment wetlands? *Water Sci. Technol.* **1997**, *35*, 11–17. [CrossRef]

314. Hammer, D.A. *Constructed Wetlands for Wastewater Treatment: Municipal, Industrial and Agricultural*; CRC Press: Boca Raton, FL, USA, 1989.

315. Marchand, L.; Mench, M.; Jacob, D.L.; Otte, M.L. Metal and metalloid removal in constructed wetlands, with emphasis on the importance of plants and standardized measurements: A review. *Environ. Pollut.* **2010**, *158*, 3447–3461. [CrossRef]

316. Baharudin, B.; Shahrel, M. Lead and Cadmium Removal in Synthetic Wastewater Using Constructed Wetland. Ph.D. Thesis, Faculty of Chemical & Natural Resources Engineering Universiti, Gambang, Malaysia, 2008.

317. Yeh, T.Y.; Chou, C.C.; Pan, C.T. Heavy metal removal within pilot-scale constructed wetlands receiving river water contaminated by confined swine operations. *Desalination* **2009**, *249*, 368–373. [CrossRef]

318. Brezinová, T.B.; Vymazal, J. Evaluation of heavy metals seasonal accumulation in *Phalaris arundinacea* in a constructed treatment wetland. *Ecol. Eng.* **2015**, *79*, 94–99. [CrossRef]

319. Vymazal, J.; Kröpfelová, L.; Švehla, J.; Štíchová, J. Can multiple harvest of aboveground biomass enhance removal of trace elements in constructed wetlands receiving municipal sewage? *Ecol. Eng.* **2010**, *36*, 939–945. [CrossRef]

320. Bonanno, G. Trace element accumulation and distribution in the organs of *Phragmites australis* (common reed) and biomonitoring applications. *Ecotoxicol. Environ. Saf.* **2011**, *74*, 1057–1064. [CrossRef] [PubMed]

321. Bhattacharya, T.; Banerjee, D.; Gopal, B. Heavy metal uptake by *Scirpus littoralis* schrad. From fly ash dosed and metal spiked soils. *Environ. Monit. Assess.* **2006**, *121*, 363–380. [CrossRef] [PubMed]

322. Rai, P.K. Heavy metals/metalloids remediation from wastewater using free floating macrophytes of a natural wetland. *Environ. Technol. Innov.* **2019**, *15*, 100393. [CrossRef]

323. Sukumaran, D. Phytoremediation of Heavy Metals from Industrial Effluent Using Constructed Wetland Technology. *Appl. Ecol. Environ. Sci.* **2013**, *1*, 92–97. [CrossRef]

324. Ladislas, S.; Gérente, C.; Chazarenc, F.; Brisson, J.; Andrès, Y. Floating treatment wetlands for heavy metal removal in highway stormwater ponds. *Ecol. Eng.* **2015**, *80*, 85–91. [CrossRef]

325. Dan, A.; Oka, M.; Fujii, Y.; Soda, S.; Ishigaki, T.; Machimura, T.; ke, M. Removal of heavy metals from synthetic landfill leachate in lab-scale vertical flow constructed wetlands. *Sci. Total Environ.* **2017**, *584*, 742–775.

326. Mustapha, H.I.A.; van Bruggen, J.J.; Lens, P.N.L. Fate of heavy metals in vertical subsurface flow constructed wetlands treating secondary treated petroleum refinery wastewater in Kaduna, Nigeria. *Int. J. Phytoreme.* **2017**, *20*, 44–53. [CrossRef]

327. Githuku, C.R.; Ndambuki, J.M.; Salim, R.W.B.; Adedayo, A. Treatment potential of Typha latifolia in removal of heavy metals from wastewater using constructed wetlands. In *Transformation towards Sustainable and Resilient Wash Services; Proceedings of the 41st WEDC International Conference, Nakuru, Kenya, 9–13 July WEDC;* Shaw, R.J., Ed.; Loughborough University United Kingdom (UK): Loughborough, UK, 2018; pp. 9–13.

328. Šíma, J.; Svoboda, L.; Šeda, M.; Krejsa, J.; Jahodová, J. The fate of selected heavy metals and arsenic in a constructed wetland. *J. Environ. Sci. Health* **2019**, *54*, 56–64. [CrossRef]

329. Vymazal, J.; Kropfelov, L.; Svehla, J.; Chrastny, V.; Stıchova, J. Trace elements in *Phragmites australis* growing in constructed wetlands for treatment of municipal wastewater. *Ecol. Eng.* **2009**, *35*, 303–309. [CrossRef]

330. Hadad, H.R.; Maine, M.A.; Bonetto, C.A. Macrophyte growth in a pilot-scale constructed wetland for industrial wastewater treatment. *Chemosphere* **2006**, *63*, 1744–1753. [CrossRef] [PubMed]

331. Maine, M.A.; Hadad, H.R.; Sánchez, G.; Caffaratti, S.; Pedro, M.C. Kinetics of Cr (III) and Cr (VI) removal from water by two floating macrophytes. *Int. J. Phytorem.* **2016**, *18*, 261–268. [CrossRef] [PubMed]

332. Xu, X.; Mills, G.L. Do constructed wetlands remove metals or increase metal bioavailability? *J. Environ. Manag.* **2018**, *218*, 245–255. [CrossRef] [PubMed]

333. Yeh, T.Y. Removal of metals in constructed wetlands. *Pract. Period. Hazard. Toxic Radioact. Waste Manag.* **2009**, *12*, 96–101. [CrossRef]

334. Cui, W.; Cheng, J. Growing duckweed for biofuel production: A review. *Plant Biol.* **2015**, *17*, 16–23. [CrossRef] [PubMed]

335. Mishima, D.; Tateda, M.; Ike, M.; Fujita, M. Comparative study on chemical pretreatments to accelerate enzymatic hydrolysis of aquatic macrophyte biomass used in water purification processes. *Bioresour. Technol.* **2006**, *97*, 2166–2172. [CrossRef]

336. Villamagna, A.M. Ecological Effects of Water Hyacinth (*Eichhornia crassipes*) on Lake Chapala, Mexico. Ph.D. Thesis, Virginia Polytechnic Institute and State University, Blacksburg, Virginia, 2009.

337. Nakphet, S.; Ritchie, R.J.; Kiriratnikom, S. Aquatic plants for bioremediation in red hybrid tilapia (*Oreochromis niloticus* × *Oreochromis mossambicus*) recirculating aquaculture. *Aquac. Int.* **2017**, *25*, 619–633. [CrossRef]

Article

A Combined System Using Lagoons and Constructed Wetlands for Swine Wastewater Treatment

Pietro Denisi, Nicola Biondo, Giuseppe Bombino, Adele Folino, Demetrio Antonio Zema * and Santo Marcello Zimbone

Department Agraria, Mediterranea University of Reggio Calabria, Località Feo di Vito, I-89122 Reggio Calabria, Italy; pietro.denisi@unirc.it (P.D.); nicolabiondo83@gmail.com (N.B.); giuseppe.bombino@unirc.it (G.B.); adelefolino@alice.it (A.F.); smzimbone@unirc.it (S.M.Z.)
* Correspondence: dzema@unirc.it

Abstract: This study evaluates the depuration efficiency of a combined system consisting of lagoons (with aerated and non-aerated tanks) and CWs (with *Typha latifolia* L.) working at pilot scale for treating SW under two recirculation rates (RRs, 4:1 and 10:1) of the CW effluent. The combined system removed about 99% of the total suspended solids and organic matter, and from 80% to 95% of the total nitrogen at both tested RRs. The lagoon system was effective as a pre-treatment of SW, particularly for nitrogen removal. It is convenient to adopt the higher RR, since nitrogen removal can be increased by approximately 20%. The irrigation of the CWs with SW did not generally determine the phyto-toxic effects on *Typha latifolia* L., except at the start of the experiment and under the lower RR. Despite the limited spatial and temporal scale of this investigation, these results provide a starting point for the use of V-SSF CWs to treat livestock wastewater with a high pollution potential (such as SW).

Keywords: livestock wastewater; *Typha latifolia* L.; V-SSF systems; total nitrogen; COD; total suspended solids

Citation: Denisi, P.; Biondo, N.; Bombino, G.; Folino, A.; Zema, D.A.; Zimbone, S.M. A Combined System Using Lagoons and Constructed Wetlands for Swine Wastewater Treatment. *Sustainability* **2021**, *13*, 12390. https://doi.org/10.3390/su132212390

Academic Editors: Muhammad Arslan, Muhammad Afzal and Naser A. Anjum

Received: 30 September 2021
Accepted: 6 November 2021
Published: 10 November 2021

Publisher's Note: MDPI stays neutral with regard to jurisdictional claims in published maps and institutional affiliations.

1. Introduction

The wastewater from livestock breeding farms has a heavy pollution potential for soil and water resources [1,2]. Swine breeding farms make a significant contribution to the production of wastewater. Swine wastewater (hereafter "SW") consists of a blend of urine (about 40–45 kg/day/1000 kg of animal live weight [3]); feces (about 90 kg/day/1000 kg live weight [3]); water; residues of undigested food; antibiotics; and pathogens, as well as water used to clean the housing sheds [4–6]. To give a rough indication, SW is characterized by a high content of solids (total solids in the range 12.6 to 42.7 g/L), organic matter (chemical oxygen demand, COD, between 16.1 and 56.2 g/L), and nutrients (total nitrogen: 1.5–5.2 g/L; ammonium: 0.9–4.3 g/L; and total phosphorus: 0.1–1.3 g/L) [1,3,7]. SW also contains other organic xenobiotic substances, such pharmaceuticals [8].

Due to these physico-chemical characteristics and large production, the management of SW can lead to severe environmental pollution [9]. Furthermore, the direct disposal of SW can contaminate surface and ground waters, cause unpleasant odor emissions in the air, and degrade soil quality [8,10]. In particular, the high nutrient loads in SW can have undesirable effects on aquatic plant proliferation and the eutrophication of water bodies, as well as direct toxicity due to the high oxygen demand in water [11].

Traditionally, SW is commonly spread to the land as fertilizer, often after lagoon retention [12], but the nutrient loads always exceed the fertilizer requirement of crops with alterations in the nutrient balance of soil [13]. The practice of SW land spreading does not respect the strict rules of many countries on SW management (e.g., the so-called "Nitrate Directive" in Europe) [14]. In other cases, the farmland for SW spreading is insufficient [12]. Several systems have been proposed as alternatives to land spreading:

(i) aerobic (e.g., sequencing batch reactors and membrane bioreactors) and anaerobic (e.g., anaerobic digesters, up-flow anaerobic sludge beds, and microbial fuel cells) biological treatments [1,15]; and (ii) other chemical, physical, and biological systems [16,17]. However, these treatments generally present a low efficiency, high costs, and process instability, mainly due to the high concentration of organic matter (OM) and toxic compounds, such as ammonia nitrogen, in SW [5,9].

Natural systems, using lagoons or constructed wetlands, may be effective for SW treatment, due to the easy construction and cheap maintenance and operation [18–20], as well as the efficiency of removing several contaminants, such as OM; nutrients; heavy metals; and pathogens [16,21,22]. In general, SW is treated in anaerobic lagoons to remove OM, while constructed wetlands are used for nutrient removal [23].

Constructed wetlands (CWs) have a low energy requirement and are compatible with typical farm operations [24]. These systems have been used for decades to treat municipal wastewater [25,26], but their capacity to depurate agro-industrial wastewater and liquid livestock effluents have been recently demonstrated [24,27,28]. The results show that CW technology is established and its application for treating these effluents is well documented [29,30]. Regarding SW treatment, different successful experiences with CWs are reported in the literature, for both free water surface (FWS) and subsurface flow (SSF) types [14,27,28,31]. Most CWs treating SW are FWS systems, and only a few belong to SSF types [32–34]. Moreover, the interest in processes occurring in vertical SSF CWs is recent [35]. However, the COD concentrations in SW are commonly much higher than in municipal wastewater, and the nitrogen (N) loads are always high [10,36]. Therefore, CWs of the SSF type have to be adapted to these high concentrations of livestock effluents [37].

CWs for livestock wastewater must always be coupled with pre-treatment strategies (e.g., filtration, lagooning, and anaerobic digestion), whose effectiveness is very important to the constructed wetlands' performance [23,37]. Before livestock effluents enter a CW, oxidation and settling treatments are required to remove much of the OM, nitrogen, phosphorous loads, and suspended solids that can clog the soils in SSF systems [31]. Often, a pre-treatment using lagoons (to remove the COD and total suspended solids) may increase the depuration efficiency of CWs. For instance, an aerated lagoon can oxidize the organic load and convert N into a nitrate, and a non-aerated lagoon can increase the settlement of the suspended solids and the oxidation of OM without energy consumption [38,39].

Some research experiences have evaluated the depuration performance of SW pre-treatments before CWs. Sievers (1997) [40] examined two types of CWs, SSF and FWS, to treat effluents from an anaerobic swine lagoon system. Hunt et al. (2002) [27] investigated the effectiveness of CWs installed downstream of an anaerobic lagoon in a swine production facility. Shappell et al. (2007) [41] evaluated the effectiveness of a lagoon–CW combined treatment on SW, for producing an effluent with a low hormonal activity. Villamar et al. (2015) [23] studied the N and phosphorus distribution in a CW fed with SW previously treated in an anaerobic lagoon.

An examination of the N content of SW shows that concentrations between 0.2 and 0.4 g/L of ammonium-N generate phyto-toxic effects on the vegetation of CWs (growth inhibition and biomass production) [7,42]. With wetland emergent plant species (such as, *Glyceria*, *Carex*, *Typha*, *Schoenoplectus/Scirpus*, and *Juncus*), the tolerance limit can be even lower (<0.1 g/L) [43,44]. The most effective process for N removal in CWs is nitrification/denitrification [7], because it converts ammonia predominantly to nitrogen [28]. Since nitrification limits the removal of N from animal wastewater, the enhancement of denitrification is expected to increase the efficiency of CW performance [28,45].

The most common method for enhancing denitrification is wastewater recirculation and the addition of partially-nitrified water [12,45]. It has been demonstrated that the recirculation of partially treated wastewater increases the total N removal in CWs from 70% to 85% [46], while water addition from a nitrifying lagoon leads to 4- to 5-fold N removal rates, compared to non-nitrified wastewater [12,47]. Moreover, effluent recirculation supplies wastewater with additional oxygen for aerobic microbial activities and a microbial biomass that is already adapted to noticeable N levels. The availability of oxygen and microbial mass in the CW is often a factor limiting the removal rates of organic and N loads from wastewater [34]. However, there is little research about the N removal in CWs for the treatment of SW [7]. Moreover, to the best of the authors' knowledge, no studies are available in literature about the combination of CW–lagoon processes and effluent recirculation in the case of SW. This leaves the depuration efficiency and usability of these methods not well understood, to date.

This study aims to address these gaps in the literature, by evaluating the depuration efficiency of a combined system consisting of lagoons (aerated and non-aerated tanks) and CWs (with *Typha latifolia* L.) working at pilot scale for treating SW under two recirculation rates (RRs) of the CW effluent. The specific aims of the research are: (i) evaluating the OM (measuring COD) and nitrogen removal rates of the system; (ii) assessing which of the two tested recirculation ratios are more effective; and (iii) identifying any phyto-toxic effects of the treated SW. The results of this study, if validated in similar environmental contexts, can contribute to a broader applicability of the studied depuration system, supporting the action of breeders to control the soil and water contamination by OM and nutrients in areas with a high pollution risk.

2. Materials and Methods

2.1. Description of the Experimental System

The present study was carried out in a swine breeding farm (geographical coordinates: 38°06′00″ N, 15°46′47″ E) in Cardeto (Calabria, Southern Italy) at 1100 m a.s.l. The climate of the area is semi-arid and belongs to the "Csb" class ("Temperate, dry summer, warm summer Mediterranean" climate), according to Koppen, ref. [48] with cold winters and temperate summers. The mean annual temperature and precipitation were 11.1 °C (max 14.6 °C and min 5.4 °C) and 1380 mm, respectively (data obtained from the weather station of Gambarie, close to Cardeto).

The farm breeds approximately 1000 animals, and its SW is stored in a 180 m^3 open pond. The pond is filled each day after stable cleaning and emptied about once a year.

The experimental pilot plant, which was installed over a flat area close to the pond, was a combination of lagoons and CW systems with two hydraulically independent treatment lines, both fed by the SW previously stored in the pond. The system operated as a batch treatment. The lagoon system of each treatment line consisted of a series of two 1000 L tanks, supplied with 40 L/d of SW. One tank was aerated (with an air flow rate of 75 L/h) by a fine bubble diffuser placed 5 cm above the tank bottom and fed by a blower. The second tank was not aerated (Figure 1). Both tanks were occasionally covered against precipitation, but the cap allowed air entry from the water surface.

The aeration system was the same as the device described by [38,39,49], in which more details can be found. Downstream of the two tanks, a third smaller tank (which was not aerated) allowed SW storage before the subsequent treatment. This tank acted as a hydraulic disconnection between the two systems (lagoon and CW). Each tank was intermittently supplied with 20 and 50 L/h of SW in eight cycles per day.

The effluent of the disconnecting tank was transferred into the CW (with a volume of 1.6 m^3 and a size of 2 m × 1 m × 0.80 m) made of LDPE. The CW system was a V-SSF type, with a distributor uniformly spreading water over the soil surface. This type of CW was chosen due to its higher capacity to transport oxygen, compared to CWs with a horizontal flow. Therefore, V-SSF CWs are more efficient in removing OM and ammonia-N from wastewater through the aerobic microbial activity [34]. The CW was filled with two layers of soil (with a total depth of 0.6 m), of which the upper layer (0.4 m) was a loamy texture, and the bottom layer (0.2 m) was composed of gravel (diameter 5–20 mm). The upper layer was mixed with sand (sand/soil ratio = 3:1) to increase the porosity (35%) and thus the infiltration of SW.

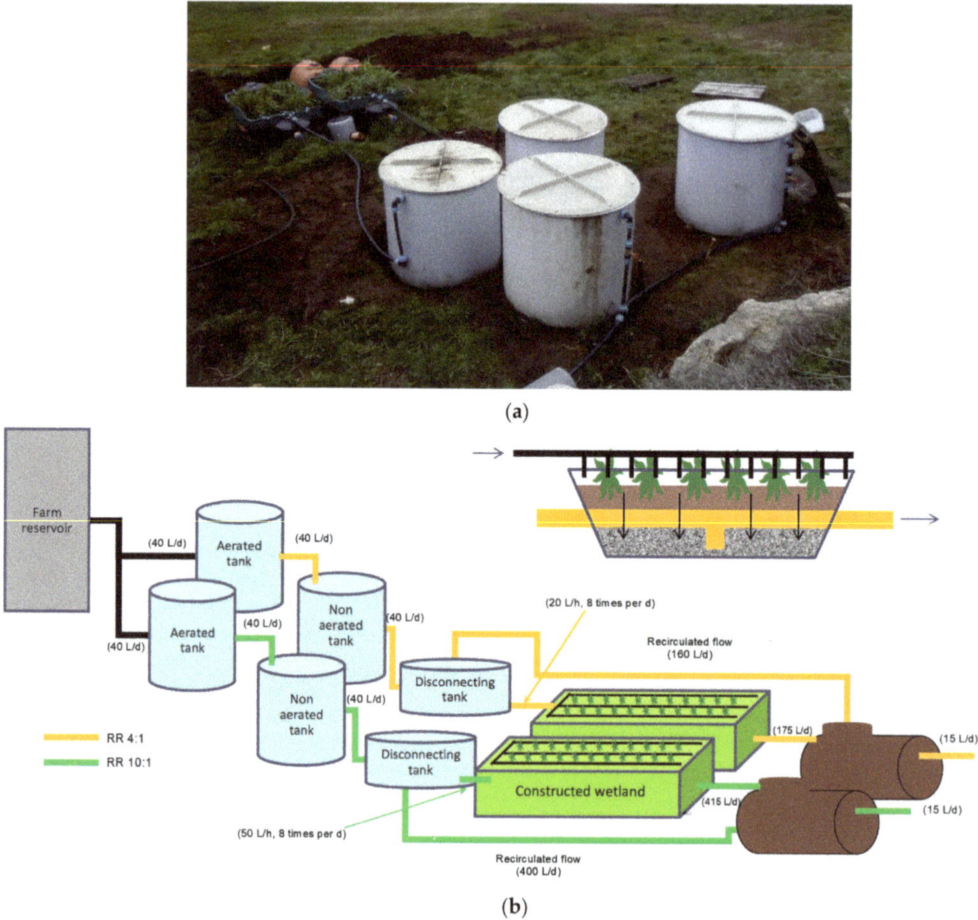

Figure 1. Photo (**a**) and scheme (**b**) of the lagoon–CW integrated system for treating swine wastewater in the experimental plant with the input/output water flow rates displayed.

The soil in the CW was planted with cattail (*Typha latifolia* L., 6 plants/m^2) in March 2017. This species is one of the most commonly used macrophytes in CWs beside bulrush (*Scirpus* spp.) and common reed (*Phragmites australis*), especially in the case of livestock water treatment [12,24].

A by-pass pipeline permitted the recirculation of the effluent in the system. More specifically, the effluent was pumped from the final tank into the disconnection tank. Two RRs were tested, one for each treatment line (Figure 1). The first RR was 4:1 (recirculated water:influent of the CW); the second was 10:1; and 160 and 400 L/h of SW were recirculated in the two lines, respectively. These ratios were selected setting up the theoretical total nitrogen (TN) concentration of the influent to about 0.2 (RR = 10:1) and 0.5 (RR = 4:1) g/L, close to the tolerance limits reported by [7,42]. The hydraulic loads were 220 and 100 L/d/m^2 for RRs 10:1 and 4:1, respectively.

A final tank receiving the effluent was located downstream of each CW. About 15 L/d was collected in this tank for both treatment lines (Figure 1).

The volume of SW to irrigate the plants in the CW was set according to the water requirement, due to the evapo-transpiration (ET) rate of the cattail. Therefore, the ET was monitored every week, measuring a third CW the water losses.

2.2. Wastewater Sampling and Characterization

Samples of SW were collected twice a month between March 2017 and March 2019, and immediately stored at 4 °C until the laboratory analysis.

The SW samples were collected: (i) in the farm storage pond (influent of the lagoon system); (ii) in the disconnection tank (effluent of lagoon system and influent of the CW), before the water flow mixing for recirculation; and (iii) in the final tank (effluent of CW).

The following characteristics were evaluated in the SW samples: the pH and electrical conductivity (EC), using a Hach Lange HQ40 multi-parameter device with dedicated probes, and the total suspended solids (TSS), after oven-drying at 105 °C for 24 h. Moreover, according to the standard methods [50], the chemical oxygen demand (COD) and total Kjeldahl nitrogen (TKN, the sum of organic nitrogen, un-ionized ammonia, and ammonium ion) were determined. The initial values of these SW parameters measured in the farm reservoir are reported in Table 1.

Table 1. Main parameters of swine wastewater initially sampled in the storage tank of the swine breeding farm.

Parameter	Value (mean $\pm$ std. dev., $n = 3$)
pH [-]	7.40 ± 0.0
TS [%]	0.96 ± 0.0
COD [g L^{-1}]	29.3 ± 1.01
TKN [g L^{-1}]	1.29 ± 0.44

Notes: TS = total solids; COD = chemical oxygen demand; and TKN = total Kjeldahl nitrogen.

2.3. Statistical Analyses

The statistical significance of differences in the main parameters of SW (farm reservoir vs. lagoon effluent vs. CW at RR = 4:1 vs. CW at RR = 10:1) and plant density in CWs (at RR = 4:1 vs. RR = 10:1) was evaluated using Kruskal–Wallis tests (a non-parametric alternative to the analysis of variance), followed by multiple pairwise comparisons using Dunn's procedure with the Bonferroni correction for the significance level for the pairwise comparisons. To differentiate the levels of significance, a *p*-level lower than 0.05 was adopted.

3. Results and Discussions

3.1. Depuration Performance of the Lagoon System

The lagoons decreased the pH (in the range of 7.48–7.86, farm reservoir, and 6.70–8.32, effluent) and EC from the influent (13.1–21.4 dS/m) to the effluent (6.09–19.7 dS/m). This parameter showed a noticeable temporal variability both in the influent and effluent. The TSS concentration decreased from values of up to 19–20 g/L measured in the farm reservoir (with a noticeable decreasing trend over time) to 0.34–1.43 g/L in the effluent (Figure 2).

After the lagoon, the COD concentration of the influent (generally with a limited temporal variability, 36.1 to 49.3 g/L) reduced, and the removal rate increased over time. The COD concentration in the effluent was between 8.56 and 36.3 g/L. The TKN decreased over time in the farm reservoir (from 1.26 to 2.60 g/L), but for this parameter the removal rate showed a low variability, reducing the N concentration from 1.04 to 1.84 g/L (Figure 2).

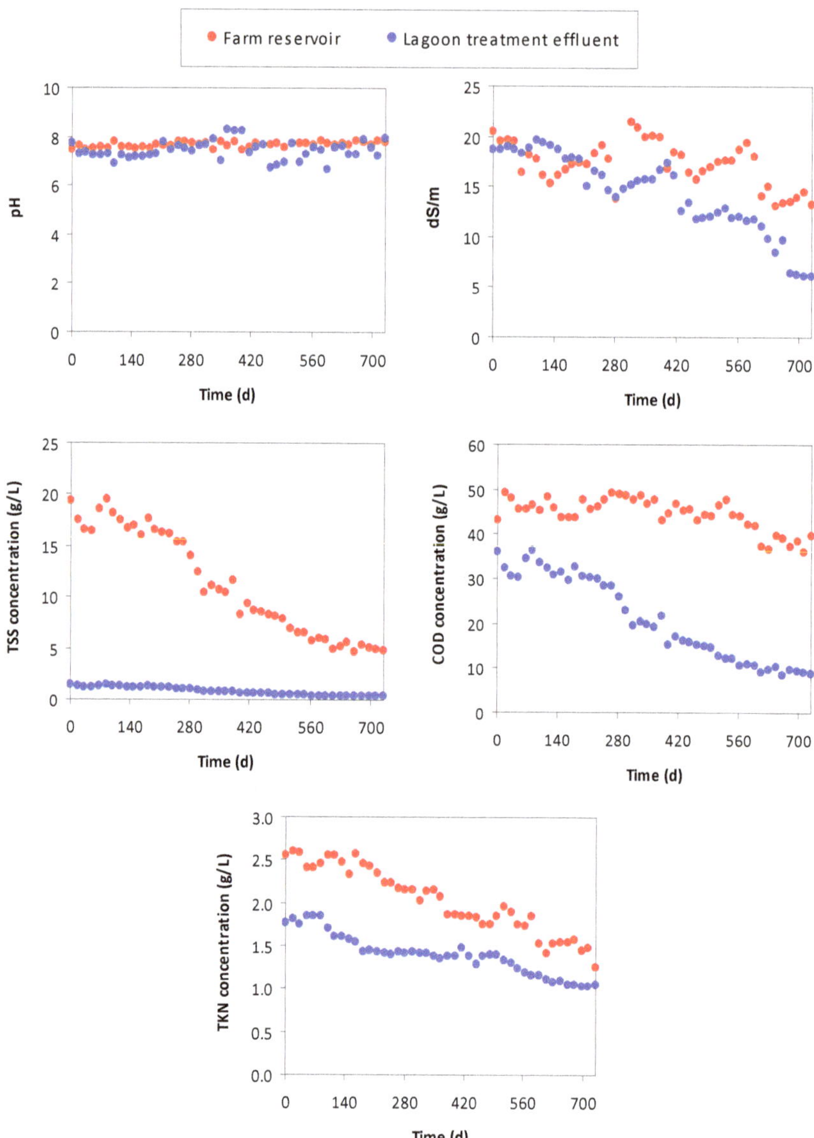

Figure 2. Variation of the main parameters over time of the lagoons of the integrated system for swine wastewater treatment (the reported values are averaged between the two lines of the lagoon system).

3.2. Depuration Performance of the Constructed Wetland

The SW treatment in CWs reduced the pH without significant differences between the tested RRs. This parameter was in the range of 6.0–8.0 for both RRs. However, the pH fluctuated in this range, according to the weather variations (Figure 3).

The EC declined in the CW under both RRs with a more noticeable effect (significant at $p < 0.05$) at the lower ratio (from 3.48 to 2.41 dS/m, RR = 4:1, and from 2.83 to 0.96 dS/m, RR = 10:1) (Figure 3).

The TSS concentration decreased (from 88.2 to 54.2 mg/L, RR = 4:1, and from 41.3 to 25.2 mg/L, RR = 10:1) for both RRs, following similar temporal trends (Figure 3), although the differences were significant. In contrast, the removal rates of both COD and TKN were significantly much higher in the CW, with an RR of 4:1. The higher RR reduced the COD from 98.2 to 40.3 mg/L in the CW with an RR of 4:1, and from 39.33 to 20.84 mg/L with an RR of 10:1. This difference was also detected for TKN, which decreased from 668 to 379 mg/L (RR = 4:1) and from 176 to 108 mg/L (RR = 10:1) (Figure 3).

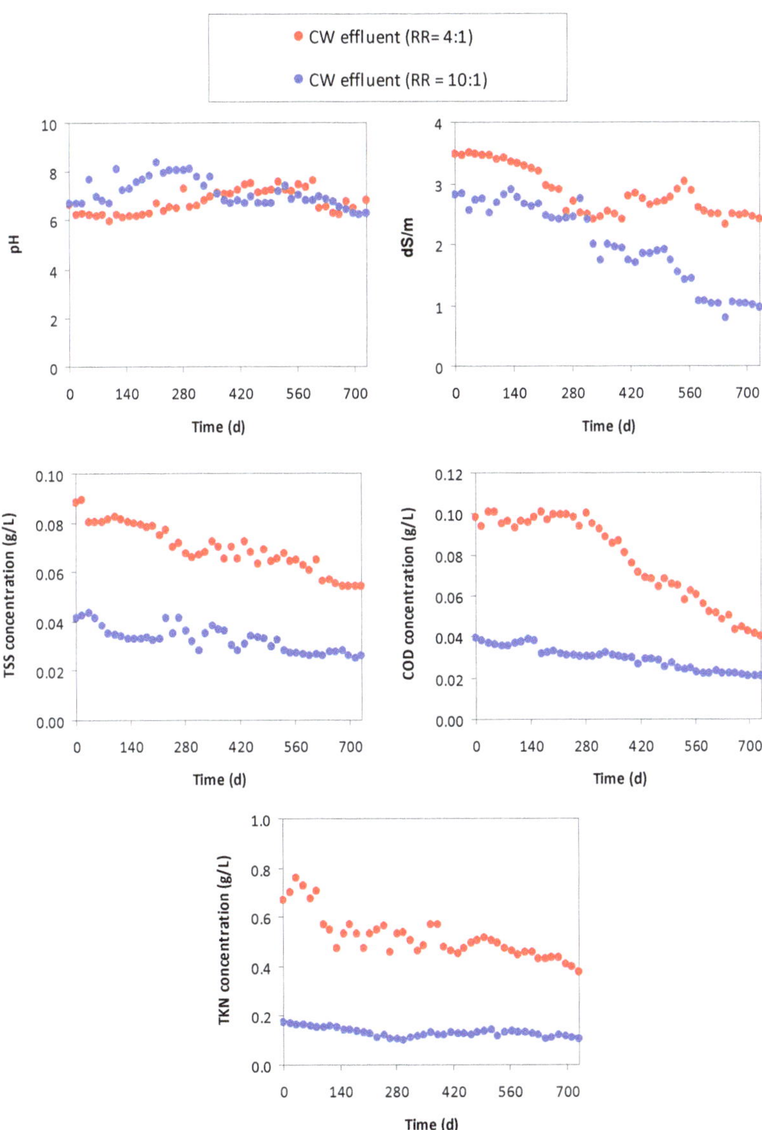

Figure 3. Variation of the main parameters over time of the constructed wetlands (at two recirculation ratios, RR) of the integrated system for swine wastewater treatment.

3.3. Variations in SW Parameters and the Removal of Pollutants

A separate analysis of the variability in the monitored parameters of SW in the two treatments (lagoon or CW) shows: (i) for the pH, slight variations in the lagoon system (by $-1.89 \pm 4.17\%$) and a greater decrease in the CW (on average 10–13%); (ii) a limited reduction in EC in the lagoon system ($-4.76 \pm 14.8\%$), and a much greater decrease in the CW system (-84.0 to -89.0%, RR = 4:1 and 10:1, respectively); (iii) an appreciable removal of TSS in the lagoon system ($42.4 \pm 35.3\%$) and a very high effectiveness in the CW (close to 100%); (iv) a limited efficiency in COD removal in the lagoon system ($2.17 \pm 4.88\%$), and, in contrast, an extremely high removal in the CW (also, in this case, close to 100%); and (v) a TKN removal by $31.0 \pm 15.6\%$ in the lagoon system, which increases to $72.7 \pm 2.48\%$ (RR = 4:1) and to $92.7 \pm 1.13\%$ (RR = 10:1) in the CW (Figure 4).

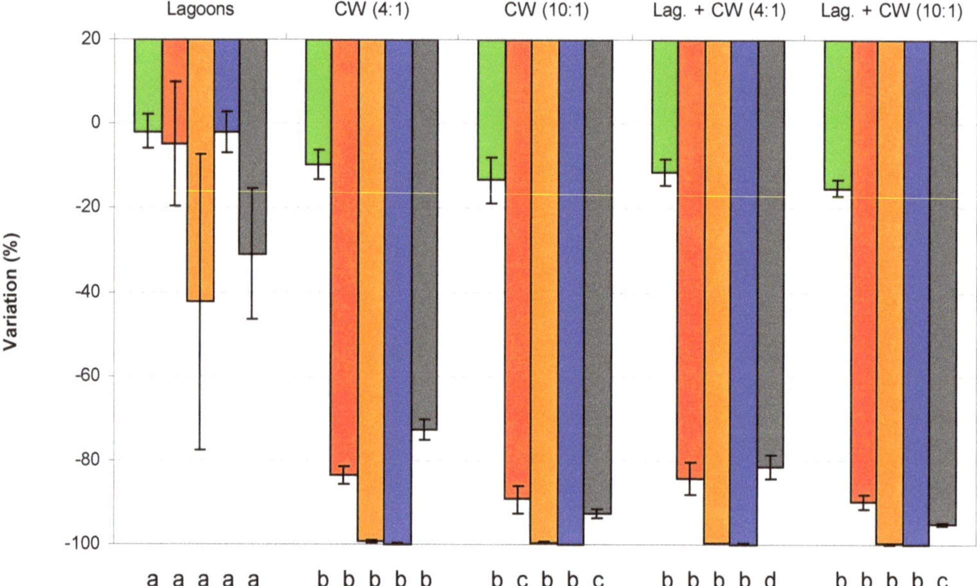

Figure 4. Variation (mean $\pm$ standard deviation) of the main parameters in the combined system using lagoons and constructed wetlands (CW, at two recirculation ratios (RR)) for swine wastewater treatment. Standard deviation is related to the parameter variability among the three survey dates (March 2017, 2018, and 2019). Letters indicate significant differences among the treatments (at $p < 0.05$).

If this analysis is carried out for the combined system (lagoon + CW under the two RRs), the following considerations can be derived: (i) the two RRs show a comparable effectiveness for pH modification ($-11.4 \pm 3.24\%$, RR = 4:1, to $-15.2 \pm 1.83\%$, RR = 10:1); (ii) a higher RR is more efficient to reduce the EC of SW ($-90 \pm 1.73\%$, RR = 10:1, against $-84.2 \pm 3.92\%$, RR = 4:1); (iii) both RRs provide an extremely high efficiency in removing TSS and COD (close to 100%); (iv) the efficiency in TKN removal increases from $81.4 \pm 2.74\%$ to $95.1 \pm 0.45\%$, when the RR is increased from 4:1 to 10:1 (Figure 4).

3.4. Effects of COD and TKN on Typha latifolia L. Plants

Compared to the initial density (6 plants/m^2), no significant effects of SW supply were noticed in the survival rates of *Typha latifolia* L. in CWs in the dry seasons (spring and summer, 18.5 ± 1.1 °C) (Figure 5). In fact, no death of plants was recorded in the CW under RR = 10:1, except 0.5 plants/m^2 in the summer of 2018. In the CW under the lower RR, the plant mortality was slightly higher in summer (especially in 2018, 1.4 plants/m^2). This plant death can be attributed to a peak in the soil N, presumably due to the excessive load of the influent (0.5–0.7 g/L) from the spring to winter seasons in 2017 (data not shown). The TKN concentrations of about 0.14–0.16 g/L in the CW with RR = 10:1 were, instead, well tolerated by the plants.

Several plants died in the cold seasons (autumn and winter), due to vegetative senescence. This plant death was particularly high in the season 2017–2018, when an extreme frost occurred in December. After this event, the killed plants were replaced by new plants, to restore the initial plant density.

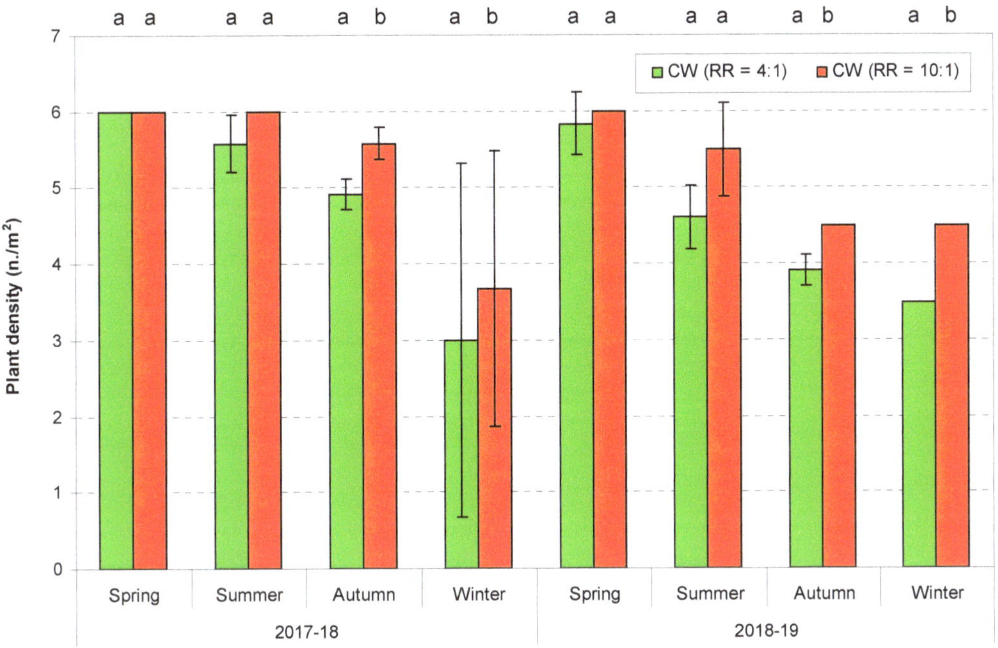

Figure 5. Plant density (mean ± standard deviation among the months) surveyed over two seasons in the constructed wetlands (CW, at two recirculation ratios, (RR)) of the integrated system for swine wastewater treatment. Letters indicate significant differences between the treatments with different RR (at $p < 0.05$).

4. Discussions

The evaluation of the main chemical properties of the SW treated in the combined system using lagoons and constructed wetlands has produced important indications in terms of depuration efficiency and, therefore, of its sustainability.

Concerning the depuration performance of the lagoon system, the decrease in the TSS concentration due to the lagoon process is well known [51]. The presence of a non-aerated tank promoted the activity of anaerobic bacteria, which degraded the organic matter concentration, and this also reduced the amount of the TSS [49,52,53]. Furthermore, the effect of aeration in the upstream tank promoted the flocculation process, due to the accelerated bacterial activity [52].

Although the reductions in the amounts of the OM and TKN, as a result of the lagoon system, were noticeable, the concentrations in the effluent noticeably exceeded the accepted amount for the discharge limits permitted by the main national rules. For example, according to the Italian environmental regulation (Legislative Decree 152 of 2006), the concentrations of nutrients and OM were about two orders of magnitude for the limits, equal to 20 mg/L for N and 160 mg/L for COD.

Concerning the depuration performance of the constructed wetland, the reduction in the pH of SW, which was close to neutrality, was in accordance with Boas et al. (2018) [54], who worked on CWs combining H-SSF and V-SSF systems, which favor the microbial activity of OM degradation and nutrient conversion.

The noticeable decline detected in the TSS concentration for both RRs is expected, since TSS is the water parameter that is strongly modified by CW treatments, as a consequence of the sedimentation, filtration, and adsorption processes that occur in CWs [33].

The COD and TKN removal in the CW were high, presumably due to the synergistic effects of both physical, chemical, and microbial processes, also under heavy loads of OM and nutrients, as is presented in this study. Organic and N compounds are removed in CWs of the SSF type by a combination of adsorption, nitrification/denitrification, volatilization, and ionic exchange [3]. Nitrification and denitrification are considered essential mechanisms for N removal [1,37], having an efficiency of more than 60% [55]. According to Vidal et al. (2018) [3], denitrification is the most effective process to remove nitrogen in CWs, and the aerated and non-aerated treatments prior to CW in the experimental plant have been beneficial for these processes. The aerated treatment has oxidized part of the TKN in SW, which was converted to nitrate. The aerated treatment should have denitrified part of this nitrate, but the remaining part was made available for denitrification in CW, which is nitrate-limited [3,27]. The nitrification of SW before the CW enhances N removal, and increases the nitrate available for denitrification [45]. Moreover, the absence of aeration should have provided anaerobic bacteria that were already adapted to the denitrification process. Denitrification is more desirable than ammonia volatilization in CWs treating the wastewater of animal origin, since ammonia is a pollutant for atmospheric, aquatic, and terrestrial environments through dry and wet deposition [45]. In our study, although not being directly measured, ammonia volatilization may have been limited, due to the pH level that was lower than eight [1,56,57], while the nitrification process should have been presumably low, due to the limited oxygen supply from the plants. Therefore, denitrification may have been the dominant process in TKN removal, in close accordance with Hunt et al. (2002) [27]. An important role in nutrient and OM removal is played by bacteria, which create a biofilm around the soil particles, allowing the catalysis of chemical reactions [33]. Effluent recirculation enabled the wastewater to flow repeatedly over this biofilm, enhancing the contact between the pollutants and microorganisms [33,34].

Plant uptake helps nitrogen removal, but its influence is lower compared to the other processes, and depends on the specific species. Plants remove ammonia nitrogen due to the stimulation of nitrifying bacteria and the uptake of nitrogen compounds [8], but these mechanisms seem to be marginal in many examples. *Typha latifolia* L. prefers slightly acidic environments, but ammonium uptake is conditioned by its toxicity (>0.2 g/L) [23] and COD concentrations of 0.6–0.8 g/L (that inhibit photo-synthesis and, consequently, nutrient incorporation) [23,58], as was evident in many stages of our study. Gonzales et al. (2009) [36] stated that the macrophyte species did not significantly contribute to the overall efficiency of V-SSF CWs in N removal, especially in the dry season. These authors attributed this minor contribution of plants to the high concentrations of contaminants. In contrast, planted CWs clearly show higher efficiencies for organic compounds, with removal efficiencies of up to 70% in wetlands planted with *Typha latifolia* L. compared to 60% of unplanted beds [36].

In relation to the variations in SW parameters and the removal of pollutants, the present study has demonstrated that the CW was more efficient in removing TSS, COD, and TKN compared to the lagoon. The lower efficiency for TSS removal in the lagoon system compared to the CW can be attributed to the great solid content of the raw SW. This low efficiency is close to the value of 25% experienced by Stone et al. (2004) [59] for SW lagoon treatment in North Carolina. To increase the system ability to remove TSS, a pre-treatment to remove further amounts of TSS in the raw SW is still necessary because it can prevent the soils of CW from being rapidly clogged. The very high efficiencies of the CW system in removing TSS are in close accordance with the values (97–99%) reported by [14,33]. In the experiences of other authors, TSS removal efficiencies between 40–50% [3] and 70–80% [35,36] were detected. Literature reports COD removal efficiencies in the range of 50–80% [33,35,36,60] with an extreme value of 99.5% detected by Masi et al. (2017) [14] working in a CW combined system (as in our study). N removal varies between 60% and 80% according to many authors [3,8,11,27,36,54], but extreme values are also reported (10–40% [33,35] to 99% [14]). In our study, the TKN removal efficiency for the CW system with RR = 4:1 is in the range reported by the majority of studies, and it is not far from the optimal value of [14] in the CW with RR = 10:1.

The analysis of the depuration efficiency of the combined system (lagoon + CW under the two RRs) suggests the adoption of an RR equal to 10:1, in order to increase the TKN removal, while the efficiency of reducing the pH and EC, and removing the TSS and COD, is comparable. Similar to the observations of Lee et al. (2006) and He et al. (2004) [33,34], the effluent recirculation in the system supplies a considerable amount of oxygen in the SW, promoting the reductions in COD and TKN. Concerning the experiences using V-SSF-CWs systems with recirculation, He et al. (2006) [34] showed that this operation strategy increased the average removal efficiencies of NH_4-N, COD, and TSS to 62%, 81%, and 77%, respectively, compared to the values of 36%, 50% and 49%, without effluent recirculation. With an RR of 100%, the average removal efficiencies were 91% for COD and 96% for TSS [61].

Regarding the effects of COD and TKN on *Typha latifolia* L., the irrigation of the CW with SW effluents from the lagoon treatment did not affect plant survival in the dry season, especially at the higher RR. In contrast, the higher plant mortality detected in the CW with the lower RR can be attributed to a peak in the nitrogen load, which exceeded the tolerance limits of *Typha latifolia* L. These limits were quantified by De los Reyes et al. (2014) [7] between 0.2 and 0.4 g/L of NH_4^+-N, which correspond to 60–80% of TKN, and therefore the expected phyto-toxic effects may have been realistic.

5. Conclusions

This study has shown that a pilot-scale system consisting of open lagoons (with or without aeration) and a constructed wetland with *Typha latifolia* L. treating raw swine wastewater has removed about 99% of the total suspended solids and COD, and from 80% to 95% of the total nitrogen in the effluent at both tested recirculation rates (4:1 and 10:1). Both the pH and the electrical conductivity (the indirect measure of wastewater salinity) were noticeably modified by the combined system, and reductions in the electrical conductivity (85–90%) were detected. The lagoon system alone showed low to moderate depuration performances (−40% of TSS, −2% of COD, and −31% of TKN removed), but represented an effective pre-treatment of CW, particularly for the nitrogen removal.

This experience has also demonstrated the suitability to increase the effluent recirculation ratio up to a value of 10:1, when high nitrogen removal rates are expected, since the higher RR allowed the removal of about 20% more of nitrogen compounds compared to an RR of 4:1. In contrast, the organic matter removal was not affected by an increased RR, given the very high depuration efficiency detected at the lower RR.

The irrigation of the CWs with SW did not generally determine the phyto-toxic effects on *Typha latifolia* L., except at the start of the experiment and under the lower effluent recirculation ratio, when a peak in the soil nitrogen killed about 25% of plants.

Although this study was carried out on pilot plants and throughout a short monitoring period (two years), the relevant results provide a starting point for the use of V-SSF CWs as a depuration solution in highly polluting livestock wastewater (such as, SW). An upscale of this preliminary investigation is suggested to verify the depuration performance of the system by real-scale experiments. A more detailed analysis of the physico-chemical and microbiological processes acting in the CW system may help to identify the most effective mechanisms for removing the polluting compounds of SW. Moreover, the incidence of each process that determines a mass loss (e.g., volatilization, denitrification, nitrogen plant uptake, and hydraulic losses) on water, soil, and plants of each sub-component of the experimental system should be quantified adopting a mass balance approach in future research.

Author Contributions: Conceptualization, G.B., D.A.Z. and S.M.Z.; methodology, N.B., G.B., P.D., A.F. and D.A.Z.; validation, G.B., D.A.Z. and S.M.Z.; formal analysis, G.B., P.D., A.F., D.A.Z. and S.M.Z.; data curation, N.B., G.B., P.D., A.F. and D.A.Z.; writing—original draft preparation, P.D. and D.A.Z.; writing—review and editing, G.B., P.D., D.A.Z. and S.M.Z.; supervision, G.B., D.A.Z. and S.M.Z.; project administration, S.M.Z.; funding acquisition, S.M.Z. All authors have read and agreed to the published version of the manuscript.

Funding: This research was funded by the National Project "PRIN—Programmi di Ricerca di Interesse Nazionale" (2015), by the Italian Ministry of Education, University and Research (M.I.U.R., Principal Investigator Attilio Toscano).

Institutional Review Board Statement: Not applicable.

Informed Consent Statement: Not applicable.

Data Availability Statement: The data presented in this study are available on request from the corresponding author.

Acknowledgments: We cordially thank the "Società Agricola Biondo dei F.lli Biondo & C." firm (Cardeto, Reggio Calabria, Italy) for hosting the experimental device and providing the manpower for its management.

Conflicts of Interest: The authors declare no conflict of interest.

References

1. Hu, H.; Li, X.; Wu, S.; Yang, C. Sustainable Livestock Wastewater Treatment via Phytoremediation: Current Status and Future Perspectives. *Bioresour. Technol.* **2020**, *315*, 123809. [CrossRef] [PubMed]
2. Algieri, A.; Andiloro, S.; Tamburino, V.; Zema, D.A. The Potential of Agricultural Residues for Energy Production in Calabria (Southern Italy). *Renew. Sustain. Energy Rev.* **2019**, *104*, 1–14. [CrossRef]
3. Vidal, G.; de Los Reyes, C.P.; Sáez, O. The Performance of Constructed Wetlands for Treating Swine Wastewater under Different Operating Conditions. In *Constructed Wetlands for Industrial Wastewater Treatment*; Alexandros, S., Ed.; John Wiley & Sons, Ltd.: Chichester, UK, 2018; pp. 203–221. ISBN 978-1-119-26837-6.
4. Cheng, D.L.; Ngo, H.H.; Guo, W.S.; Liu, Y.W.; Zhou, J.L.; Chang, S.W.; Nguyen, D.D.; Bui, X.T.; Zhang, X.B. Bioprocessing for Elimination Antibiotics and Hormones from Swine Wastewater. *Sci. Total Environ.* **2018**, *621*, 1664–1682. [CrossRef] [PubMed]
5. Folino, A.; Calabrò, P.S.; Zema, D.A. Effects of Ammonia Stripping and Other Physico-Chemical Pretreatments on Anaerobic Digestion of Swine Wastewater. *Energies* **2020**, *13*, 3413. [CrossRef]
6. Viancelli, A.; Kunz, A.; Steinmetz, R.L.R.; Kich, J.D.; Souza, C.K.; Canal, C.W.; Coldebella, A.; Esteves, P.A.; Barardi, C.R.M. Performance of Two Swine Manure Treatment Systems on Chemical Composition and on the Reduction of Pathogens. *Chemosphere* **2013**, *90*, 1539–1544. [CrossRef] [PubMed]
7. De Los Reyes, C.P.; Pozo, G.; Vidal, G. Nitrogen Behavior in a Free Water Surface Constructed Wetland Used as Posttreatment for Anaerobically Treated Swine wastewater effluent. *J. Environ. Sci. Health-Part A Toxic/Hazard. Subst. Environ. Eng.* **2014**, *49*, 218–227. [CrossRef] [PubMed]
8. Dordio, A.; Carvalho, A.J.P. Constructed Wetlands with Light Expanded Clay Aggregates for Agricultural Wastewater Treatment. *Sci. Total Environ.* **2013**, *463–464*, 454–461. [CrossRef]
9. Folino, A.; Zema, D.A.; Calabrò, P.S. Environmental and Economic Sustainability of Swine Wastewater Treatments Using Ammonia Stripping and Anaerobic Digestion: A Short Review. *Sustainability* **2020**, *12*, 4971. [CrossRef]
10. Folino, A.; Zema, D.A.; Calabrò, P.S. Organic Matter Removal and Ammonia Recovery by Optimised Treatments of Swine Wastewater. *J. Environ. Manag.* **2020**, *270*, 110692. [CrossRef]
11. Tanner, C.C.; Clayton, J.S.; Upsdell, M.P. Effect of Loading Rate and Planting on Treatment of Dairy Farm Wastewaters in Constructed Wetlands—II. Removal of Nitrogen and Phosphorus. *Water Res.* **1995**, *29*, 27–34. [CrossRef]
12. Harrington, C.; Scholz, M. Assessment of Pre-Digested Piggery Wastewater Treatment Operations with Surface Flow Integrated Constructed Wetland Systems. *Bioresour. Technol.* **2010**, *101*, 7713–7723. [CrossRef]
13. Qian, X.; Wang, Z.; Shen, G.; Chen, X.; Tang, Z.; Guo, C.; Gu, H.; Fu, K. Heavy Metals Accumulation in Soil after 4 Years of Continuous Land Application of Swine Manure: A Field-Scale Monitoring and Modeling Estimation. *Chemosphere* **2018**, *210*, 1029–1034. [CrossRef]
14. Masi, F.; Rizzo, A.; Martinuzzi, N.; Wallace, S.D.; Van Oirschot, D.; Salazzari, P.; Meers, E.; Bresciani, R. Upflow Anaerobic Sludge Blanket and Aerated Constructed Wetlands for Swine Wastewater Treatment: A Pilot Study. *Water Sci. Technol.* **2017**, *76*, 68–78. [CrossRef] [PubMed]
15. Zema, D.A. Planning the Optimal Site, Size, and Feed of Biogas Plants in Agricultural Districts. *Biofuels Bioprod. Bioref.* **2017**, *11*, 454–471. [CrossRef]
16. Kim, J.-H.; Chen, M.; Kishida, N.; Sudo, R. Integrated Real-Time Control Strategy for Nitrogen Removal in Swine Wastewater Treatment Using Sequencing Batch Reactors. *Water Res.* **2004**, *38*, 3340–3348. [CrossRef] [PubMed]

17. Mehta, C.M.; Khunjar, W.O.; Nguyen, V.; Tait, S.; Batstone, D.J. Technologies to Recover Nutrients from Waste Streams: A Critical Review. *Crit. Rev. Environ. Sci. Technol.* **2015**, *45*, 385–427. [CrossRef]

18. Hamza, R.A.; Iorhemen, O.T.; Tay, J.H. Anaerobic-Aerobic Granular System for High-Strength Wastewater Treatment in Lagoons. *Adv. Environ. Res.* **2016**, *5*, 169–178. [CrossRef]

19. Gikas, G.D.; Tsakmakis, I.D.; Tsihrintzis, V.A. Hybrid Natural Systems for Treatment of Olive Mill Wastewater. *J. Chem. Technol. Biotechnol.* **2018**, *93*, 800–809. [CrossRef]

20. Molari, G.; Milani, M.; Toscano, A.; Borin, M.; Taglioli, G.; Villani, G.; Zema, D.A. Energy Characterisation of Herbaceous Biomasses Irrigated with Marginal Waters. *Biomass Bioenergy* **2014**, *70*, 392–399. [CrossRef]

21. Pishgar, R.; Hamza, R.A.; Tay, J.H. Augmenting Lagoon Process Using Reactivated Freeze-Dried Biogranules. *Appl. Biochem. Biotechnol.* **2017**, *183*, 137–154. [CrossRef]

22. Van Kessel, M.A.H.J.; Speth, D.R.; Albertsen, M.; Nielsen, P.H.; Op Den Camp, H.J.M.; Kartal, B.; Jetten, M.S.M.; Lücker, S. Complete Nitrification by a Single Microorganism. *Nature* **2015**, *528*, 555–559. [CrossRef] [PubMed]

23. Villamar, C.A.; Rivera, D.; Neubauer, M.E.; Vidal, G. Nitrogen and Phosphorus Distribution in a Constructed Wetland Fed with Treated Swine Slurry from an Anaerobic Lagoon. *J. Environ. Sci. Health-Part A Toxic/Hazard. Subst. Environ. Eng.* **2015**, *50*, 60–71. [CrossRef] [PubMed]

24. Knight, R.L.; Payne, V.W.E.; Borer, R.E.; Clarke, R.A.; Pries, J.H. Constructed Wetlands for Livestock Wastewater Management. *Ecol. Eng.* **2000**, *15*, 41–55. [CrossRef]

25. Cirelli, G.L.; Consoli, S.; Di Grande, V.; Milani, M.; Toscano, A. Subsurface Constructed Wetlands for Wastewater Treatment and Reuse in Agriculture: Five Years of Experiences in Sicily, Italy. *Water Sci. Technol.* **2007**, *56*, 183–191. [CrossRef] [PubMed]

26. Toscano, A.; Langergraber, G.; Consoli, S.; Cirelli, G.L. Modelling Pollutant Removal in a Pilot-Scale Two-Stage Subsurface Flow Constructed Wetlands. *Ecol. Eng.* **2009**, *35*, 281–289. [CrossRef]

27. Hunt, P.G.; Szögi, A.A.; Humenik, F.J.; Rice, J.M.; Matheny, T.A.; Stone, K.C. Constructed Wetlands for Treatment of Swine Wastewater from an Anaerobic Lagoon. *Trans. ASAE* **2002**, *45*, 639.

28. Poach, M.E.; Hunt, P.G.; Reddy, G.B.; Stone, K.C.; Johnson, M.H.; Grubbs, A. Swine Wastewater Treatment by Marsh–Pond–Marsh Constructed Wetlands under Varying Nitrogen Loads. *Ecol. Eng.* **2004**, *23*, 165–175. [CrossRef]

29. Han, Z.; Dong, J.; Shen, Z.; Mou, R.; Zhou, Y.; Chen, X.; Fu, X.; Yang, C. Nitrogen Removal of Anaerobically Digested Swine Wastewater by Pilot-Scale Tidal Flow Constructed Wetland Based on in-Situ Biological Regeneration of Zeolite. *Chemosphere* **2019**, *217*, 364–373. [CrossRef]

30. Zhang, M.; Luo, P.; Liu, F.; Li, H.; Zhang, S.; Xiao, R.; Yin, L.; Zhou, J.; Wu, J. Nitrogen Removal and Distribution of Ammonia-Oxidizing and Denitrifying Genes in an Integrated Constructed Wetland for Swine Wastewater Treatment. *Ecol. Eng.* **2017**, *104*, 30–38. [CrossRef]

31. Meers, E.; Tack, F.M.G.; Tolpe, I.; Michels, E. Application of a Full-Scale Constructed Wetland for Tertiary Treatment of Piggery Manure: Monitoring Results. *Water Air Soil Pollut.* **2008**, *193*, 15–24. [CrossRef]

32. Hunt, P.G.; Poach, M.E. State of the Art for Animal Wastewater Treatment in Constructed Wetlands. *Water Sci. Technol.* **2001**, *44*, 19–25. [CrossRef] [PubMed]

33. Lee, C.Y.; Lee, C.C.; Lee, F.Y.; Tseng, S.K.; Liao, C.J. Performance of Subsurface Flow Constructed Wetland Taking Pretreated Swine Effluent under Heavy Loads. *Bioresour. Technol.* **2004**, *92*, 173–179. [CrossRef]

34. He, L.S.; Liu, H.L.; Xi, B.D.; Zhu, Y.B. Effects of Effluent Recirculation in Vertical-Flow Constructed Wetland on Treatment Efficiency of Livestock Wastewater. *Water Sci. Technol.* **2006**, *54*, 137–146. [CrossRef]

35. Sarmento, A.P.; Borges, A.C.; Matos, A.T. Evaluation of Vertical-Flow Constructed Wetlands for Swine Wastewater Treatment. *Water Air Soil Pollut.* **2012**, *223*, 1065–1071. [CrossRef]

36. González, F.T.; Vallejos, G.G.; Silveira, J.H.; Franco, C.Q.; García, J.; Puigagut, J. Treatment of Swine Wastewater with Subsurface-Flow Constructed Wetlands in Yucatán, Mexico: Infuence of Plant Species and Contact Time. *Water SA* **2009**, *35*, 335–342. [CrossRef]

37. Cronk, J.K. Constructed Wetlands to Treat Wastewater from Dairy and Swine Operations: A Review. *Agric. Ecosyst. Environ.* **1996**, *58*, 97–114. [CrossRef]

38. Zema, D.A.; Andiloro, S.; Bombino, G.; Caridi, A.; Sidari, R.; Tamburino, V. Comparing Different Schemes of Agricultural Wastewater Lagooning: Depuration Performance and Microbiological Characteristics. *Water Air Soil Pollut.* **2016**, *227*, 439. [CrossRef]

39. Zema, D.A.; Andiloro, S.; Bombino, G.; Tamburino, V.; Sidari, R.; Caridi, A. Depuration in Aerated Ponds of Citrus Processing Wastewater with a High Concentration of Essential Oils. *Environ. Technol.* **2012**, *33*, 1255–1260. [CrossRef] [PubMed]

40. Sievers, D.M. Performance of Four Constructed Wetlands Treating Anaerobic Swine Lagoon Effluents. *Trans. ASAE* **1997**, *40*, 769–775. [CrossRef]

41. Shappell, N.W.; Billey, L.O.; Forbes, D.; Matheny, T.A.; Poach, M.E.; Reddy, G.B.; Hunt, P.G. Estrogenic Activity and Steroid Hormones in Swine Wastewater through a Lagoon Constructed-Wetland System. *Environ. Sci. Technol.* **2007**, *41*, 444–450. [CrossRef]

42. Clarke, E.; Baldwin, A.H. Responses of Wetland Plants to Ammonia and Water Level. *Ecol. Eng.* **2002**, *18*, 257–264. [CrossRef]

43. Harrington, R.; Dunne, E.J.; Carroll, P.; Keohane, J.; Ryder, C. The Concept, Design and Performance of Integrated Constructed Wetlands for the Treatment of Farmyard Dirty Water. In *Nutrient Management in Agricultural Watersheds: A Wetlands Solution*; Wageningen Academic Publishers: Wageningen, The Netherlands, 2005; pp. 179–188.

44. Harrington, R.; McInnes, R. Integrated Constructed Wetlands (ICW) for Livestock Wastewater Management. *Bioresour. Technol.* **2009**, *100*, 5498–5505. [CrossRef] [PubMed]

45. Poach, M.E.; Hunt, P.G.; Vanotti, M.B.; Stone, K.C.; Matheny, T.A.; Johnson, M.H.; Sadler, E.J. Improved Nitrogen Treatment by Constructed Wetlands Receiving Partially Nitrified Liquid Swine Manure. *Ecol. Eng.* **2003**, *20*, 183–197. [CrossRef]

46. Kantawanichkul, S.; Neamkam, P.; Shutes, R.B.E. Nitrogen Removal in a Combined System: Vertical Vegetated Bed over Horizontal Flow Sand Bed. *Water Sci. Technol.* **2001**, *44*, 137–142. [CrossRef] [PubMed]

47. Humenik, F.J.; Szogi, A.A.; Hunt, P.G.; Broome, S.; Rice, M. Wastewater Utilization: A Place for Managed Wetlands-Review. *Asian-Australas. J. Anim. Sci.* **1999**, *12*, 629–632. [CrossRef]

48. Kottek, M.; Grieser, J.; Beck, C.; Rudolf, B.; Rubel, F. World Map of the Köppen-Geiger Climate Classification Updated. *Meteorol. Z.* **2006**, *15*, 259–263. [CrossRef]

49. Andiloro, S.; Bombino, G.; Tamburino, V.; Zema, D.A.; Zimbone, S.M. Aerated Lagooning of Agro-Industrial Wastewater: Depuration Performance and Energy Requirements. *J. Agric. Eng.* **2013**, *44*, 827–832. [CrossRef]

50. APHA. *Standard Methods for the Examination of Water and Wastewater*; American Public Health Association: Washington, DC, USA, 2012.

51. Liu, J.; Ge, Y.; Cheng, H.; Wu, L.; Tian, G. Aerated Swine Lagoon Wastewater: A Promising Alternative Medium for Botryococcus Braunii Cultivation in Open System. *Bioresour. Technol.* **2013**, *139*, 190–194. [CrossRef]

52. Bartsch, E.H.; Randall, C.W. Aerated Lagoons—A Report on the State of the Art. *J. Water Pollut. Control Fed.* **1971**, *43*, 699–708.

53. Osada, T.; Haga, K.; Harada, Y. Removal of Nitrogen and Phosphorus from Swine Wastewater by the Activated Sludge Units with the Intermittent Aeration Process. *Water Res.* **1991**, *25*, 1377–1388. [CrossRef]

54. Bôas, R.B.V.; Fia, R.; Fia, F.R.L.; Campos, A.T.; de Souza, G.R. Nutrient Removal From Swine Wastewater in A Combined Vertical and Horizontal Flow Constructed Wetland System. *Eng. Agríc.* **2018**, *38*, 411–416. [CrossRef]

55. Liu, F.; Zhang, S.; Wang, Y.; Li, Y.; Xiao, R.; Li, H.; He, Y.; Zhang, M.; Wang, D.; Li, X. Nitrogen Removal and Mass Balance in Newly-Formed Myriophyllum Aquaticum Mesocosm during a Single 28-Day Incubation with Swine Wastewater Treatment. *J. Environ. Manag.* **2016**, *166*, 596–604. [CrossRef] [PubMed]

56. Vymazal, J. Constructed Wetlands for Wastewater Treatment: Five Decades of Experience. *Environ. Sci. Technol.* **2011**, *45*, 61–69. [CrossRef]

57. Vymazal, J. Constructed Wetlands for Wastewater Treatment. *Water* **2010**, *2*, 530–549. [CrossRef]

58. Xu, J.; Li, C.; Yang, F.; Dong, Z.; Zhang, J.; Zhao, Y.; Qi, P.; Hu, Z. Typha Angustifolia Stress Tolerance to Wastewater with Different Levels of Chemical Oxygen Demand. *Desalination* **2011**, *280*, 58–62. [CrossRef]

59. Stone, K.C.; Poach, M.E.; Hunt, P.G.; Reddy, G.B. Marsh-Pond-Marsh Constructed Wetland Design Analysis for Swine Lagoon Wastewater Treatment. *Ecol. Eng.* **2004**, *23*, 127–133. [CrossRef]

60. Klomjek, P. Swine Wastewater Treatment Using Vertical Subsurface Flow Constructed Wetland Planted with Napier Grass. *Sustain. Environ. Res.* **2016**, *26*, 217–223. [CrossRef]

61. Ayaz, S.Ç.; Findik, N.; Akça, L.; Erdoğan, N.; Kınacı, C. Effect of Recirculation on Organic Matter Removal in a Hybrid Constructed Wetland System. *Water Sci. Technol.* **2011**, *63*, 2360–2366. [CrossRef]

 *sustainability*

Article

Bacterial Augmented Floating Treatment Wetlands for Efficient Treatment of Synthetic Textile Dye Wastewater

Neeha Nawaz [1], Shafaqat Ali [1,2,*], Ghulam Shabir [3], Muhammad Rizwan [1],
Muhammad Bilal Shakoor [1], Munazzam Jawad Shahid [1], Muhammad Afzal [3,*],
Muhammad Arslan [4], Abeer Hashem [5,6], Elsayed Fathi Abd_Allah [7],
Mohammed Nasser Alyemeni [5] and Parvaiz Ahmad [5,8]

[1]   Department of Environmental Sciences and Engineering, Government College University, Faisalabad 38000,
      Pakistan; neehanawaz66@gmail.com (N.N.); mrazi1532@yahoo.com (M.R.);
      bilalshakoor88@gmail.com (M.B.S.); munazzam01@gmail.com (M.J.S.)
[2]   Department of Biological Sciences and Technology, China Medical University, Taichung 40402, Taiwan
[3]   National Institute of Biotechnology and Genetic Engineering, Soil and Environmental Biotechnology
      Division, Faisalabad 38000, Pakistan; gshabirnlbge@yahoo.com
[4]   Department of Civil and Environmental Engineering, University of Alberta, Edmonton, AB T6G 2R3,
      Canada; marslan@ualberta.com
[5]   Botany and Microbiology Department, College of Science, King Saud University, P.O. Box. 2460, Riyadh
      11451, Saudi Arabia; habeer@ksu.edu.sa (A.H.); mnalyemeni@gmail.com (M.N.A.);
      parvaizbot@yahoo.com (P.A.)
[6]   Mycology and Plant Disease Survey Department, Plant Pathology Research Institute, ARC, Giza 12511, Egypt
[7]   Plant Production Department, College of Food and Agricultural Sciences, King Saud University, P.O. Box.
      2460, Riyadh 11451, Saudi Arabia; eabdallah@ksu.edu.sa
[8]   Department of Botany, S.P. College, Srinagar, Jammu and Kashmir 180001, India
*   Correspondence: shafaqataligill@yahoo.com or shafaqataligill@gcuf.edu.pk (S.A.); manibge@yahoo.com or
      afzal@nibge.org (M.A.)

Received: 12 March 2020; Accepted: 23 April 2020; Published: 4 May 2020

Abstract: Floating treatment wetland (FTW) is an innovative, cost effective and environmentally friendly option for wastewater treatment. The dyes in textile wastewater degrade water quality and pose harmful effects to living organisms. In this study, FTWs, vegetated with *Phragmites australis* and augmented with specific bacteria, were used to treat dye-enriched synthetic effluent. Three different types of textile wastewater were synthesized by adding three different dyes in tap water separately. The FTWs were augmented with three pollutants degrading and plant growth promoting bacterial strains (i.e., *Acinetobacter junii* strain NT-15, *Rhodococcus sp.* strain NT-39, and *Pseudomonas indoloxydans* strain NT-38). The water samples were analyzed for pH, electrical conductivity (EC), total dissolved solid (TDS), total suspended solids (TSS), chemical oxygen demand (COD), biological oxygen demand (BOD), color, bacterial survival and heavy metals (Cr, Ni, Mn, Zn, Pb and Fe). The results indicated that the FTWs removed pollutants and color from the treated water; however, the inoculated bacteria in combination with plants further enhanced the remediation potential of floating wetlands. In FTWs with *P. australis* and augmented with bacterial inoculum, pH, EC, TDS, TSS, COD, BOD and color of dyes were significantly reduced as compared to only vegetated and non-vegetated floating treatment wetlands without bacterial inoculation. Similarly, the FTWs application successfully removed the heavy metal from the treated dye-enriched wastewater, predominately by FTWs inoculated with bacterial strains. The bacterial augmented vegetated FTWs, in the case of dye 1, reduced the concentration of Cu, Ni, Zn, Fe, Mn and Pb by 75%, 73.3%, 86.9%, 75%, 70% and 76.7%, respectively. Similarly, the bacterial inoculation to plants in the case of dye 2 achieved 77.5% (Cu), 73.3% (Ni), 83.3% (Zn), 77.5% (Fe), 66.7% (Mn) and 73.3% (Pb) removal rates. Likewise in the case of dye 3, which was treated with plants and inoculated bacteria, the metals removal rates were 77.5%, 73.3%, 89.7%,

81.0%, 70% and 65.5% for Cu, Ni, Zn, Fe, Mn and Pb, respectively. The inoculated bacteria showed persistence in water, in roots and in shoots of the inoculated plants. The bacteria also reduced the dye-induced toxicity and promoted plant growth for all three dyes. The overall results suggested that FTW could be a promising technology for the treatment of dye-enriched textile effluent. Further research is needed in this regard before making it commercially applicable.

Keywords: floating treatment wetlands; bio-augmentation; dye degradation; bacteria; *Phragmites australis*

---

1. Introduction

Industrialization is a main source of water pollution. The negative impact of polluted water is more severe in developing countries as compare to developed nations [1]. Textile wastewaters contain dyes, and these dyes are one of the worst polluters of our environment [2]. Almost 17% to 20% of industrial water pollution is due to textile dyeing and finishing treatments given to fabrics [3]. Many dyes are derived from heavy metals such as copper (Cu), lead (Pb) and cadmium (Cd). The uses of these metal-complex dyes is a source of heavy metals contamination in water bodies [4]. The release of textile wastewater into open waters causes oxygen level depletion. Dyes block the sunlight in water bodies, thus stopping photosynthesis [5]. These textile contaminants are also carcinogenic and mutagenic for all life forms [3].

Some plants have the capacity to take up pollutants from the environment into themselves [6]. In the past, many plant species have proved to remove or degrade dyes, such as *Sesuvium portulacastrum* that removed Green HE4B, *Portulaca grandiflora* that removed Navy blue HD2R, *Brassica juncea* that removed methyl orange and *Glandularia pulchella* that removed green HE4B [7–9]. Bacteria has the potential to remove dyes from wastewater [10]. Bacteria can also degrade synthetic dyes and use them as a sole source of carbon and energy [11]. There are many examples like degradation of crystal violet by *Enterobacter sp. CV-S1* [12].

Wetland technology has emerged as a sustainable approach for wastewater treatment as compared to conventional treatment processes [13–15]. Floating treatment wetland (FTW) is a variant of pond and wetland land technology (Figure 1), that has been proven as an innovative tool for wastewater treatment [16]. In FTWs, plants are vegetated on an artificial floating mat, such that their roots are submerged in the contaminated water and the aerial parts of the plants remain above the water [13]. The mat can be made of PVC pipes, polyethylene or any other suitable material that can support plants on a water surface [13,17]. Roots play an integral role in and provide space for biofilm formation [16]. Organic matter and other pollutants like heavy metals are taken up by the plants' roots and eventually degraded by bacteria inside the plants and on the roots' surface [11,18]. The roots of plants also act as biological filters as they help in filtration, sedimentation and adsorption of organic matter and suspended particles, as well as other pollutants [19]. In contrast to conventional wetlands, floating wetlands can be installed on any aquatic pond without digging, earth moving and additional land acquisition [13].

The application of specific microorganisms in combination with macrophytes in FTW systems is a recent approach to enhance the pollutant removal efficiency of the system [20,21]. Naturally occurring bacteria and fungi reside inside and outside the plant roots and water, and contribute to pollutants removal process [22]. However, these microorganisms may have limited potential to degrade and remove toxic pollutants [23]. To overcome this concern, FTWs can be restorative by appropriate plant–microbe partnerships [24,25]. This plant–bacteria association may be plant–rhizospheric and or plant–endophytic, depending upon the nature of the bacteria and macrophytes [26,27].

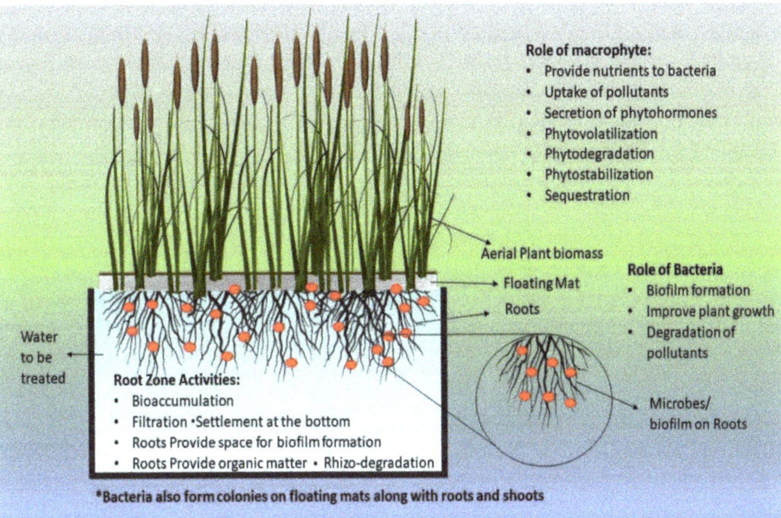

Figure 1. Schematic representation of floating treatment wetland and associated pollutant removal process.

Floating wetlands have been widely used for the treatment of wastewater from different sources [28,29]. However, the potential of FTWs composed of *Phragmites australis* in combination with inoculated bacteria has not been fully explored for the treatment of dye-enriched textile effluent. This study was carried out to analyze the potential of *P. australis* and selected bacteria in the degradation of dyes, pollutants reduction and the ultimate alleviation of toxicity of dye enriched water. Further, the focus of this study was on the persistence and survival of inoculated bacteria within the floating wetland system.

2. Materials and Methods

2.1. Synthesis of Textile Effluent

Three different types of textile effluent were synthesized in the laboratory by mixing three different dyes (500 g) in tap water separately. The first type of effluent contained Bemaplex Navy Blue DRD (D1), the second type of effluent contained Bemaplex Rubine DB (D2) and third one contained Bemaplex Black DRKP Bezma (D3). The concentration of these dyes was 500 mg L^{-1} in each type of synthetic textile effluent. These dyes were selected because of their common use in the textile industry and the high concentration of these toxic dyes and associated degraded products in textile effluent [30]. The experiments were performed individually on each type of effluent.

2.2. Macrophytes

The *Phragmites australis* commonly known as common reed was used to carry out this research. It was selected because it has previously proven its effectiveness in reducing the toxicity of polluted wastewater in different studies [11,25,31]. The *P. australis* has an extensive root and shoot system that helps in better oxygen supply to the root zone, thereby enhancing the bacterial propagation and increased pollutants degradation [32].

2.3. Endophytic Bacterial Strains

In this study a consortium of three bacterial strains was applied, namely *Rhodococcus sp.* (NCBI Accession: MF326802), *Pseudomonas indoloxydans* (NCBI Accession: MF478985) and *Acinetobacter junii*

(NCBI Accession: MF478980) [25,30]. The strain *P. indoloxydans* was endophyte because it was isolated from the root interior of *Polygonum aviculare*. The strain *Rhodococcus sp.* was rhizospheric as it was isolated from the rhizosphere of *Poa labillardierei*, and the strain *A. junii* was isolated from activated sludge [28]. These specific bacterial strains were chosen due to their potential to reduce textile dyes and assist the macrophytes to alleviate pollutant-induced toxicity without compromising plant growth and development.

The bacterial strains were cultured as separate cultures at 30 °C for 24 h in Luria–Bertani (LB) broth. The bacterial cell pellets were isolated by centrifugation at 4 °C, followed by resuspension in 0.9% NaCl solution [25]. The optical density of each bacterial inoculum was adjusted to 0.9 at 600 nm according to the guidelines of the turbid metric method [33]. The bacterial consortium (10^8 colony forming units (CFU) mL^{-1}) was prepared by mixing all bacterial inoculum together in equal proportion. This bacterial consortium was used as an inoculum to inoculate the floating treatment wetlands.

2.4. Fabrication of FTWs and Experimental Setup

The macrocosms experimental setup was comprised of nine tanks with 1000 liter capacity each, and the dimensions were 1.2 m (L × W × H). The tanks were painted black form all sides to minimize the algal growth. The floating mats were fabricated from expanded polystyrene (EPS)-based sheets manufactured by Diamond® Foam Private Ltd., Pakistan [11,34–36]. EPS sheets are rigid, have low thermal conductivity, are moisture resistant and consist of non-porous closed cell foam [37]. The size of the floating mats was adjusted so that they could float in each tank with >95% coverage on the water surface. All four sides of the floating mats were wrapped with aluminum foil to protect the sheets from sun and water damage. In each floating sheet eight equidistant holes, equal in diameter, were made for the plantation of macrophytes on the floating mats. Each hole was planted with three healthy seedlings of *P. australis*, thus having 24 seedlings in each mat. Each seedling weighed 45 to 65 g and their length was 55–65 cm. The seedlings were supported by coconut shavings and soil in the floating mat. The seedlings were allowed to grow in fresh water for one month to gain optimum growth of roots and shoots. After one month, the average height of the plants was about 145 cm, and the fresh water in tanks was replaced with the synthetic textile effluent enriched with dyes. The experiment was run in triplicate with the subsequent treatment design:

T1D1, T1D2, T1D3: Only dye;

T2D1, T2D2, T2D3: Containing dye and plants;

T3D1, T3D2, T3D3: Dye, plants and bacterial consortium;

T4: Fresh water and plants.

(D1: Bemaplex Navy Blue DRD, D2: Bemaplex Rubine DB, D3: Bemaplex Black DRKP Bezma).

The treatments T3D1, T3D2, T3D3 were inoculated by pouring one liter of inoculum into each tank. The experiment lasted for 20 days until a maximum of dye and pollutants were removed from treated water. One liter of sample was collected from each tank every 5 days starting from day 0 using a sequencing fill-and-draw batch mode method (for convenience, data of only the 0, 10th and 20th day are presented in results). The samples were stored in a cool and dry place for further analysis [38]. The collected water samples were analyzed for pH, electrical conductivity (EC), total dissolved solids (TDS), total suspended solids (TSS), dye concentration, chemical oxygen demand (COD), biological oxygen demand (BOD), colony forming unit (CFU) and metal concentration (Cu, Fe, Mn, Ni, Zn and Pb) according to standard methods [38]. The evapo-transpiration losses were recovered by pouring fresh water in treatment tanks up to the level of 1000 L in each tank [34]. In case of rain, the tanks were covered with plastic sheets.

2.5. Persistence of Inoculated Bacteria in Treated Water and Plants

The persistence of bacteria in water, root and shoot samples were periodically analyzed during the experiment using the cultivation-dependent plate count method [24,25]. The collected roots and shoots samples were surface sterilized by 70% ethanol and 2% sodium hypochlorite solution. Then these roots and shoots were homogenized in a 0.9% NaCl solution and serial dilution of these suspensions was spread on LB agar plates. Similarly, the collected water samples from all treatments were spread on LB agar plates and these plates were incubated at 37 °C for 48 h for CFU analysis [35,39].

2.6. Plant Biomass

In order to determine the effect of bacterial inoculation and dye-induced toxicity on plants growth and development, the data about plants agronomic parameters (root and shoot length and dry biomass) were noted at the end of the experiment. The root and shoot length was measured manually by a measuring scale. The root and shoots were harvested near the surface of the floating mat and oven dried at the 80 °C for 72 h until a constant weight was achieved [11,34].

2.7. Statistical Analysis

The results of physicochemical parameters (pH, EC, TDS, TSS, BOD, COD, color and heavy metals), bacterial persistence and plant biomass were evaluated by the SPSS software package. The comparison between treatments was executed by analysis of variance (ANOVA) followed by a Post-Hoc Tukey test ($p \leq 0.05$) [40]. The alphabet labels over the values show the significant/non-significant differences among treatments.

3. Results

3.1. Changes in Physicochemical Parameters of Treated Textile Effluent

The graphs in Figures 2–4 represent the changes in the physicochemical parameters of the dye-enriched tap water treated by floating treatment wetlands. The floating wetlands had a positive impact and predominately reduced the pH, EC, TSS, TDS, COD, BOD and color within the retention period of 20 days. All of the above-mentioned pollutants were reduced sharply in vegetated treatments (T2D1, T2D2, T2D3 and T3D1, T3D2, T3D3) as compared to non-vegetated treatments. However, the vegetated treatments inoculated with bacterial consortium (T3D1, T3D2, T3D3) achieved highest pollutants removal rate, outperforming all other treatments in all three types of dyes.

In the treatment containing dye 1, *P. australis* and bacterial consortium (T3D1), maximum pollutants removal efficiency was achieved. In this treatment, pH was reduced to 6.7 from 8.5, EC was reduced from 6.13 to 1.00 mS cm^{-1}, TDS was reduced from 400 to 60 mg L^{-1}, TSS was reduced from 92 to 19 mg L^{-1}, COD was reduced from 310 to 30 mg L^{-1}, BOD was reduced from 121 to 20 mg L^{-1} and color was reduced from 40.0 to 6.0 m^{-1}.

Similarly, in the case of dye 2, maximum pollutant removal efficiency was obtained from T3D2, in which pH was reduced to 6.8 from 8.5, EC was reduced from 6.13 to 1.02 mS cm^{-1}, TDS was reduced from 400 to 63 mg L^{-1}, TSS was reduced from 92 to 21 mg L^{-1}, COD was reduced from 308 to 33 mg L^{-1}, BOD was reduced from 121 to 18 mg L^{-1} and color was reduced from 40.0 to 6.7 m^{-1}.

As in the case of dye 1 and dye 2, the maximum pollutant removal rate was achieved by T3D3 containing dye 3, *P. australis* and bacterial consortium. In this treatment, pH was reduced to 6.7 from 8.5, EC was reduced from 6.15 to 1.05 mS cm^{-1}, TDS was reduced from 401 to 62 mg L^{-1}, TSS was reduced from 91 to 24 mg L^{-1}, COD was reduced from 309 to 31 mg L^{-1}, BOD was reduced from 120 to 19 mg L^{-1} and color was reduced from 40.0 to 6.4 m^{-1}.

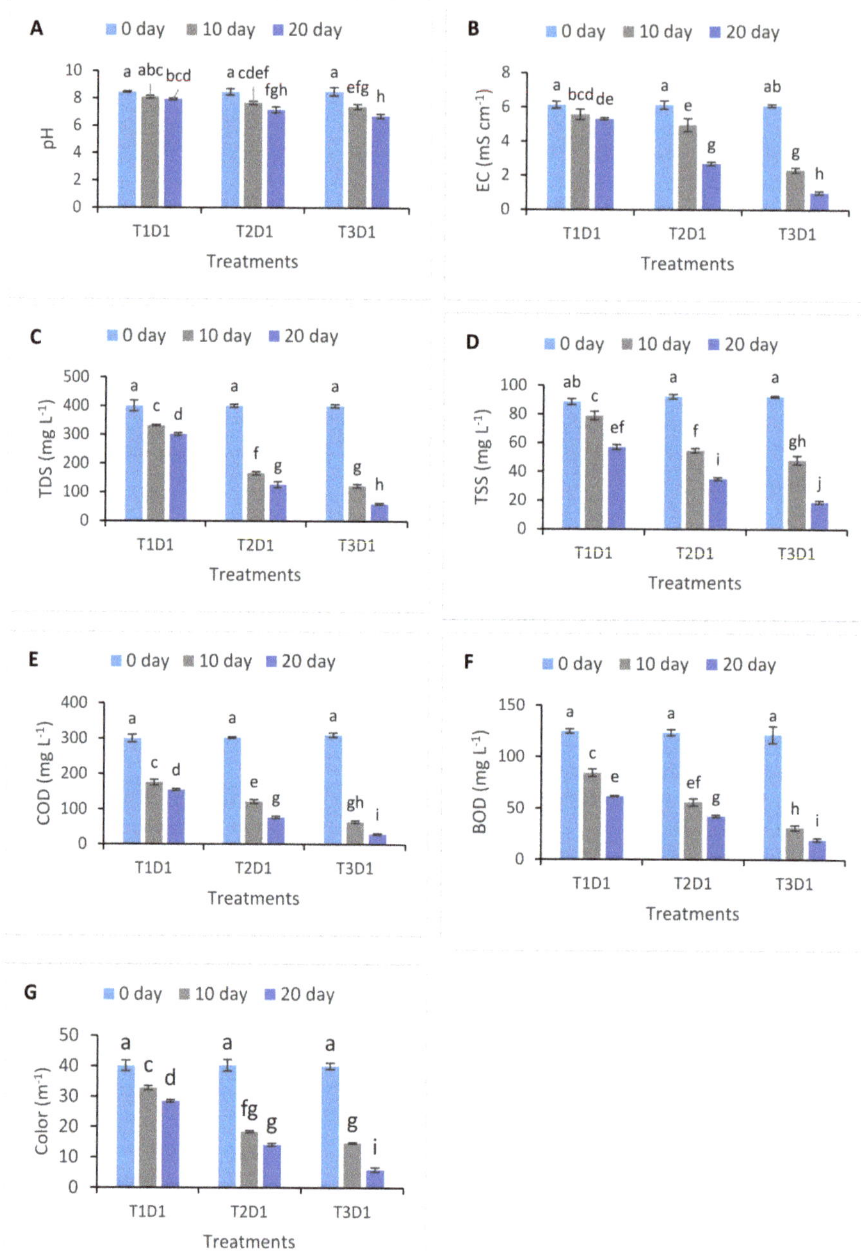

Figure 2. Effect of floating treatment wetlands on pH (**A**), EC (**B**), TDS (**C**), TSS (**D**), COD (**E**), BOD (**F**) and color (**G**) after 20 days of retention time. D1: Bemaplex Navy Blue DRD. Each value is a mean of three replicates and error bars represent the standard deviation. Lettering shows that various treatments are significantly different at $p \leq 0.05$.

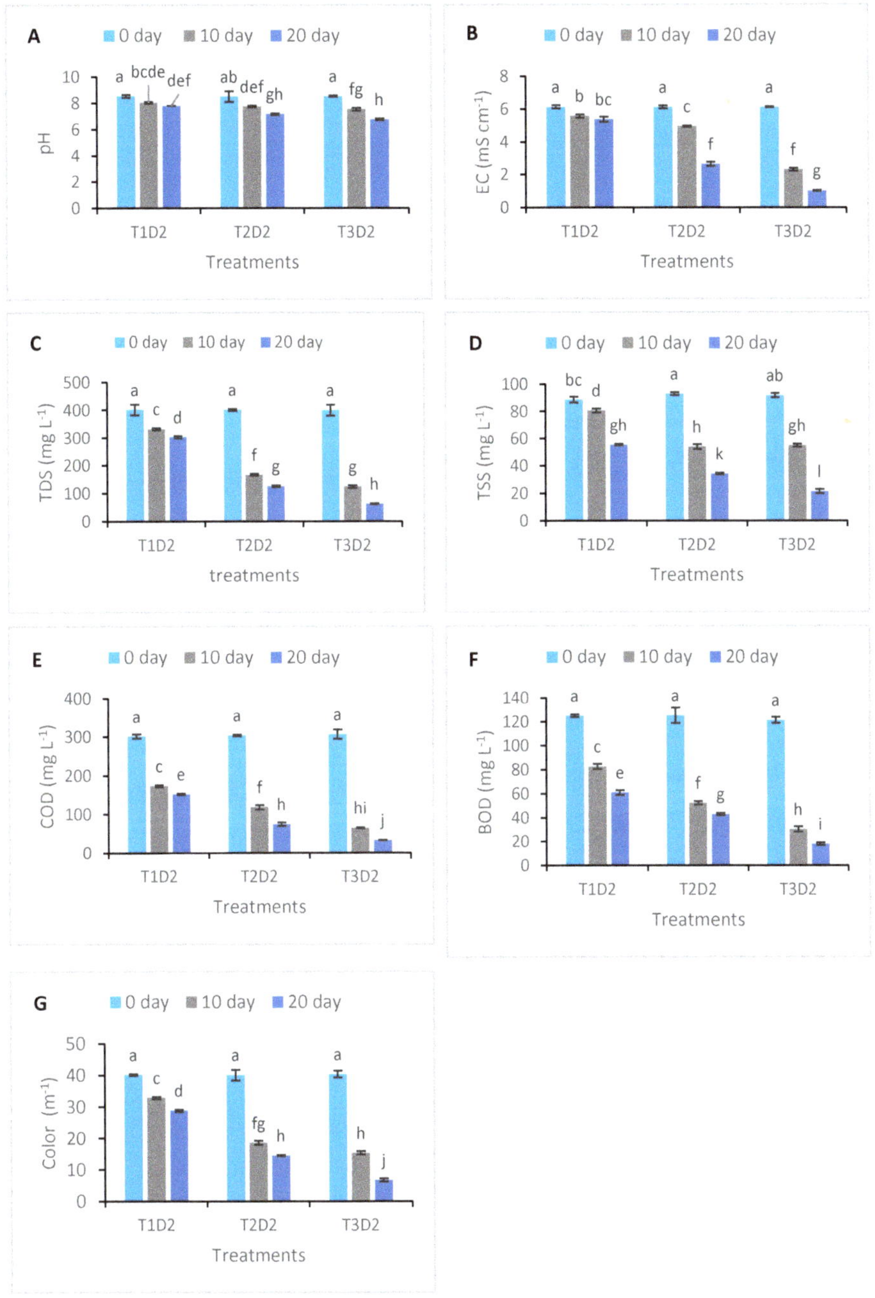

Figure 3. Effect of floating treatment wetlands on pH (**A**), EC (**B**), TDS (**C**), TSS (**D**), COD (**E**), BOD (**F**) and color (**G**) after 20 days of retention time. D2: Bemaplex Rubine DB. Each value is a mean of three replicates and error bars represent the standard deviation. Lettering shows that various treatments are significantly different at $p \leq 0.05$.

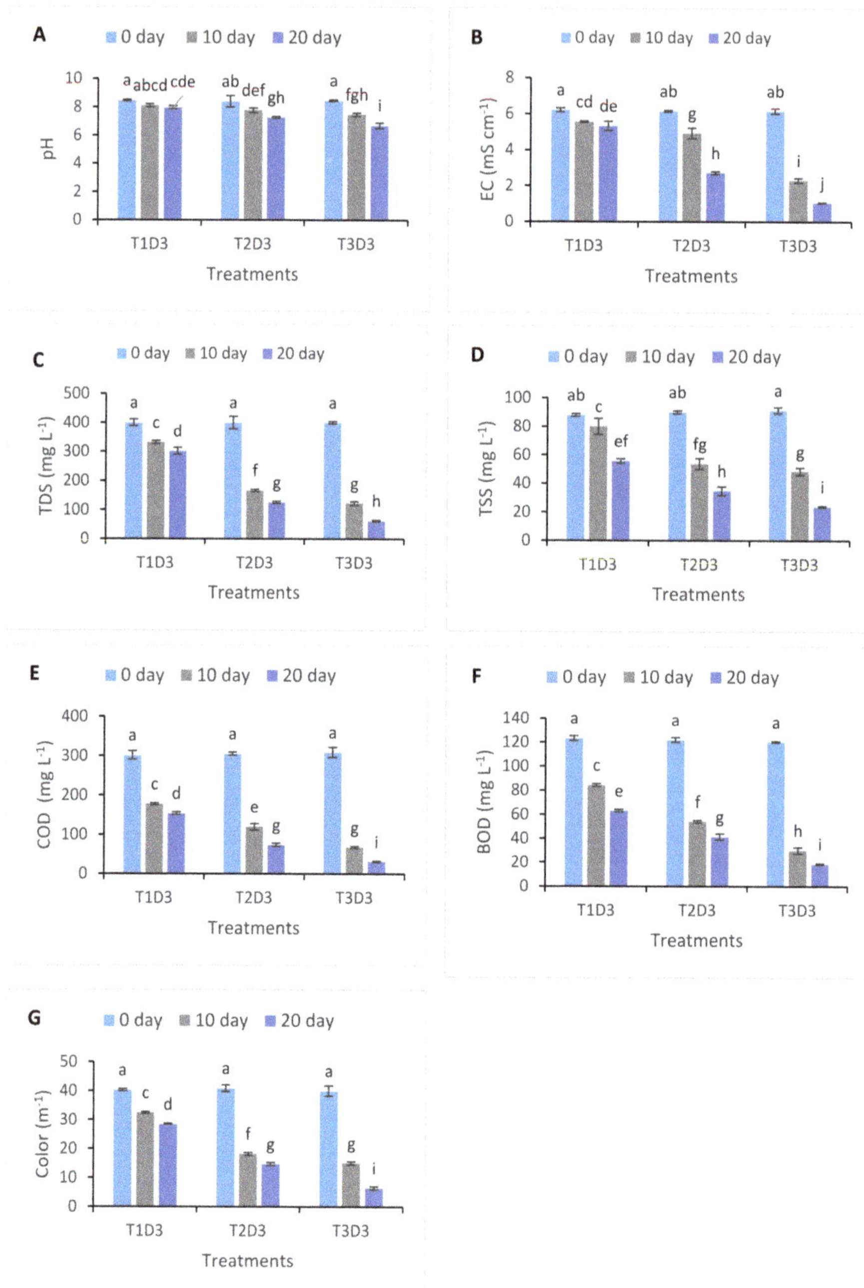

Figure 4. Effect of floating treatment wetlands on pH (**A**), EC (**B**), TDS (**C**), TSS (**D**), COD (**E**), BOD (**F**) and color (**G**) after 20 days of retention time. D3: Bemaplex Black DRKP Bezma. Each value is a mean of three replicates and error bars represent the standard deviation. Lettering shows that various treatments are significantly different at $p \leq 0.05$.

3.2. Removal of Heavy Metals from Water

The concentration of six heavy metals (Cu, Fe, Mn, Ni, Zn and Pb) considerably reduced in the FTWs-treated water samples (Table 1). All vegetated treatments (T2 and T3) showed significantly better removal of trace metals from the dye-polluted water (D1, D2 and D3) as compared to the non-vegetated treatments (T1D1, T1D2 and T1D3). Next, the efficiency of bacterial augmented treatments (T3D1, T3D2 and T3D3) was significantly better than non-inoculated vegetated treatments (T2D1, T2D2 and T2D3). In treatment T3D1, the metal concentrations for Cu, Ni, Zn, Fe, Mn and Pb were reduced by up to 75%, 73.3%, 86.9%, 75%, 70% and 76.7%, respectively, in the 20 days retention time. Similar results were achieved for dye 2 and dye 3 in the case of treatment T3, in which bacterial inoculation efficiently removed the metals from dye water as compared to non-inoculated vegetated treatments (T2) and un-vegetated non-inoculated treatments (T1).

Table 1. Percentage (%) reduction in concentration of metals with time by floating treatment wetlands.

Treatment		T1 Only Dye		T2 Dye + Plant		T3 Dye + Plant + Bacteria	
Dye	Metals	10 Days	20 Days	10 Days	20 Days	10 Days	20 Days
D1	Cu	20.0 [b,c] (0.0)	30.0 [c,d] (0.0)	58.5 [e] (0.0)	65.9 [e,f,g] (0.0)	67.5 [e,f,g] (0.0)	75.0 [g] (0.0)
	Ni	19.4 [b,c] (0.0)	32.3 [d,e] (0.0)	40.0 [e,f] (0.0)	60.0 [g,h] (0.0)	60.0 [g,h] (0.0)	73.3 [h] (0.0)
	Zn	8.8 [c] (0.0)	21.1 [d] (0.0)	60.0 [e,f] (0.0)	66.7 [f] (0.0)	75.4 [g] (0.0)	86.9 [h] (0.0)
	Fe	7.5 [b,c] (0.0)	12.5 [c] (0.0)	48.8 [e] (0.0)	65.9 [g] (0.0)	62.5 [f,g] (0.0)	75.0 [h] (0.0)
	Mn	13.3 [b,c] (0.0)	23.3 [c,d] (0.0)	34.5 [d,e] (0.0)	48.3 [e,f] (0.0)	56.7 [f,g] (0.0)	70.0 [h] (0.0)
	Pb	20.0 [b,c] (0.0)	26.7 [c,d] (0.0)	40.0 [d,e,f] (0.0)	60.0 [g] (0.0)	46.7 [f] (0.0)	76.7 [h] (0.0)
D2	Cu	20.0 [c] (0.0)	27.5 [c] (0.0)	55.0 [d,e,f] (0.0)	65.0 [f,g,h] (0.0)	70.0 [g,h,i] (0.0)	77.5 [i] (0.0)
	Ni	16.7 [b,c] (0.0)	30.0 [c] (0.0)	46.7 [d] (0.0)	60.0 [d,e,f] (0.0)	60.0 [d,e,f] (0.0)	73.3 [f] (0.0)
	Zn	13.3 [b,c] (0.0)	33.3 [c] (0.0)	56.9 [d,e] (0.0)	65.5 [e,f] (0.0)	75.0 [f,g] (0.0)	83.3 [g] (0.0)
	Fe	5.1 [a,b] (0.0)	12.8 [b,c] (0.0)	48.8 [c,d] (0.0)	65.9 [e,f] (0.0)	62.5 [e] (0.0)	77.5 [f] (0.0)
	Mn	22.6 [b] (0.0)	25.8 [b,c] (0.0)	40.0 [c,d,e] (0.0)	50.0 [e,f,g] (0.0)	60.0 [f,g] (0.0)	66.7 [g] (0.0)
	Pb	22.6 [b] (0.0)	29.0 [b,c] (0.0)	43.3 [c,d,e] (0.0)	56.7 [e,f,g] (0.0)	50.0 [e,f] (0.0)	73.3 [g] (0.0)
D3	Cu	20.0 [a,b,c] (0.0)	30.0 [a,b,c,d,e] (0.0)	55.0 [d,e,f,h] (0.0)	65.0 [f,g] (0.0)	67.5 [e,f,g] (0.0)	77.5 [g] (0.0)
	Ni	20.0 [c,d] (0.0)	30.0 [d,e] (0.0)	43.8 [e,f] (0.0)	59.4 [g,h] (0.0)	60.0 [g,h,i] (0.0)	73.3 [i] (0.0)
	Zn	10.5 [c] (0.0)	24.6 [d] (0.0)	58.3 [e] (0.0)	70.0 [f,g,h] (0.0)	75.9 [g,h] (0.0)	89.7 [i] (0.0)
	Fe	14.6 [b,c] (0.0)	19.5 [c] (0.0)	52.4 [d,e] (0.0)	69.0 [f,g] (0.0)	66.7 [e,f] (0.0)	81.0 [g] (0.0)
	Mn	10.0 [a,b,c] (0.0)	26.7 [c,d,e] (0.0)	35.7 [d,e,f] (0.0)	46.4 [f,g,h] (0.0)	63.3 [g,h] (0.0)	70.0 [h] (0.0)
	Pb	23.3 [bc] (0.0)	30.0 [cd] (0.0)	43.3 [ef] (0.0)	56.7 [hi] (0.0)	44.8 [fg] (0.0)	65.5 [i] (0.0)

T symbolizes treatments (T1, T2, T3) and D symbolizes dye (D1: Bemaplex Navy Blue DRD, D2: Bemaplex Rubine DB, D3: Bemaplex Black DRKP Bezma). Values represent the means of three replicates and standard deviations are presented in parenthesis. Lettering shows that various treatments are significantly different at $p \leq 0.05$.

3.3. Bacterial Persistence in Roots, Shoots and Water

The presence of a significantly high population of bacteria in water (Table 2), roots and shoots (Table 3) in the bacterial inoculated treatment (T3) as compared to non-inoculated treatments (T1 and T2) confirmed the persistence of inoculated bacteria during the treatment process in inoculated treatments for all three dyes. The bacteria showed the highest population in wastewater compared to roots and shoots. On the other hand, the count of bacteria was found higher in roots than shoots.

Table 2. Average concentration of bacteria in water (colony forming unit (CFU) mL^{-1}).

Treatment	Days	Dye 1	Dye 2	Dye 3
Only Dye (T1)	5	1.5×10^{3} a (0.3)	1.6×10^{3} a (0.2)	1.5×10^{3} a (0.4)
	10	1.6×10^{3} a (0.4)	1.6×10^{3} a (0.5)	1.7×10^{3} a (0.5)
	15	1.9×10^{3} a (0.6)	1.8×10^{3} a (0.7)	1.8×10^{3} a (0.6)
	20	1.8×10^{3} a (0.6)	1.8×10^{3} a (0.8)	2.0×10^{3} a (0.9)
Dye + Plant (T2)	5	2.1×10^{5} b (1.0)	2.3×10^{5} b (1.0)	2.2×10^{5} b (0.9)
	10	2.7×10^{5} b (1.1)	2.9×10^{5} b (1.2)	2.5×10^{5} b (1.1)
	15	3×10^{5} b,c (0.9)	3.3×10^{5} b,c (0.8)	3.4×10^{5} b,c (0.9)
	20	3.7×10^{5} c (1.1)	3.5×10^{5} c (1.1)	3.6×10^{5} c (1.1)
Dye + Plant + Bacteria (T3)	5	9.6×10^{8} d (0.5)	9.8×10^{8} d (0.6)	9.9×10^{8} d (0.5)
	10	7.1×10^{9} e (0.7)	7.2×10^{9} e (0.5)	7.1×10^{9} e (0.4)
	15	6.4×10^{9} f (0.2)	6.6×10^{9} f (0.6)	6.6×10^{9} f (0.6)
	20	5.0×10^{8} g (0.6)	5.1×10^{8} g (0.6)	4.9×10^{8} g (0.7)

Dye 1: Bemaplex Navy Blue DRD, Dye 2: Bemaplex Rubine DB, Dye 3: Bemaplex Black DRKP Bezma. Values represent the means of three replicates and standard deviations are presented in parenthesis. Lettering shows that various treatments are significantly different at $p \leq 0.05$.

3.4. Plant Growth in Response to Bacterial Inoculation

It is well established that the presence of toxic pollutants in water inhibits plant growth and ultimately phytoremediation efficiency. The root and shoot length (Table 4) and root and shoot dry mass (Table 5) were noted at the end of the experiment and it was found that the plants grown in dye water inoculated with bacteria (T3) showed more growth as compared to the plants grown only in dye water. The plants grown in only tap water with no dye showed maximum growth out of all treatments. The dye water hindered the growth of plants and root and shoot length were reduced in case of all three dyes. Similarly, the plants grown in dye water inoculated with bacteria gained high shoot and root dry biomass due to good growth as compared to plants grown in dye water without bacterial inoculation. These results showed that despite the toxic effect of dyes, the inoculation of bacteria to dye water predominantly increased the length and dry weight of shoot and root of *P. australis*.

Table 3. Average concentration of bacteria in roots and shoots (CFU mL^{-1}).

Root/Shoot	Treatment	Days	Dye 1	Dye 2	Dye 3
Root	Dye + Plant (T2)	5	2×10^2 [a] (0.8)	2.2×10^2 [a] (0.9)	2.2×10^2 [a] (0.8)
		10	3.2×10^2 [b] (1.0)	3.3×10^2 [b] (1.0)	3.3×10^2 [b] (1.0)
		15	3.7×10^2 [b,c] (0.8)	3.7×10^2 [b,c] (0.8)	3.7×10^2 [b,c] (0.8)
		20	4.1×10^2 [c] (0.9)	4.0×10^2 [c] (0.9)	4.1×10^2 [c] (0.9)
	Dye + Plant + Bacteria (T3)	5	4×10^3 [d] (1.2)	4.4×10^3 [d] (1.3)	4.3×10^3 [d] (0.9)
		10	11.9×10^3 [e] (1.1)	12.0×10^3 [e] (1.3)	11.6×10^3 [e] (1.1)
		15	17.2×10^3 [f] (1.1)	17.6×10^3 [f] (1.3)	18.5×10^3 [f] (0.9)
		20	22.8×10^3 [g] (1.1)	23.1×10^3 [g] (1.1)	23.4×10^3 [g] (1.1)
Shoot	Dye + Plant (T2)	5	1.1×10^2 [a] (0.2)	1.2×10^2 [a] (0.2)	1.2×10^2 [a] (0.3)
		10	1.2×10^2 [a] (0.4)	1.0×10^2 [a] (0.5)	1.2×10^2 [a] (0.3)
		15	1.3×10^2 [a] (0.2)	1.3×10^2 [a] (0.2)	1.2×10^2 [a] (0.7)
		20	1.3×10^2 [a] (0.2)	1.2×10^2 [a] (0.3)	1.3×10^2 [a] (0.5)
	Dye + Plant + Bacteria (T3)	5	1.4×10^3 [b] (0.3)	1.6×10^3 [b] (0.2)	1.5×10^3 [b] (0.4)
		10	6.2×10^3 [c] (0.7)	6.0×10^3 [c] (0.7)	6.2×10^3 [c] (0.8)
		15	10.5×10^3 [d] (0.4)	10.1×10^3 [d] (0.9)	11.2×10^3 [d] (1.3)
		20	14.3×10^3 [e] (2.1)	13.9×10^3 [e] (2.3)	14.0×10^3 [e] (2.5)

Dye 1: Bemaplex Navy Blue DRD, Dye 2: Bemaplex Rubine DB, Dye 3: Bemaplex Black DRKP Bezma. Values represent the means of three replicates and standard deviations are presented in parenthesis. Lettering shows that various treatments are significantly different at $p \leq 0.05$.

Table 4. Comparison between shoot lengths and root lengths in different treatments.

Treatments	Shoot Length (cm)			Root Length (cm)		
	Dye 1	Dye 2	Dye 3	Dye 1	Dye 2	Dye 3
Dye + Plants (T2)	187.7 [d] (24.9)	197.7 [c,d] (7.8)	202.7 [b,c,d] (3.8)	29.7 [c] (0.58)	30.7 [c] (1.2)	31.0 [c] (0.0)
Dye + Plants + Bacteria (T3)	222.0 [a,b,c] (10.8)	228.0 [a,b] (2.6)	224.3 [a,b,c] (9.3)	38.0 [b] (1.0)	39.0 [b] (1.0)	38.3 [b] (1.2)
Fresh water + Plants (T4)	233.3 [a] (3.1)	232.3 [a] (2.5)	230.0 [a,b] (2.6)	43.0 [a] (1.0)	44.3 [a] (0.58)	43.7 [a] (2.1)

Dye 1: Bemaplex Navy Blue DRD, Dye 2: Bemaplex Rubine DB, Dye 3: Bemaplex Black DRKP Bezma. Values represent the means of three replicates and standard deviations are presented in parenthesis. Lettering shows that various treatments are significantly different at $p \leq 0.05$.

Table 5. Shoot and root dry weight of the plants.

Treatments	Shoot Dry Weight (g)			Root Dry Weight (g)		
	Dye 1	**Dye 2**	**Dye 3**	**Dye 1**	**Dye 2**	**Dye 3**
Dye + Plants (T2)	492.0 [c] (65.3)	517.7 [b,c] (20.3)	533.0 [a,b,c] (9.6)	62.3 [c] (1.2)	64.3 [c] (3.2)	64.7 [c] (0.6)
Dye + Plants+ Bacteria (T3)	572.7 [a,b] (29.3)	591.3 [a,b] (7.1)	581.3 [a,b] (21.4)	79.3 [a,b,c] (2.1)	81.0 [b,c] (1.7)	80.3 [a,b,c] (2.1)
Fresh water + Plants (T4)	605.0 [a] (8.2)	609.0 [a] (9.5)	603.7 [a] (8.3)	89.3 [a,b] (3.1)	93.0 [a] (1.7)	90.0 [a,b] (1.7)

Dye 1: Bemaplex Navy Blue DRD, Dye 2: Bemaplex Rubine DB, Dye 3: Bemaplex Black DRKP Bezma. Values represent the means of three replicates and standard deviations are presented in parenthesis. Lettering shows that various treatments are significantly different at $p \le 0.05$.

4. Discussion

In this study, pH, EC, TDS, TSS, COD, BOD and color of the dye-enriched water and heavy metals contents were significantly decreased in the vegetated and vegetated-inoculated floating treatment wetlands. The reductions in pollutants load in treated dye-contaminated water emphasize the prominent role of vegetation and bacteria in floating wetlands.

The pH might be decreased due to the release of organic acids by the roots of the plants as reported in earlier studies [31,41]. The decrease in EC might be associated with the uptake of nutrients by plants and the biological and physicochemical binding of pollutants to roots and soil particles [13,36]. The pH and EC reduction was highest in treatment vegetated with plants and augmented with bacteria. This suggests the key role of plants and bacteria in pH and EC reduction through the release of organic acids and the uptake of nutrients by plants and bacteria [25,34,35]. The TDS and TSS loads were reduced due to the combination of physical and biological processes supported by floating wetlands [41]. The suspended particles in the water are trapped in the biofilm of the roots of macrophytes, and there they either precipitate at the bottom or adsorb on biofilm where they might be degraded [42]. Physical entrapment in roots, sorption and settlement at the bottom might contribute to the removal of TDS and TSS from treated water [16,19,43]. Further, the roots of plants act as physical filters and provide appropriate organic matter that acts as a bio-sorbent and contributed to the removal of particulate matter [11,21].

Roots allow microbial communities to assimilate carbon compounds and reduce BOD and COD [44]. In this study, the high removal of BOD in wetland systems might be attributed to the deposition and filtration of organic compounds that can be settled. The speedy and high removal rate in bacterial augmented FTWs could be attributed to the biofilm on roots, which contributes to the removal of organic matter by decomposing it into simple nutrients, thus aiding in the direct uptake by the plant [20,45]. Uptake by plants' roots is an important method of nutrient removal [42]. The nutrients in the wastewater might be taken up by the roots of the plants. There, they can either accumulate in the plant biomass or be degraded by endophytic bacteria present inside the plants [25,46]. The similar findings have been reported by earlier studies, where plants and bacterial combinations enhanced the removal of organic pollutants from highly polluted wastewater [31,47,48].

Color was also removed to a great extent in this study by the vegetated treatment and the vegetated-inoculated treatment. It has been well reported that COD, a measure of oxidizable contaminants, has a positive correlation with color in textile wastewater [11]. Correspondingly, in this study color was reduced with the reduction in COD. However, the rate of decolorization was high in vegetated-inoculated floating wetlands. This could be associated with the combined action of plant and bacteria in the degradation of dyes and removal of color [11]. This emphasized the key role of bacteria in the decolorization of dye from textile effluent. The previous studies also showed that many bacteria are helpful in the removal of dyes, and that bacteria have the ability to degrade dyes by aerobic as well as anaerobic mechanisms [11,49].

In this study, the concentration of six heavy metals (Cr, Ni, Zn, Fe, Pb and Mn) was decreased significantly in the treated dye-containing wastewater. The unique potential of *P. australis* to remove heavy metals has been reported by many researchers [25,34]. In the previous studies, *P. australis* showed similar pattern of removal of heavy metals from industrial effluent [11,28,36]. These previous studies also demonstrated that the heavy metals from wastewater were taken up by the *P. australis* in its roots and shoots [41,50,51]. The maximum concentrations of heavy metals were found in the roots of the plant, meaning that the root has most potential to uptake heavy metals [50].

In the case of inoculation of *P. australis* with bacteria, the heavy metal removal capacity was further enhanced. The improved performance of bacterially augmented FTWs emphasized the role of bacteria in the removal of heavy metals from polluted water. The inoculated bacteria reduced the metals load in polluted water by their bioaccumulation potential [31]. These bacteria might contribute to reducing metal-induced toxicity and increase the bioavailability and metals uptake of plants [27]. It is well reported that in FTWs the inoculated bacteria may boost the metals removal process by entrapment of metals in root biofilms, sorbing of metallic ions on the bacterial cell wall and oxidation of metal ions [52,53]. Further, the plaque formation by the combined action of plant and bacteria on plant roots may increase the Fe, Mn, Cu and Zn binding in roots biofilms [13,54]. This emphasizes that *P.australis* and inoculated bacterial combined role, which contributes to metals removal from treated dye-contaminated wastewater. The significantly substantial removal of metals from bacterial inoculated treatments relative to non-inoculated vegetated treatments could be attributed to a high population of bacteria in the inoculated treatment.

The inoculated bacteria showed persistence in polluted water being treated by floating treatment wetlands. The periodic analysis of water from all three treatments showed the high population of bacteria in inoculated treatments as compared to the non-inoculated treatments. The higher population of bacteria in the water of inoculated treatments confirmed that inoculated bacteria showed persistence and were responsible for dye removal and pollutant removal. This could be due to the fact that the inoculated bacteria successfully made mutualistic relationships with plants, which supported the survival of inoculated bacterial [55]. This finding is consistent with previous studies in which inoculated bacteria improved the pollutant removal process [24,34]. The survival of inoculated bacteria depends upon the nutrient supply, pH, temperature and the interaction with the host [56,57]. In this study, the bacterial population in the roots and shoots of inoculated plants were found to be higher as compared to non-inoculated vegetated treatment. This could be due to the preferential survival of bacteria in roots and shoots of *P. australis* in inoculated treatments, as reported in previous studies [28,58]. Further, these bacteria were initially isolated from the roots and shoots of the plants; hence these bacteria possibly have an adaptive mechanism to survive and grow in these parts of the plant in this hostile environment [18,27]. In order to make FTWs a potential wastewater treatment method, periodic inoculation of bacteria should be performed in order to overcome the problem of decreasing bacteria with time in inoculated water [57,59].

Toxic pollutants in the environment inhibit plant growth [27]. Dyes containing toxic chemicals and potentially toxic heavy metals also inhibit plant growth [28]. In this study, the *P. australis*, synergistic with bacteria, achieved high root and shoot growth as compared to plants without inoculation. The control tank having only water and plants with no added dye showed maximum growth of roots and shots of plants due to the absence of any toxic pollutant. The bacteria present in the system can promote plant growth by decreasing biotic and abiotic stress [60]. Bacteria also positively affect plant growth by releasing phyto-hormones and by the solubilisation of essential nutrients [61]. Pollutant-degrading rhizospheric and endophytic bacteria have been proven as effective to enhance plant growth development and phytoremediation efficacy [53]. Similar results have been reported by previous studies where inoculated bacteria promoted plant growth by alleviating pollutant-induced toxicity and improved plant nutrition, health and disease resistance [34,62].

5. Conclusions

The present study evaluated the potential of *P. australis* in FTWs along with three inoculated bacterial strains to remove dye as well as organic and inorganic pollutants from dye-enriched water. The results clearly indicated that *P. australis* along with inoculated strains have a great potential to remove different types of dyes and pollutants, including potentially toxic metals, from textile effluent. The floating wetlands are capable of efficiently decreasing the levels of pH, EC, TDS, TSS, BOD, COD, color and toxic metals from dye-polluted wastewater. The high rate of pollutants removal by vegetated-inoculated FTWs validates the potential role of bacteria in FTWs. The bacteria showed high persistence in water as well as in the roots and shoots of the inoculated plants. It suggests that bacteria have the ability to make a mutualistic relationship with *P. australis* in FTWs system to collectively remove pollutants form the water body. These plant growth-promoting rhizospheric and endophytic bacteria also increased the plants' ability to tolerate pollutant-induced toxicity and alleviate the toxicity of textile effluent. We conclude that the FTWs can be a promising technology to treat textile effluent and can be a propitious substitute for conventional wastewater technology for the treatment of textile effluent. The pollutant removal efficiency of already existing water retention ponds can be enhanced by installing floating wetland systems. However, there is a need for conducting meticulous research about the careful and objective-based selection of plants and bacteria, which can further enhance the efficiency of the FTWs system.

Author Contributions: Conceptualization, N.N., S.A., G.S., M.A. (Muhammad Afzal), M.A. (Muhammad Arslan) and M.N.A.; Data curation, N.N. and G.S.; Formal analysis, N.N., M.R. and M.J.S.; Funding acquisition, A.H. and P.A.; Investigation, S.A., M.A. (Muhammad Arslan) and P.A.; Methodology, N.N., G.S., M.B.S., M.J.S., M.A. (Muhammad Afzal) and A.H.; Project administration, S.A., M.R. and M.A. (Muhammad Afzal); Resources, A.H., M.N.A., E.F.A., A.H. and P.A.; Software, E.F.A., G.S. and M.B.S.; Supervision, S.A., M.R., M.B.S., M.A. (Muhammad Afzal) and A.H.; Validation, M.J.S. and A.H.; Writing—Original draft, N.N., A.H. and P.A.; Writing—Review and editing, S.A., M.A. (Muhammad Arslan), A.H., M.N.A. and P.A. All authors have read and agreed to the published version of the manuscript.

Funding: This research was supported by the Higher Education Commission, Pakistan (grant number 20-3854/R&D/HEC/14). The authors would like to extend their sincere appreciation to the Researchers Supporting Project Number (RSP-2019/116), King Saud University, Riyadh, Saudi Arabia.

Acknowledgments: The authors would like to appreciate the Government College University, Faisalabad, Pakistan for support. The authors would like to extend their sincere appreciation to the Researchers Supporting Project Number (RSP-2019/116), King Saud University, Riyadh, Saudi Arabia.

Conflicts of Interest: The authors declare that they have no conflict of interest.

References

1. Evans, A.E.; Mateo-Sagasta, J.; Qadir, M.; Boelee, E.; Ippolito, A. Agricultural water pollution: Key knowledge gaps and research needs. *Curr. Opin. Environ. Sustain.* **2019**, *36*, 20–27. [CrossRef]
2. Madhav, S.; Ahamad, A.; Singh, P.; Mishra, P.K. A review of textile industry: Wet processing, environmental impacts, and effluent treatment methods. *Environ. Qual. Manag.* **2018**, *27*, 31–41. [CrossRef]
3. Khan, S.; Malik, A. Environmental and health effects of textile industry wastewater. In *Environmental Deterioration and Human Health*; Springer: Berlin/Heidelberg, Germany, 2014; pp. 55–71.
4. Carmen, Z.; Daniela, S. *Textile Organic Dyes—Characteristics, Polluting Effects and Separation/Elimination Procedures from Industrial Effluents—A Critical Overview, Organic Pollutants Ten Years after the Stockholm Convention-Environmental and Analytical Update*; InTech: Rijeka, Croatia, 2012; p. 31.
5. Chandanshive, V.V.; Kadam, S.K.; Khandare, R.V.; Kurade, M.B.; Jeon, B.-H.; Jadhav, J.P.; Govindwar, S.P. In situ phytoremediation of dyes from textile wastewater using garden ornamental plants, effect on soil quality and plant growth. *Chemosphere* **2018**, *210*, 968–976. [CrossRef] [PubMed]
6. Bello, A.O.; Tawabini, B.S.; Khalil, A.B.; Boland, C.R.; Saleh, T.A. Phytoremediation of cadmium-, lead-and nickel-contaminated water by Phragmites australis in hydroponic systems. *Ecol. Eng.* **2018**, *120*, 126–133. [CrossRef]

7. Khandare, R.V.; Kabra, A.N.; Kurade, M.B.; Govindwar, S.P. Phytoremediation potential of Portulaca grandiflora Hook. (Moss-Rose) in degrading a sulfonated diazo reactive dye Navy Blue HE2R (Reactive Blue 172). *Bioresour. Technol.* **2011**, *102*, 6774–6777. [CrossRef]

8. Patil, A.V.; Lokhande, V.H.; Suprasanna, P.; Bapat, V.A.; Jadhav, J.P. *Sesuvium portulacastrum* (L.) L.: A potential halophyte for the degradation of toxic textile dye, Green HE4B. *Planta* **2012**, *235*, 1051–1063. [CrossRef]

9. Telke, A.A.; Kagalkar, A.N.; Jagtap, U.B.; Desai, N.S.; Bapat, V.A.; Govindwar, S.P. Biochemical characterization of laccase from hairy root culture of *Brassica juncea* L. and role of redox mediators to enhance its potential for the decolorization of textile dyes. *Planta* **2011**, *234*, 1137–1149. [CrossRef]

10. Khan, M.U.; Sessitsch, A.; Harris, M.; Fatima, K.; Imran, A.; Arslan, M.; Shabir, G.; Khan, Q.M.; Afzal, M. Cr-resistant rhizo-and endophytic bacteria associated with *Prosopis juliflora* and their potential as phytoremediation enhancing agents in metal-degraded soils. *Front. Plant Sci.* **2015**, *5*, 755. [CrossRef]

11. Tara, N.; Arslan, M.; Hussain, Z.; Iqbal, M.; Khan, Q.M.; Afzal, M. On-site performance of floating treatment wetland macrocosms augmented with dye-degrading bacteria for the remediation of textile industry wastewater. *J. Clean. Prod.* **2019**, *217*, 541–548. [CrossRef]

12. Roy, D.C.; Biswas, S.K.; Saha, A.K.; Sikdar, B.; Rahman, M.; Roy, A.K.; Prodhan, Z.H.; Tang, S.-S. Biodegradation of Crystal Violet dye by bacteria isolated from textile industry effluents. *PeerJ* **2018**, *6*, e5015. [CrossRef]

13. Shahid, M.J.; Arslan, M.; Ali, S.; Siddique, M.; Afzal, M. Floating Wetlands: A Sustainable Tool for Wastewater Treatment. *CLEAN-Soil Air Water* **2018**, *46*, 1800120. [CrossRef]

14. Wu, S.; Vymazal, J.; Brix, H. Critical Review: Biogeochemical Networking of Iron, Is It Important in Constructed Wetlands for Wastewater Treatment? *Environ. Sci. Technol.* **2019**. [CrossRef] [PubMed]

15. He, Y.; Nurul, S.; Schmitt, H.; Sutton, N.B.; Murk, T.A.; Blokland, M.H.; Rijnaarts, H.H.; Langenhoff, A.A. Evaluation of attenuation of pharmaceuticals, toxic potency, and antibiotic resistance genes in constructed wetlands treating wastewater effluents. *Sci. Total Environ.* **2018**, *631*, 1572–1581. [CrossRef] [PubMed]

16. Tanner, C.C.; Headley, T.R. Components of floating emergent macrophyte treatment wetlands influencing removal of stormwater pollutants. *Ecol. Eng.* **2011**, *37*, 474–486. [CrossRef]

17. White, S.A.; Cousins, M.M. Floating treatment wetland aided remediation of nitrogen and phosphorus from simulated stormwater runoff. *Ecol. Eng.* **2013**, *61*, 207–215. [CrossRef]

18. Yeh, N.; Yeh, P.; Chang, Y.-H. Artificial floating islands for environmental improvement. *Renew. Sustain. Energy Rev.* **2015**, *47*, 616–622. [CrossRef]

19. Karstens, S.; Nazzari, C.; Bâlon, C.; Bielecka, M.; Grigaitis, Ž.; Schumacher, J.; Stybel, N.; Razinkovas-Baziukas, A. Floating wetlands for nutrient removal in eutrophicated coastal lagoons: Decision support for site selection and permit process. *Mar. Policy* **2018**, *97*, 51–60. [CrossRef]

20. Keizer-Vlek, H.E.; Verdonschot, P.F.; Verdonschot, R.C.; Dekkers, D. The contribution of plant uptake to nutrient removal by floating treatment wetlands. *Ecol. Eng.* **2014**, *73*, 684–690. [CrossRef]

21. Ladislas, S.; Gerente, C.; Chazarenc, F.; Brisson, J.; Andres, Y. Floating treatment wetlands for heavy metal removal in highway stormwater ponds. *Ecol. Eng.* **2015**, *80*, 85–91. [CrossRef]

22. Ijaz, A.; Iqbal, Z.; Afzal, M. Remediation of sewage and industrial effluent using bacterially assisted floating treatment wetlands vegetated with *Typha domingensis*. *Water Sci. Technol.* **2016**, *74*, 2192–2201. [CrossRef]

23. Emenike, C.U.; Jayanthi, B.; Agamuthu, P.; Fauziah, S. Biotransformation and removal of heavy metals: A review of phytoremediation and microbial remediation assessment on contaminated soil. *Environ. Rev.* **2018**, *26*, 156–168. [CrossRef]

24. Ijaz, A.; Shabir, G.; Khan, Q.M.; Afzal, M. Enhanced remediation of sewage effluent by endophyte-assisted floating treatment wetlands. *Ecol. Eng.* **2015**, *84*, 58–66. [CrossRef]

25. Shahid, M.J.; Arslan, M.; Siddique, M.; Ali, S.; Tahseen, R.; Afzal, M. Potentialities of floating wetlands for the treatment of polluted water of river Ravi, Pakistan. *Ecol. Eng.* **2019**, *133*, 167–176. [CrossRef]

26. Compant, S.; Clément, C.; Sessitsch, A. Plant growth-promoting bacteria in the rhizo-and endosphere of plants: Their role, colonization, mechanisms involved and prospects for utilization. *Soil Biol. Biochem.* **2010**, *42*, 669–678. [CrossRef]

27. Afzal, M.; Khan, Q.M.; Sessitsch, A. Endophytic bacteria: Prospects and applications for the phytoremediation of organic pollutants. *Chemosphere* **2014**, *117*, 232–242. [CrossRef] [PubMed]

28. Tara, N.; Iqbal, M.; Mahmood Khan, Q.; Afzal, M. Bioaugmentation of floating treatment wetlands for the remediation of textile effluent. *Water Environ. J.* **2019**, *33*, 124–134. [CrossRef]

29. Wang, C.-Y.; Sample, D.J.; Day, S.D.; Grizzard, T.J. Floating treatment wetland nutrient removal through vegetation harvest and observations from a field study. *Ecol. Eng.* **2015**, *78*, 15–26. [CrossRef]

30. Shahid, A.; Singh, J.; Bisht, S.; Teotia, P.; Kumar, V. Biodegradation of textile dyes by fungi isolated from north Indian field soil. *EnvironmentAsia* **2013**, *6*, 51–57.

31. Arslan, M.; Imran, A.; Khan, Q.M.; Afzal, M. Plant–bacteria partnerships for the remediation of persistent organic pollutants. *Environ. Sci. Pollut. Res.* **2017**, *24*, 4322–4336. [CrossRef]

32. Saleem, H.; Arslan, M.; Rehman, K.; Tahseen, R.; Afzal, M. *Phragmites australis*—A helophytic grass—Can establish successful partnership with phenol-degrading bacteria in a floating treatment wetland. *Saudi J. Biol. Sci.* **2019**, *26*, 1179–1186. [CrossRef]

33. Sutton, S. Measurement of microbial cells by optical density. *J. Valid. Technol.* **2011**, *17*, 46–49.

34. Rehman, K.; Imran, A.; Amin, I.; Afzal, M. Inoculation with bacteria in floating treatment wetlands positively modulates the phytoremediation of oil field wastewater. *J. Hazard. Mater.* **2018**, *349*, 242–251. [CrossRef] [PubMed]

35. Saleem, H.; Rehman, K.; Arslan, M.; Afzal, M. Enhanced degradation of phenol in floating treatment wetlands by plant-bacterial synergism. *Int. J. Phytorem.* **2018**, *20*, 692–698. [CrossRef] [PubMed]

36. Shahid, M.J.; Tahseen, R.; Siddique, M.; Ali, S.; Iqbal, S.; Afzal, M. Remediation of polluted river water by floating treatment wetlands. *Water Supply* **2019**, *19*, 967–977. [CrossRef]

37. Chen, W.; Hao, H.; Hughes, D.; Shi, Y.; Cui, J.; Li, Z.-X. Static and dynamic mechanical properties of expanded polystyrene. *Mater. Des.* **2015**, *69*, 170–180. [CrossRef]

38. Apha, A. *Standard Methods for the Examination of Water and Wastewater*; WEF: Cologny, Switzerland, 2012; Volume 22.

39. Kämpfer, P.; Erhart, R.; Beimfohr, C.; Böhringer, J.; Wagner, M.; Amann, R. Characterization of bacterial communities from activated sludge: Culture-dependent numerical identification versus in situ identification using group-and genus-specific rRNA-targeted oligonucleotide probes. *Microb. Ecol.* **1996**, *32*, 101–121. [CrossRef]

40. Chen, T.; Xu, M.; Tu, J.; Wang, H.; Niu, X. Relationship between Omnibus and Post-hoc Tests: An Investigation of performance of the F test in ANOVA. *Shanghai Arch. Psychiatry* **2018**, *30*, 60–64.

41. Rehman, K.; Imran, A.; Amin, I.; Afzal, M. Enhancement of oil field-produced wastewater remediation by bacterially-augmented floating treatment wetlands. *Chemosphere* **2019**, *217*, 576–583. [CrossRef]

42. Borne, K.E.; Fassman, E.A.; Tanner, C.C. Floating treatment wetland retrofit to improve stormwater pond performance for suspended solids, copper and zinc. *Ecol. Eng.* **2013**, *54*, 173–182. [CrossRef]

43. Borne, K.E. Floating treatment wetland influences on the fate and removal performance of phosphorus in stormwater retention ponds. *Ecol. Eng.* **2014**, *69*, 76–82. [CrossRef]

44. Ma, Y.; Rajkumar, M.; Oliveira, R.S.; Zhang, C.; Freitas, H. Potential of plant beneficial bacteria and arbuscular mycorrhizal fungi in phytoremediation of metal-contaminated saline soils. *J. Hazard. Mater.* **2019**, *379*, 120813. [CrossRef] [PubMed]

45. Billore, S.; Sharma, J. Treatment performance of artificial floating reed beds in an experimental mesocosm to improve the water quality of river Kshipra. *Water Sci. Technol.* **2009**, *60*, 2851–2859. [CrossRef] [PubMed]

46. Shehzadi, M.; Fatima, K.; Imran, A.; Mirza, M.; Khan, Q.; Afzal, M. Ecology of bacterial endophytes associated with wetland plants growing in textile effluent for pollutant-degradation and plant growth-promotion potentials. *Plant. Biosyst.* **2016**, *150*, 1261–1270. [CrossRef]

47. Shehzadi, M.; Afzal, M.; Khan, M.U.; Islam, E.; Mobin, A.; Anwar, S.; Khan, Q.M. Enhanced degradation of textile effluent in constructed wetland system using Typha domingensis and textile effluent-degrading endophytic bacteria. *Water Res.* **2014**, *58*, 152–159. [CrossRef] [PubMed]

48. Winston, R.J.; Hunt, W.F.; Kennedy, S.G.; Merriman, L.S.; Chandler, J.; Brown, D. Evaluation of floating treatment wetlands as retrofits to existing stormwater retention ponds. *Ecol. Eng.* **2013**, *54*, 254–265. [CrossRef]

49. Joshi, P.; Jaybhaye, S.; Mhatre, K. Biodegradation of dyes using consortium of bacterial strains isolated from textile effluent. *Eur. J. Exp. Biol.* **2015**, *5*, 36–40.

50. Manjate, E.; Ramos, S.; Almeida, C.M.R. Potential interferences of microplastics in the phytoremediation of Cd and Cu by the salt marsh plant Phragmites australis. *J. Environ. Chem. Eng.* **2020**, *8*, 103658. [CrossRef]

51. Maxwell, B.; Winter, D.; Birgand, F. Floating treatment wetland retrofit in a stormwater wet pond provides limited water quality improvements. *Ecol. Eng.* **2020**, *149*, 105784. [CrossRef]

52. Kong, Z.; Wu, Z.; Glick, B.R.; He, S.; Huang, C.; Wu, L. Co-occurrence patterns of microbial communities affected by inoculants of plant growth-promoting bacteria during phytoremediation of heavy metal-contaminated soils. *Ecotoxicol. Environ. Saf.* **2019**, *183*, 109504. [CrossRef]

53. Afzal, M.; Shabir, G.; Tahseen, R.; Islam, E.; Iqbal, S.; Khan, Q.M.; Khalid, Z.M. Endophytic Burkholderia sp. strain PsJN improves plant growth and phytoremediation of soil irrigated with textile effluent. *CLEAN–Soil Air Water* **2014**, *42*, 1304–1310. [CrossRef]

54. Kong, Z.; Glick, B.R. The role of plant growth-promoting bacteria in metal phytoremediation. *Adv. Microb. Physiol.* **2017**, *71*, 97–132.

55. Zhang, N.; Wang, D.; Liu, Y.; Li, S.; Shen, Q.; Zhang, R. Effects of different plant root exudates and their organic acid components on chemotaxis, biofilm formation and colonization by beneficial rhizosphere-associated bacterial strains. *Plant Soil* **2014**, *374*, 689–700. [CrossRef]

56. Ojuederie, O.; Babalola, O. Microbial and plant-assisted bioremediation of heavy metal polluted environments: A review. *Int. J. Environ. Res. Public Health* **2017**, *14*, 1504. [CrossRef] [PubMed]

57. Afzal, M.; Rehman, K.; Shabir, G.; Tahseen, R.; Ijaz, A.; Hashmat, A.J.; Brix, H. Large-scale remediation of oil-contaminated water using floating treatment wetlands. *NPJ Clean Water* **2019**, *2*, 3. [CrossRef]

58. Ali, H.; Khan, E.; Sajad, M.A. Phytoremediation of heavy metals—Concepts and applications. *Chemosphere* **2013**, *91*, 869–881. [CrossRef]

59. Ashraf, S.; Afzal, M.; Naveed, M.; Shahid, M.; Ahmad Zahir, Z. Endophytic bacteria enhance remediation of tannery effluent in constructed wetlands vegetated with Leptochloa fusca. *Int. J. Phytorem.* **2018**, *20*, 121–128. [CrossRef]

60. Afridi, M.S.; Mahmood, T.; Salam, A.; Mukhtar, T.; Mehmood, S.; Ali, J.; Khatoon, Z.; Bibi, M.; Javed, M.T.; Sultan, T. Induction of tolerance to salinity in wheat genotypes by plant growth promoting endophytes: Involvement of ACC deaminase and antioxidant enzymes. *Plant Physiol. Biochem.* **2019**, *139*, 569–577. [CrossRef] [PubMed]

61. Glick, B.R. Plant growth-promoting bacteria: Mechanisms and applications. *Scientifica* **2012**, *2012*, 963401. [CrossRef]

62. Fahid, M.; Arslan, M.; Shabir, G.; Younus, S.; Yasmeen, T.; Rizwan, M.; Siddique, K.; Ahmad, S.R.; Tahseen, R.; Iqbal, S. *Phragmites australis* in combination with hydrocarbons degrading bacteria is a suitable option for remediation of diesel-contaminated water in floating wetlands. *Chemosphere* **2020**, *240*, 124890. [CrossRef]

Article

Cyperus laevigatus L. Enhances Diesel Oil Remediation in Synergism with Bacterial Inoculation in Floating Treatment Wetlands

Muhammad Fahid [1,2], Shafaqat Ali [1,3,*], Ghulam Shabir [2], Sajid Rashid Ahmad [4], Tahira Yasmeen [1], Muhammad Afzal [2,*], Muhammad Arslan [5], Afzal Hussain [1], Abeer Hashem [6,7], Elsayed Fathi Abd Allah [8], Mohammed Nasser Alyemeni [6] and Parvaiz Ahmad [6,9]

[1] Department of Environmental Sciences and Engineering, Government College University Faisalabad, Faisalabad 38000, Pakistan; fahadceespu@gmail.com (M.F.); rida_akash@hotmail.com (T.Y.); afzaalh345@gmail.com (A.H.)

[2] Soil and Environmental Biotechnology Division, National Institute for Biotechnology and Genetic Engineering (NIBGE), Faisalabad 38000, Pakistan; gshabirnibge@yahoo.com

[3] Department of Biological Sciences and Technology, China Medical University, Taichung 40402, Taiwan

[4] Colleges of Earth and Environmental Sciences, University of the Punjab, Lahore 54000, Pakistan; sajidpu@yahoo.com

[5] Department of Civil and Environmental Engineering, University of Alberta, Edmonton, AB T6G 2R3, Canada; marslan@ualberta.ca

[6] Botany and Microbiology Department, College of Science, King Saud University, P.O. Box. 2460, Riyadh 11451, Saudi Arabia; habeer@ksu.edu.sa (A.H.); mnalyemeni@gmail.com (M.N.A.); parvaizbot@yahoo.com (P.A.)

[7] Mycology and Plant Disease Survey Department, Plant Pathology Research Institute, ARC, Giza 12511, Egypt

[8] Plant Production Department, College of Food and Agricultural Sciences, King Saud University, P.O. Box. 2460, Riyadh 11451, Saudi Arabia; eabdallah@ksu.edu.sa

[9] Department of Botany, S.P. College, Kothi Bagh, Srinagar, Jammu and Kashmir 190001, India

[*] Correspondence: shafaqataligill@yahoo.com (S.A.); manibge@yahoo.com (M.A.)

Received: 20 January 2020; Accepted: 3 March 2020; Published: 18 March 2020

Abstract: Diesel oil is considered a very hazardous fuel due to its adverse effect on the aquatic ecosystem, so its remediation has become the focus of much attention. Taking this into consideration, the current study was conducted to explore the synergistic applications of both plant and bacteria for cleaning up of diesel oil contaminated water. We examined that the application of floating treatment wetlands (FTWs) is an economical and superlative choice for the treatment of diesel oil contaminated water. In this study, a pilot scale floating treatment wetlands system having diesel oil contaminated water (1% w/v), was adopted using *Cyperus laevigatus* L and a mixture of hydrocarbons degrading bacterial strains; viz., *Acinetobacter* sp.61KJ620863, *Bacillus megaterium* 65 KF478214, and *Acinetobacter* sp.82 KF478231. It was observed that consortium of hydrocarbons degrading bacteria improved the remediation of diesel oil in combination with *Cyperus laevigatus* L. Moreover, the performance of the FTWs was enhanced by colonization of bacterial strains in the root and shoot of *Cyperus laevigatus* L. Independently, the bacterial consortium and *Cyperus laevigatus* L exhibited 37.46% and 56.57% reduction in diesel oil, respectively, while 73.48% reduction in hydrocarbons was exhibited by the joint application of both plant and bacteria in FTWs. Furthermore, microbial inoculation improved the fresh biomass (11.62%), dry biomass (33.33%), and height (18.05%) of plants. Fish toxicity assay evaluated the effectiveness of FTWs by showing the extent of improvement in the water quality to a level that became safe for living organisms. The study therefore concluded that *Cyperus laevigatus* L augmented with hydrocarbons degrading bacterial consortium exhibited a remarkable ability to decontaminate the diesel oil from water and could enhance the FTWs performance.

Keywords: floating treatment wetlands; *Cyperus laevigatus* L; diesel oil; plant-bacteria synergism; toxicity

1. Introduction

Discharge of petroleum hydrocarbons into the environment whether unintentionally or due to anthropogenic sources is the main cause of surface water and ground water pollution. Because of this alarming situation and the hazardous effect of hydrocarbons on the aquatic ecosystem, much attention is being focused on the remediation of hydrocarbons [1]. Moreover, during the petroleum refining process, raw crude oil is converted into various useful end products such as gasoline, diesel fuel, kerosene, and fuel oils. Purification of crude oil utilizes huge volumes of water which causes the generation of wastewater enriched with toxic organic compounds [2,3]. Treatment of such wastewater is very essential to achieve safe environmental water quality standards before discharge into any water body [4,5]. Traditional wastewater treatment processes such as electrochemical oxidation, membrane filtration, coagulation, flocculation [6,7] require highly skilled man power and high operational and maintenance costs. Moreover, application of these techniques generates toxic waste that further needs treatment before disposal [8,9].

It has been reported by many studies that various types of domestic and industrial wastewaters are widely treated using floating treatment wetlands [10–14]. This innovative technology has low installation, operational and maintenance costs, along with aesthetic value and environmentally friendly quality [15,16]. In floating treatment wetlands, plants are grown on a floating mat, whereas roots are hanged in the water column [17]. The extended roots in the water body offer plants the ability to create a direct contact between contaminants and the roots-associated microbial community. In addition, the suspended roots in water accelerate the sedimentation process by trapping suspended particles and reducing the water turbulence [18]. The roots grow horizontally and vertically to provide a large surface area for nutrient uptake and biofilm enlargement [13]. The associated microbial community degrades complex organic matter into simple components which are removed through the combined action of plants and microbes [19,20]. Floating treatment wetlands in assistance with bacterial consortium can be a promising alternative and green technology for remediation of oil refinery effluent. Many bacterial genera have been reported to degrade the hydrocarbons by their metabolic process [21]. Bacteria enhance the solubility, bioavailability, biodegradation, and uptake of hydrophobic compounds by production of biosurfactants which also facilitate microbial growth and hydrocarbon emulsification [22]. Plants provide nutrients, metabolites, phyto-hormones, and habitat for bacteria [23]. In FTWs, the plants augmented with hydrocarbon degrading bacteria could be an effective methodology for the remediation of diesel oil from water. Floating treatment wetlands augmented with *Cyperus laevigatus* L and hydrocarbon-degrading bacteria have not been widely tested for treatment of water polluted by diesel oil. So, considering this, the current study was undertaken with the main objective to assess the synergistic potential of *Cyperus laevigatus* L and hydrocarbons degrading bacterial strains in the FTWs system to remediate diesel oil from water.

2. Methodology

2.1. Diesel Oil

The diesel oil used in this study was purchased from a local filling station. The diesel oil was filter sterilized via syringe filters and was used as such without any further analysis for detail composition of the entire compounds.

2.2. Preparation of Mixed Bacterial Culture

Three hydrocarbons degrading bacterial strains viz.; Acinetobacter sp.61KJ620863, Bacillus megaterium 65 KF478214, and Acinetobacter sp. 82 KF478231 already isolated and characterized [24] were applied in the current experiment. All the hydrocarbons degrading bacterial strains were grown separately in M9 solution containing diesel oil 1% (w/v) as a sole carbon source. After growth in M9 solution, the bacterial culture was centrifuged then resuspended in normal saline solution. To achieve 10^8 cells mL^{-1}, the optical density of each microbial suspension was adjusted finally using normal saline solution. The bacterial suspension was spread on the LB agar plates to count the number of colonies. The cell suspension of each bacterial strain was mixed in equal proportion and the consortium (50 mL) was inoculated in FTWs microcosms.

2.3. Manufacturing of Floating Treatment Wetlands

The FTWs were fabricated at the National Institute for Biotechnology and Genetic Engineering, (NIBGE) Faisalabad Pakistan. The experiment was conducted for 90 days beginning in the month of April 2018. In the FTWs microcosms, the tanks having 20 L volume capacity were made of poly ethylene material, whereas the floating mats were preapared using Jumbolon sheets of Diamond Foam Company, Pvt. Ltd., Pakistan. The Jumbolon foam piece, which was used as a floating mat, comprises the dimension of 50.8 cm (length) × 38.1 cm (width) × 7.62 cm (thickness) and five holes in each mat which were drilled at equal distance for growing healthy seedlings of *Cyperus laevigatus* L.

The seedlings of selected plant were allowed to grow for thirty days duration in the FTWs microcosms containing tap water without any treatment. However, Hoagland solution was applied periodically to stabilize the process of root-establishment. After thirty days acclimatization of plant roots, the outer surface of roots was sterilized using 5% NaOCl (sodium hypochlorite) solution. The established microcosms were spiked with 1% diesel oil (w/v). The concentration of diesel oil was selected based on our previous studies of diesel oil biodegradation (0.5%, 1.0%, and 1.5% diesel oil) in M9 media by shake flask experimentation.

2.4. Experimental Design

The experimental setup of floating treatment wetlands is shown in Figure 1. The floating treatment wetland microcosms were prepared in triplicates and different types of treatments were as follows:

Control-1 (C1): Microcosm consist of diesel oil polluted water and no plants.
Control-2 (C2): Microcosm consist of tap water and plants.
Treatment-1 (T1): Microcosm consist of diesel oil polluted water and bacterial consortium.
Treatment-2 (T2): Microcosm consist of diesel oil polluted water and plants.
Treatment-3 (T3): Microcosm consist of diesel oil polluted water, plants, and bacterial consortium.

Figure 1. Experimental setup showing the floating treatment wetlands system for the cleanup of diesel oil polluted water.

2.5. Plant Biomass

After 3 months of the experimental period, the plant roots and shoots were cropped 2 cm down and overhead the floating mat. For both roots and shoots, the fresh and dry biomass was determined to check the influence of microbial inoculation and plant growth on diesel oil remediation. Fresh roots and shoots samples were placed in an oven at 70 °C and dry biomass was determined [25].

2.6. Hydrocarbons Assessment

The residual amount of hydrocarbons present in the treated water sample was extracted as an extracting solvent using the organochloride compound dichloromethane. Briefly, water sample having amount 25 ml was shaken with 15 mL of dichloromethane in a glass separatory funnel for 15 min. After 30 sec agitation and 3 min settling time, the water layer was discarded. The procedure was repeated thrice until the entire water sample was completely extracted. The obtained extract was dried using 5 g anhydrous sodium sulphate. The extract was then transferred to a Teflon-capped glass tube. The extracted hydrocarbons were analyzed by Spectrum Two Fourier transform infrared (FTIR) spectrometer [26].

2.7. Water Quality Parameter Analyses

Water samples were collected at different time intervals. These water samples were tested for pH, electrical conductivity (EC), dissolved oxygen (DO), total suspended solids (TSS), total solids (TS), total dissolved solids (TDS), chemical oxygen demand (COD), biochemical oxygen demand (BOD), and total organic carbon (TOC) using established standard protocols [27].

2.8. Persistence of Bacterial Culture

Treated water, roots, and shoots samples collected at different time intervals were analyzed for the survival of the hydrocarbons degrading bacteria in water, plant rhizosphere, and endosphere by plate count method. By following the established protocols, the surface sterilization was applied to the roots and shoots for the isolation of endophytic bacteria [28]. Briefly, plant roots and shoots samples were washed by using autoclaved distilled water, followed by ethanol (70%) and sodium hypochlorite solution (2%). Finally, the roots and shoots samples were also wash away with autoclaved distilled water. The surface sterilized roots and shoots were ground (5 g) by using pestle and mortar and were mixed using 10 mL NaCl solution (0.9% w/v) to make a suspension. The suspension was serially diluted up to 10^{-6}. A 100 µL of the suspension was spread on the M9 agar media containing diesel oil (50 mg/L) as a sole carbon source by spreading plate methodology. For the determination of total hydrocarbons degrading bacteria, the petri dishes were incubated at 37 °C for 48 h.

2.9. Evaluation of Toxicity

After the completion of the experimentation, the treated water samples were tested for toxicity using fish toxicity assay. Glass tanks were filled with treated water from each treatment. In each tank, ten fish *Labeo rohita* of equal size and weight were added. The fish toxicity experiment was conducted for a duration of 96 h. After every 24 h of regular interval, the number of fish survival was recorded [29,30].

2.10. Statistical Analyses

Water pollution parameters, residual hydrocarbon concentration, perseverance of hydrocarbons degrading bacteria in water, root and shoot, plant biomass, and reduction in toxicity level were analyzed through Statistics 8.1. Factorial Analysis of Variance (ANOVA) to make the comparison between independent variable. Further all pairwise comparisons between time into treatments were analyzed by Tukey HSD ($\alpha = 0.05$). The alphabets on values represent the significant/non-significant difference among the treatments.

3. Results and Discussion

3.1. Hydrocarbons Degradation

Discharges of petroleum oil during transportation of oil tankers, refining of crude oil, and leakage in underground storage tanks is the main cause of environmental contamination and ultimately damage of the ecosystem [31]. Diesel oil has been widely reported as a very harmful petroleum product that is composed of a complex mixture of hydrocarbons. These hydrocarbons pose severe threats to human health due to their mutagenic, carcinogenic, and immune toxic behavior [32]. Consequently, serious attention has been focused on remediating the adverse effect of hydrocarbons on the water quality. Figure 2 shows gradual reduction of hydrocarbons concentrations in diesel oil contaminated water during the 90 day experiment under different treatments. It was noticed that in T3 treatment consisting of *Cyperus laevigatus* L and bacterial consortium, the removal of hydrocarbons in FTWs microcosms was maximum (73.48%). It may be due to the combined effect of both plant and hydrocarbon bacteria. It has been reported that in the presence of the microorganism, plants get enough support in severe conditions and can perform better organic pollutant degradation [33,34]. It has also been noted that the root exudates secreted by the plant, boost up the growth and activity of rhizosphere bacterial communities [35].

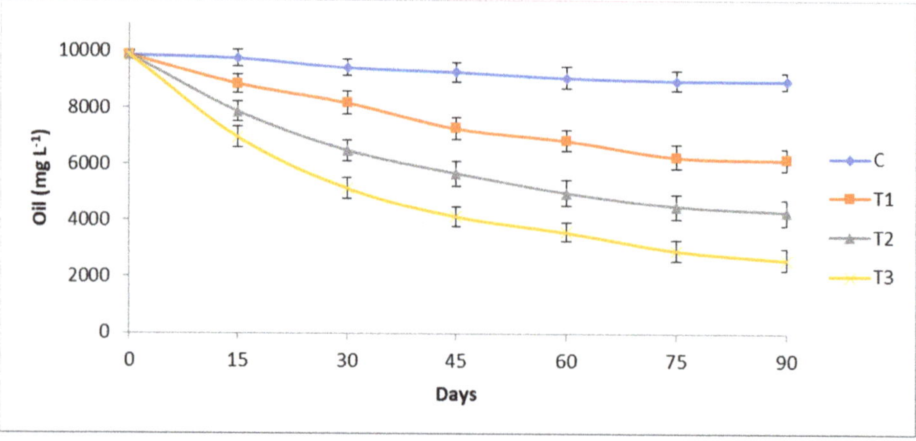

Figure 2. Diesel oil removal from water by floating treatment wetlands. **C:** Microcosm containing diesel oil polluted water and no plants; **T1:** Microcosm containing diesel oil polluted water and bacterial consortium; **T2:** Microcosm containing diesel oil polluted water and plants; **T3:** Microcosm containing diesel oil polluted water, plants and bacterial consortium. Each value is a mean of triplicate determination. Error bars represent the standard deviations among all three replicates.

In un-vegetated treatment with bacterial consortium (T1), only 37.46% reduction of hydrocarbons was observed. It has been described that bacterial populations have the tendency to mineralize hydrocarbons present in diesel oil [36]. However, it was observed that in T1 treatment, hydrocarbon reduction was 2 times lower than T3 treatment. It may be due to the reason that in T1 treatment, the growth of the microorganism is suppressed due to the presence of a higher amount of toxic hydrocarbons in absence of the plants that resulted in lower reduction in this treatment [37,38]. Relatively higher reduction in hydrocarbons (56.56%) was detected in T2 treatment, vegetated with the plant but deprived of bacterial consortium rather than un-vegetated treatment (T1). Due to absorption of easily degradable hydrocarbons in the plant roots, the superior hydrocarbons removal was observed during the initial 30 days of the experiment in T2 treatment. These results are in agreement with the finding of previous research [39,40]. The literature study revealed that in spite of microorganisms, the degradation of hydrocarbons is also assisted by plants that play a fundamental role by taking up the hydrocarbons in their roots and shoots and change them into less harmful substances [41,42]. In control without the plant (C), the hydrocarbons content also decreased up to 9.30% which may be due to evaporation of volatile hydrocarbons present in the diesel oil and/or due to the presence of indigenous bacteria in water or photolysis in the unplanted control [41,43,44].

3.2. Chemical and Biological Oxygen Demand

Reduction in COD and BOD is illustrated in Figure 3; Figure 4, respectively. In T3 treatment vegetated with *Cyperus laevigatus* and bacteria, COD and BOD were reduced up to 52.18% and 72.28%, respectively. These results are in agreement with our previous findings that *growths* of plants with bacterial consortium *improve* the remediation potential of organic components *present* in wastewater [45]. It is stated that bacterial consortium emulsifies the hydrocarbons in water resulting in lowering of COD and BOD values. Relatively lower reduction in COD (36.61%) and BOD (56.68%) was noticed in T2 treatment. It was described that the higher the number of plants, the more reduction in the COD and BOD [11,26]. However, relatively minimum reduction of COD (26.54%) and BOD (39.98%) was observed in T1 treatment. Control exhibited very less reduction in COD (10.33%) and BOD (12.2%).

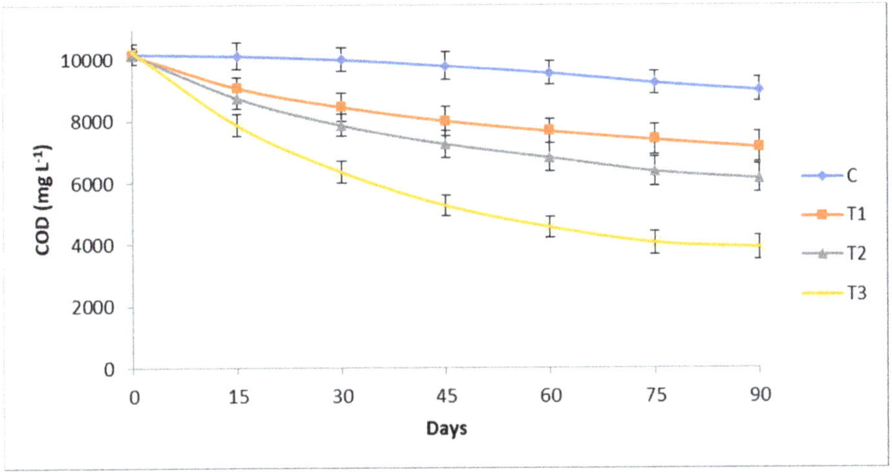

Figure 3. Chemical oxygen demand (COD) removal from water by floating treatment wetlands. **C:** Microcosm containing diesel oil polluted water and no plants; **T1:** Microcosm containing diesel oil polluted water and bacterial consortium; **T2:** Microcosm containing diesel oil polluted water and plants; **T3:** Microcosm containing diesel oil polluted water, plants and bacterial consortium. Each value is a mean of triplicate determination. Error bars represent the standard deviations among all three replicates.

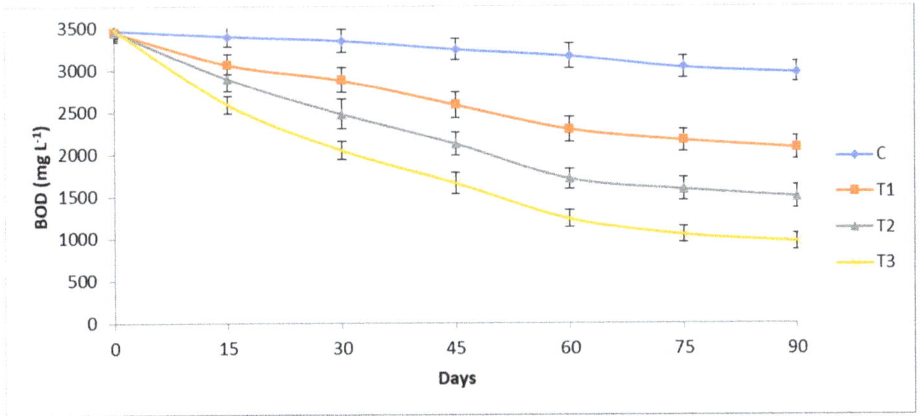

Figure 4. Biochemicalical oxygen demand (BOD) removal from water by floating treatment wetlands. **C:** Microcosm containing diesel oil polluted water and no plants; **T1:** Microcosm containing diesel oil polluted water and bacterial consortium; **T2:** Microcosm containing diesel oil polluted water and plants; **T3:** Microcosm containing diesel oil polluted water, plants and bacterial consortium. Each value is a mean of triplicate determination. Error bars represent the standard deviations among all three replicates.

3.3. Total Organic Carbon and Phenol Reduction

Total organic carbon (TOC) reduction is shown in Figure 5. Higher TOC reduction (91.71%) was observed in T3 treatment as compared to T2 treatment that exhibited lower TOC reduction (76.96%).

However, minimum TOC reduction (67.70%) was recorded in T1 treatment among all the treatments. Non-significant results for TOC reduction (17.36%) were seen in control. It has been revealed that the growth of plants by utilizing the organic matter as a source of nutrients is supported by the cluster of microbes availability in the plants roots, which is the main reason for the higher reduction in TOC [46,47].

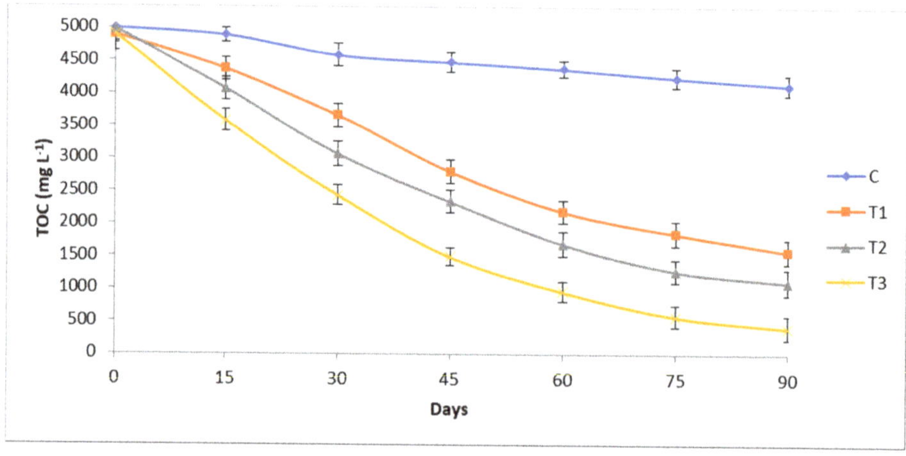

Figure 5. Total organic carbon (TOC) removal from water by floating treatment wetlands. **C:** Microcosm containing diesel oil polluted water and no plants; **T1:** Microcosm containing diesel oil polluted water and bacterial consortium; **T2:** Microcosm containing diesel oil polluted water and plants; **T3:** Microcosm containing diesel oil polluted water, plants and bacterial consortium. Each value is a mean of triplicate determination. Error bars represent the standard deviations among all three replicates.

Higher reduction (94.88%) in phenol was examined in T3 treatment, which is significant as compared to other treatments (T1 and T2). It was also observed that effective reduction (93.44%) in phenol was seen in T2 treatment. However, lower removal of phenol (77.14%) was detected in the treatment T1 (Table 1). Our results are in agreement with the earlier study who reported the effectiveness of bacterial augmentation in phenol removal [48].

3.4. Removal of Solids

Table 1 illustrates the removal of solids from diesel oil contaminated water. Consequently, highest reduction occurred in TS (52.19%), TSS (75.56%), TDS (49.63%), and EC (74.09%) in bacterially-augmented treatment (T3). Apparently, it was observed that the FTWs showed efficiency to improve the quality of water by reducing the pH value that ranged from 8.5 to 7.5, which is authenticated by previous findings [45,49]. It was observed that due the presence of the *Cyperus laevigatus* plant in T2 treatment, in comparison to T1 treatment (without plants), higher concentrations of nutrients were removed from the wastewater. This finding was within the permissible range reported earlier [50].

Table 1. Inoculation bacterial effect on remediation of diesel oil polluted water in floating treatment wetlands microcosms vegetated with *Cyperus laevigatus* L.

Treatment	Days	Parameters						
		pH	EC (ms/cm)	TDS (mg/L)	TS (mg/L)	TSS (mg/L)	DO (mg/L)	Phenol (mg/L)
Control	0	8.7^{ab}(0.13)	3.3^{ab}(0.01)	1918^{a}(175)	2274^{a}(232)	356^{b}(13)	5.5^{ef}(0.13)	0.35^{a}(0.01)
	15	8.5^{abc}(0.12)	3.3^{ab}(0.01)	1912^{ab}(197)	2270^{a}(215)	350^{bcd}(25)	5.2^{fg}(0.14)	0.34^{b}(0.01)
	30	8.3^{def}(0.16)	3.2^{ab}(0.03)	1908^{ab}(178)	2268^{a}(234)	346^{bcde}(16)	5.0^{ghi}(0.12)	0.32^{c}(0.01)
	45	8.2^{efg}(0.15)	3.2^{ab}(0.07)	1906^{ab}(183)	2255^{a}(196)	336^{cdef}(24)	5.1^{gh}(0.13)	0.32^{cd}(0.06)
	60	8.1^{fgh}(0.11)	3.2^{b}(0.01)	1903^{ab}(192)	2245^{a}(234)	331^{def}(26)	4.9^{hij}(0.14)	0.32^{cd}(0.01)
	75	8.0^{fgh}(0.19)	3.2^{b}(0.08)	1902^{ab}(125)	2239^{a}(266)	329^{ef}(16)	4.7^{ij}(0.16)	0.32^{cd}(0.01)
	90	8.0^{fgh}(0.14)	3.1^{b}(0.01)	1900^{ab}(147)	2225^{a}(297)	317^{f}(24)	4.5^{jk}(0.15)	0.31^{d}(0.07)
T1	0	8.8^{a}(0.14)	3.3^{ab}(0.01)	1921^{a}(115)	2276^{a}(195)	355^{bc}(15)	5.3^{fg}(0.15)	0.35^{ab}(0.01)
	15	8.7^{ab}(0.15)	2.3^{c}(0.04)	1615^{d}(156)	1811^{c}(176)	296^{g}(18)	5.1^{gh}(0.14)	0.17^{f}(0.03)
	30	8.6^{abc}(0.12)	2.1^{d}(0.05)	1312^{g}(169)	1509^{g}(197)	197^{ijk}(17)	4.8^{jk}(0.13)	0.13^{i}(0.00)
	45	8.6^{abc}(0.13)	1.2^{fghij}(0.05)	1208^{h}(149)	1321^{h}(144)	143^{l}(15)	4.3^{kl}(0.15)	0.10^{j}(0.00)
	60	8.5^{bcd}(0.15)	1.1^{ghij}(0.03)	1129^{i}(124)	1217^{ij}(105)	117^{m}(17)	4.0^{lm}(0.16)	0.09^{k}(0.00)
	75	8.4^{cde}(0.17)	1.1^{hij}(0.04)	1078^{i}(156)	1161^{jk}(108)	103^{mn}(18)	3.8^{mn}(0.17)	0.08^{k}(0.00)
	90	8.2^{def}(0.19)	1.0^{ij}(0.04)	1011^{k}(108)	1117^{kl}(145)	92^{n}(15)	3.5^{n}(0.13)	0.08^{k}(0.00)
T2	0	8.5^{bcd}(0.13)	3.2^{ab}(0.06)	1924^{a}(254)	2275^{a}(196)	359^{b}(15)	5.6^{de}(0.14)	0.35^{ab}(0.04)
	15	8.2^{efg}(0.11)	2.9^{b}(0.04)	1880^{b}(147)	1972^{b}(162)	388^{a}(22)	5.3^{fg}(0.15)	0.19^{e}(0.01)
	30	8.1^{fgh}(0.16)	2.3^{c}(0.06)	1505^{e}(138)	1681^{d}(144)	258^{h}(23)	5.9^{cd}(0.16)	0.13^{h}(0.00)
	45	8.0^{gh}(0.17)	1.5^{defgh}(0.05)	1008^{k}(148)	1514^{f}(177)	198^{i}(12)	6.3^{bc}(0.14)	0.03^{n}(0.00)
	60	7.5^{i}(0.14)	1.3^{efghij}(0.06)	937^{lm}(168)	1475^{fg}(176)	184^{ij}(14)	6.1^{c}(0.16)	0.03^{no}(0.00)
	75	7.2^{j}(0.13)	1.4^{efghi}(0.04)	929^{m}(159)	1357^{h}(156)	145^{l}(15)	5.9^{cd}(0.13)	0.02^{no}(0.00)
	90	7.1^{j}(0.15)	1.3^{efghij}(0.01)	916^{m}(148)	1253^{i}(145)	86^{n}(15)	6.3^{bc}(0.16)	0.02^{o}(0.00)
T3	0	8.5^{cd}(0.17)	3.6^{a}(0.05)	1922^{a}(138)	2278^{a}(184)	356^{b}(16)	5.6^{de}(0.17)	0.35^{ab}(0.01)
	15	8.1^{fgh}(0.14)	1.9^{de}(0.04)	1718^{c}(129)	1921^{b}(192)	317^{f}(18)	5.7^{de}(0.14)	0.14^{g}(0.00)
	30	7.8^{h}(0.16)	1.6^{def}(0.05)	1408^{f}(127)	1611^{e}(147)	193^{i}(15)	6.5^{b}(0.16)	0.12^{hi}(0.00)
	45	7.5^{i}(0.15)	1.5^{defg}(0.05)	1211^{h}(148)	1438^{g}(174)	173^{jk}(19)	6.9^{a}(0.15)	0.06^{l}(0.00)
	60	7.2^{j}(0.17)	1.2^{fghij}(0.01)	1179^{h}(136)	1338^{h}(134)	160^{kl}(14)	7.1^{a}(0.16)	0.05^{m}(0.00)
	75	7.1^{j}(0.14)	1.0^{ij}(0.01)	1029^{k}(148)	1138^{kl}(136)	112^{m}(17)	7.0^{a}(0.14)	0.02^{no}(0.00)
	90	6.9^{k}(0.18)	0.9^{j}(0.01)	968^{l}(149)	1089^{l}(142)	87^{n}(13)	7.1^{a}(0.15)	0.02^{o}(0.00)

Control: Microcosm containing diesel oil polluted water and no plants; **T1:** Microcosm containing diesel oil polluted water and bacterial consortium; **T2:** Microcosm having diesel oil polluted water and plants; **T3:** Microcosm having diesel oil polluted water, plants and bacterial consortium. Each value is a mean of triplicate determination. Standard deviations among three replicates are presented in parenthesis and the alphabets represent the significant/non-significant difference among the treatments.

3.5. Persistence of Microbial Population

Mostly plant-associated microbes mineralize the organic pollutants. It has been explored that the effectiveness of the FTWs technique is truly related to the biodegradation of organic pollutants and persistence of inoculated bacterial in the water investigated for remediation [49,51]. Persistence of bacterial population in the root interior, shoot interior, water, and in the rhizoplane is shown in Table 2. Higher level of bacterial colonization in the plant roots and shoots and also in hydrocarbon contaminated water was observed in this study. The greater number of survival of inoculated bacteria (1.01×10^6 cfu/mL) was recorded after 90 days of the experiment in the treated water. In different plant compartments, bacterial survival follows the order as: rhizoplane (4.5×10^6) > root interior (4.0×10^6) > shoot interior (1.2×10^6). Higher numbers of bacterial populations were counted in the T3 treatment; it was due to the efficient plant bacterial partnership. As reported by the previous study that plant roots provide residency and nutrients for proliferation of bacterial community present in outer and inner part of the tissues of the plant [52]. It has also been reported that rhizo bacteria that are involved in plant growth are probably to be present in inner tissue of plant (endophyte bacteria) at particular phase of their lifecycle so microbes can effectively penetrate in exposed plant parts especially in root and process of colonization of bacteria occurs by a dynamic mechanism for pollutant removal [23,53].

Table 2. Enumeration of total microbial loads in the water and tissues of *Cyperus laevigatus* L augmented with bacterial consortium (T3) during different sampling time.

Treatments	Cfu × 10^5						
	0 d	15 d	30 d	45 d	60 d	75 d	90 d
Water (Cfu/mL)	27.8^f(1.5)	26.3^h(1.3)	17.8^j(0.9)	16.5^k(1.8)	14.4^l(1.6)	12.3^m(1.5)	10.1^o(0.8)
Rhizoplane (Cfu/g)	8.7^p(0.8)	12.3^m(1.1)	19.7^i(1.2)	27.5^{fg}(1.9)	34.2^d(2.1)	39.6^c(2.4)	45.8^a(2.7)
Root (Cfu/g)	0.4^v(0.05)	5.4^s(1.5)	11.7^n(1.8)	19.9^i(1.5)	27.3^g(1.3)	33.8^e(1.4)	40.5^b(1.3)
Shoot (Cfu/g)	0.1^v(0.01)	1.2^u(1.4)	3.5^t(0.1)	5.9^r(0.3)	7.0^q(1.1)	9.8^o(1.3)	12.3^m(1.6)

Each value is a mean of triplicate determination. Standard deviations amongst three replicates are existing in parenthesis and the alphabets represent the significant/non-significant difference among the treatments.

It has now recently demonstrated that growth of microbial population in roots and shoots of plants and decrease of their survival in water is due to presence of enormous amount of carbon in hydrocarbons contaminated water that provides a source of energy during the microbial proliferation [54].

3.6. Plant Height and Biomass

Effectiveness of phytoremediation is of great importance and correlated with selection of a particular plant species, their survival and tolerance in hydrocarbons contaminated water. Due to different interaction of roots of the plants with hydrocarbons, the contaminants are absorbed and transported in the shoot of plants, ultimately affecting the growth and biomass of plants. Roots of plants offer a large surface area for microbial population and act as a modified place for every microbe endorsing the constant source of nutrients [55]. To check the effect of bacterial inoculation for hydrocarbon degradation and growth of the *Cyperus laevigatus* L plant, both fresh and dry biomass of this plant were recorded (Table 3). In FTWs, *Cyperus laevigatus* L planted in microcosms that contain diesel oil (T2) displayed lesser root length (54.14%), shoot length (49.11%), fresh (61.77%), and dry (77.06%), biomasses in comparison to the plants that were vegetated in tap water (Control 2). It has been reported by earlier studies that hydrocarbon pollution significantly affected the growth of plants during rhizoremediation of petroleum hydrocarbons [56,57].

Table 3. Inoculated bacterial effect on biomass, root length, and shoot length of *Cyperus laevigatus* in using floating treatment wetlands.

Treatments	Fresh Biomass (g)		Dry Biomass (g)		Length (cm)	
	Root	Shoot	Root	Shoot	Root	Shoot
C2	302^b(7.6)	454^a(33.7)	191^b(7.8)	306^a(11.8)	31.4^c(1.1)	56.2^a(2.1)
T2	91^f(3.8)	198^d(6.3)	46^f(1.3)	68^e(3.1)	14.4^f(0.8)	28.6^d(1.8)
T3	109^e(6.7)	218^c(13.8)	78^d(4.9)	93^c(4.2)	18.5^e(1.2)	34.9^b(2.2)

C2: Microcosm containing tap water and plants; T2: Microcosm containing diesel oil polluted water and plants; T3: Microcosm containing diesel oil polluted water, plants and bacterial consortium. All value is a mean of triplicate determination. Standard deviation among three replicates is existing in parenthesis and the alphabets represent the significant/non-significant difference among the treatments.

It has been proposed that by the absorption of toxic hydrocarbons by plants, the reduction occurs in uptake of water and growth of plants, which are ascribed to chlorosis, oxygen depletion, and dryness in vegetated plants [49,58]. However, the treatment containing *Cyperus laevigatus* L and bacterial consortium (T3) exhibited lesser root length (41.08%), shoot length (37.90%), fresh biomass (56.74%), and dry biomass (65.59%) in the context of the control irrigated with tap water.

It has been described earlier that specific bacteria, especially those involved in hydrocarbon degradation, have the capability to decrease the toxicity of organic pollutant in hydrocarbons contaminated water, which is directly attributed to effective growth of plants and their biomass [59]. Existence of plant growth stimulating and hydrocarbons degrading bacteria, exist interior and exterior of the plant tissues, emulsify the hydrocarbons, and make their availability easy for bacteria to degrade them into compounds that can be utilized by the plants, ultimately diminishing the toxic effect of hydrocarbons for better growth and biomass production of plants [60,61].

3.7. Reduction of Toxicity

After the completion of the experiment, the level of remediation of hydrocarbons was further confirmed by the exposure of fish to treated water (Table 4). Fish toxicity testing was performed in order to evaluate the effectiveness of FTWs in improving water quality to a level that also becomes safe for living organisms. In the FTWs system, the treated water with *Cyperus laevigatus* L and the bacterial consortium (T3) showed less toxification. In treated water of T3 treatment, only two fish died out of 10 exposed to hydrocarbon contaminated water.

Table 4. Evaluation of toxicity of diesel oil contaminated water detoxified by floating treatment wetlands.

Treatments	Fish Death Time				Total Death	Detoxification Position
	24 h	48 h	72 h	96 h		
Control	10	0	0	0	10/10[a]	Negligible
T1	2	1	1	1	5/10[b]	Partial
T2	1	1	1	1	3/10[c]	Partial
T3	1	1	0	0	2/10[d]	Complete

Control: Microcosm containing diesel oil contaminated water and no plants; T1: Microcosm containing diesel oil polluted water and bacterial consortium; T2: Microcosm containing diesel oil contaminated water and plants; T3: Microcosm containing diesel oil polluted water and bacterial consortium. The alphabets represent the significant/non-significant difference among the treatments.

For assessment of toxicity level among different treatments, it was observed that T1 and T2 treatments exhibited death of 5 and 3 fish out of 10 fish, respectively; nevertheless, after 24 h duration, fish were entirely dead in the control. The survival of fish in T2 and T3 treatments indicated the detoxification and pollutant reduction in the hydrocarbon contaminated wastewater. Besides, it was observed that presence of hydrocarbon degrading bacterial strains in FTWs excellently assisted in enhancement of water quality and decrease in toxicity of the contaminated water. Similar investigations have been reported in earlier studies that combined use of plant and bacteria is a more active methodology in the detoxification of the polluted water than individual use of plant and bacteria [30,62]. Due to the presence of a higher concentration of hydrocarbons in the control, oxidative stress increased which resulted in chronic cellular DNA damage, so a number of fish died in the untreated wastewater [63].

4. Conclusions

The FTWs augmented with bacteria is proven efficient among all the phytoremediation techniques. Hydrocarbons degrading bacterial strains and the *Cyperus laevigatus* L plant improved the diesel oil remediation in FTWs. This study investigates and justifies a way to remediate the diesel oil pollution in water. *Cyperus laevigatus* L plant can be the appropriate choice in FTWs for phytoremediation of diesel oil contaminated water. The performance and effectiveness of the developed FTWs were proven by successful reduction in hydrocarbons, COD, BOD, and TOC of diesel oil contaminated water. Though the experiment was completed in FTWs consisting of a microcosm set up, it displayed an effective bacterial association together with wetland plants. The study highlights a very supportable application of FTWs for conceivable removal of hydrocarbons from contaminated wastewater. This study further offers the prospects of evaluating the effectiveness of FTWs and is a promising option in the wastewater treatment at pilot and field scale levels. Furthermore, FTWs technology involves natural ways of treating wastewater and minimum energy is required for its operational cost. Application of this FTWs technology in developing countries like Pakistan is more feasible and cost effective in contrast to expensive technologies used for wastewater treatment worldwide.

Author Contributions: Conceptualization, M.F., S.A., S.R.A., M.A. (Muhammad Afzal), M.A. (Muhammad Arslan), E.F.A.A., M.N.A. and P.A.; Data curation, M.F., G.S. and M.A. (Muhammad Afzal); Formal analysis, M.F., G.S., S.R.A., T.Y., A.H. (Afzal Hussain) and M.A. (Muhammad Afzal); Funding acquisition, M.A. (Muhammad Afzal), A.H. (Abeer Hashem), E.F.A.A. and P.A.; Investigation, S.A., S.R.A., M.A. (Muhammad Arslan), M.N.A. and P.A.; Methodology, M.F., T.Y., A.H. (Abeer Hashem), A.H. (Afzal Hussain) and P.A.; Project administration, S.A., T.Y. and E.F.A.A.; Resources, M.A. (Muhammad Afzal), M.A. (Muhammad Arslan), A.H.(Abeer Hashem), E.F.A.A., M.N.A. and P.A.; Software, G.S. and A.H. (Afzal Hussain); Supervision, S.A., T.Y. and M.A. (Muhammad Afzal); Validation, G.S.; Visualization, M.N.A.; Writing—original draft, M.F.; Writing—review & editing, S.A., S.R.A., M.A. (Muhammad Afzal) and M.A. (Muhammad Arslan). All authors have read and agreed to the published version of the manuscript.

Funding: This research was supported by the Higher Education Commission, Pakistan (grant number 20-3854/R&D/HEC/14). The authors would like to extend their sincere appreciation to the Researchers Supporting Project Number (RSP-2019/116), King Saud University, Riyadh, Saudi Arabia.

Acknowledgments: This research was supported by the Higher Education Commission, Pakistan (grant number 20-3854/R&D/HEC/14). The authors would like to extend their sincere appreciation to the Researchers Supporting Project Number (RSP-2019/116), King Saud University, Riyadh, Saudi Arabia.

Conflicts of Interest: The authors declare no conflicts of interest.

References

1. Li, H.; Hao, H.; Yang, X.; Xiang, L.; Zhao, F.; Jiang, H.; He, Z. Purification of refinery wastewater by different perennial grasses growing in a floating bed. *J. Plant Nutr.* **2012**, *35*, 93–110. [CrossRef]

2. Tony, M.A.; Purcell, P.J.; Zhao, Y. Oil refinery wastewater treatment using physicochemical, Fenton and Photo-Fenton oxidation processes. *J. Environ. Sci. Health* **2012**, *47*, 435–440. [CrossRef]

3. Daflon, S.D.A.; Guerra, I.L.; Reynier, M.V.; Botta, C.R.; Campos, J.C. Toxicity identification and evaluation of a refinery wastewater from Brazil (Phase I). *Ecotoxicol. Environ. Contam.* **2015**, *10*, 41–45. [CrossRef]

4. Shahriari, T.; Karbassi, A.; Reyhani, M. Treatment of oil refinery wastewater by electrocoagulation–flocculation (Case Study: Shazand Oil Refinery of Arak). *Int. J. Environ. Sci. Technol.* **2018**, *16*, 4159–4166. [CrossRef]

5. Hernández-Francisco, E.; Peral, J.; Blanco-Jerez, L. Removal of phenolic compounds from oil refinery wastewater by electrocoagulation and Fenton/photo-Fenton processes. *J. Water Proc. Eng.* **2017**, *19*, 96–100. [CrossRef]

6. Agarry, S.E.; Oghenejoboh, K.M.; Latinwo, G.K.; Owabor, C.N. Biotreatment of petroleum refinery wastewater in vertical surface-flow constructed wetland vegetated with *Eichhornia crassipes*: Lab-scale experimental and kinetic modelling. *Environ. Technol.* **2018**, 1–21. [CrossRef]

7. Mandri, T.; Lin, J. Isolation and characterization of engine oil degrading indigenous microrganisms in Kwazulu-Natal, South Africa. *Afr. J. Biotechnol.* **2007**, *7*, 1927–1932.

8. Mustapha, H.I.; Van Bruggen, J.; Lens, P. Vertical subsurface flow constructed wetlands for polishing secondary Kaduna refinery wastewater in Nigeria. *Ecol. Eng.* **2015**, *84*, 588–595. [CrossRef]

9. Fakhru'l-Razi, A.; Pendashteh, A.; Abdullah, L.C.; Biak, D.R.A.; Madaeni, S.S.; Abidin, Z.Z. Review of technologies for oil and gas produced water treatment. *J. Hazard. Mater.* **2009**, *170*, 530–551. [CrossRef]

10. Chen, Z.; Reiche, N.; Vymazal, J.; Kuschk, P. Treatment of water contaminated by volatile organic compounds in hydroponic root mats. *Ecol. Eng.* **2017**, *98*, 339–345. [CrossRef]

11. Shahid, M.J.; Arslan, M.; Siddique, M.; Ali, S.; Tahseen, R.; Afzal, M. Potentialities of floating wetlands for the treatment of polluted water of river Ravi, Pakistan. *Ecol. Eng.* **2019**, *133*, 167–176. [CrossRef]

12. Tara, N.; Arslan, M.; Hussain, Z.; Iqbal, M.; Khan, Q.M.; Afzal, M. On-site performance of floating treatment wetland macrocosms augmented with dye-degrading bacteria for the remediation of textile industry wastewater. *J. Clean. Prod.* **2019**, *217*, 541–548. [CrossRef]

13. Chen, Z.; Kuschk, P.; Reiche, N.; Borsdorf, H.; Kästner, M.; Köser, H. Comparative evaluation of pilot scale horizontal subsurface-flow constructed wetlands and plant root mats for treating groundwater contaminated with benzene and MTBE. *J. Hazard. Mater.* **2012**, *209*, 510–515. [CrossRef] [PubMed]

14. Wang, C.-Y.; Sample, D.J. Assessment of the nutrient removal effectiveness of floating treatment wetlands applied to urban retention ponds. *J. Environ. Manag.* **2014**, *137*, 23–35. [CrossRef] [PubMed]

15. Headley, T.R.; Tanner, C.C. Constructed wetlands with floating emergent macrophytes: An innovative stormwater treatment technology. *Crit. Rev. Environ. Sci. Technol.* **2012**, *42*, 2261–2310. [CrossRef]

16. Tanner, C.C.; Headley, T.R. Components of floating emergent macrophyte treatment wetlands influencing removal of stormwater pollutants. *Ecol. Eng.* **2011**, *37*, 474–486. [CrossRef]

17. Chen, Z.; Cuervo, D.P.; Müller, J.A.; Wiessner, A.; Köser, H.; Vymazal, J.; Kästner, M.; Kuschk, P. Hydroponic root mats for wastewater treatment—A review. *Environ. Sci. Pollut. Res.* **2016**, *23*, 15911–15928. [CrossRef]

18. Faulwetter, J.; Burr, M.D.; Cunningham, A.B.; Stewart, F.M.; Camper, A.K.; Stein, O.R. Floating treatment wetlands for domestic wastewater treatment. *Water Sci. Technol.* **2011**, *64*, 2089–2095. [CrossRef]

19. Ghosh, A.; Dastidar, M.G.; Sreekrishnan, T. Bioremediation of Chromium Complex Dye by Growing Aspergillus flavus. In *Water Quality Management*; Springer: New York, NY, USA, 2018; pp. 81–92.

20. Sessitsch, A.; Mitter, B.; Compant, S.; Trognitz, F.; Brader, G. The plant microbiome: Ecology and functioning of bacterial endophytes and how plants can benefit. In Proceedings of the EGU General Assembly Conference Abstracts, Vienna, Austria, 8–13 April 2018; p. 2201.

21. Gomes, M.; Gonzales-Limache, E.; Sousa, S.; Dellagnezze, B.; Sartoratto, A.; Silva, L.; Gieg, L.; Valoni, E.; Souza, R.; Torres, A. Exploring the potential of halophilic bacteria from oil terminal environments for biosurfactant production and hydrocarbon degradation under high-salinity conditions. *Int. Biodeterior. Biodegrad.* **2018**, *126*, 231–242. [CrossRef]

22. Lopes, P.R.M.; Montagnolli, R.N.; Cruz, J.M.; Claro, E.M.T.; Bidoia, E.D. Biosurfactants in Improving Bioremediation Effectiveness in Environmental Contamination by Hydrocarbons. In *Microbial Action on Hydrocarbons*; Springer: New York, NY, USA, 2018; pp. 21–34.

23. Afzal, M.; Khan, Q.M.; Sessitsch, A. Endophytic bacteria: Prospects and applications for the phytoremediation of organic pollutants. *Chemosphere* **2014**, *117*, 232–242. [CrossRef]

24. Fatima, K.; Afzal, M.; Imran, A.; Khan, Q.M. Bacterial rhizosphere and endosphere populations associated with grasses and trees to be used for phytoremediation of crude oil contaminated soil. *Bull. Environ. Contam. Toxicol.* **2015**, *94*, 314–320. [CrossRef]

25. Arslan, M.; Afzal, M.; Amin, I.; Iqbal, S.; Khan, Q.M. Nutrients can enhance the abundance and expression of alkane hydroxylase CYP153 gene in the rhizosphere of ryegrass planted in hydrocarbon-polluted soil. *PLoS ONE* **2014**, *9*, e111208. [CrossRef]

26. Rehman, K.; Imran, A.; Amin, I.; Afzal, M. Enhancement of oil field-produced wastewater remediation by bacterially-augmented floating treatment wetlands. *Chemosphere* **2019**, *217*, 576–583. [CrossRef] [PubMed]

27. APHA; AWWA; WEF. *Standard Methods for the Examination of Water and Wastewater*; American Public Health Association: Washington, DC, USA, 2012.

28. Afzal, M.; Yousaf, S.; Reichenauer, T.G.; Sessitsch, A. The inoculation method affects colonization and performance of bacterial inoculant strains in the phytoremediation of soil contaminated with diesel oil. *Int. J. Phytoremediat.* **2012**, *14*, 35–47. [CrossRef] [PubMed]

29. Shahid, M.J.; Tahseen, R.; Siddique, M.; Ali, S.; Iqbal, S.; Afzal, M. Remediation of polluted river water by floating treatment wetlands. *Water Supply* **2019**, *19*, 967–977. [CrossRef]

30. Shehzadi, M.; Afzal, M.; Khan, M.U.; Islam, E.; Mobin, A.; Anwar, S.; Khan, Q.M. Enhanced degradation of textile effluent in constructed wetland system using Typha domingensis and textile effluent-degrading endophytic bacteria. *Water Res.* **2014**, *58*, 152–159. [CrossRef] [PubMed]

31. Geetha, S.; Banat, I.M.; Joshi, S.J. Biosurfactants: Production and potential applications in microbial enhanced oil recovery (MEOR). *Biocatal. Agric. Biotechnol.* **2018**, *14*, 23–32.

32. Varjani, S.J.; Joshi, R.R.; Kumar, P.S.; Srivastava, V.K.; Kumar, V.; Banerjee, C.; Kumar, R.P. Polycyclic aromatic hydrocarbons from petroleum oil industry activities: Effect on human health and their biodegradation. In *Waste Bioremediation*; Springer: New York, NY, USA, 2018; pp. 185–199.

33. Jambon, I.; Thijs, S.; Weyens, N.; Vangronsveld, J. Harnessing plant-bacteria-fungi interactions to improve plant growth and degradation of organic pollutants. *J. Plant Interact.* **2018**, *13*, 119–130. [CrossRef]

34. Singh, P.; Borthakur, A. A review on biodegradation and photocatalytic degradation of organic pollutants: A bibliometric and comparative analysis. *J. Clean. Prod.* **2018**, *196*, 1669–1680. [CrossRef]

35. Turkovskaya, O.; Muratova, A. Plant–Bacterial Degradation of Polyaromatic Hydrocarbons in the Rhizosphere. *Trends Biotechnol.* **2019**, *37*, 926–930. [CrossRef]

36. Kumari, S.; Regar, R.K.; Manickam, N. Improved polycyclic aromatic hydrocarbon degradation in a crude oil by individual and a consortium of bacteria. *Bioresour. Technol.* **2018**, *254*, 174–179. [CrossRef] [PubMed]

37. Zhang, L.; Zhao, J.; Cui, N.; Dai, Y.; Kong, L.; Wu, J.; Cheng, S. Enhancing the water purification efficiency of a floating treatment wetland using a biofilm carrier. *Environ. Sci. Pollut. Res.* **2016**, *23*, 7437–7443. [CrossRef] [PubMed]

38. Adeboye, P.T.; Bettiga, M.; Olsson, L. The chemical nature of phenolic compounds determines their toxicity and induces distinct physiological responses in *Saccharomyces cerevisiae* in lignocellulose hydrolysates. *Amb Express* **2014**, *4*, 46. [CrossRef]

39. Dhote, M.; Kumar, A.; Jajoo, A.; Juwarkar, A. Assessment of hydrocarbon degradation potentials in a plant–microbe interaction system with oil sludge contamination: A sustainable solution. *Int. J. Phytoremediat.* **2017**, *19*, 1085–1092. [CrossRef]

40. Pawlik, M.; Cania, B.; Thijs, S.; Vangronsveld, J.; Piotrowska-Seget, Z. Hydrocarbon degradation potential and plant growth-promoting activity of culturable endophytic bacteria of *Lotus corniculatus* and *Oenothera biennis* from a long-term polluted site. *Environ. Sci. Pollut. Res.* **2017**, *24*, 19640–19652. [CrossRef]

41. Kösesakal, T.; Ünal, M.; Kulen, O.; Memon, A.; Yüksel, B. Phytoremediation of petroleum hydrocarbons by using a freshwater fern species *Azolla filiculoides* Lam. *Int. J. Phytoremediat.* **2016**, *18*, 467–476. [CrossRef]

42. Darajeh, N.; Idris, A.; Truong, P.; Abdul Aziz, A.; Abu Bakar, R.; Che Man, H. Phytoremediation potential of vetiver system technology for improving the quality of palm oil mill effluent. *Adv. Mater. Sci. Eng.* **2014**, *2014*. [CrossRef]

43. Zhang, C.-B.; Liu, W.-L.; Pan, X.-C.; Guan, M.; Liu, S.-Y.; Ge, Y.; Chang, J. Comparison of effects of plant and biofilm bacterial community parameters on removal performances of pollutants in floating island systems. *Ecol. Eng.* **2014**, *73*, 58–63. [CrossRef]

44. Mitter, E.K.; Kataoka, R.; de Freitas, J.R.; Germida, J.J. Potential use of endophytic root bacteria and host plants to degrade hydrocarbons. *Int. J. Phytoremediat.* **2019**, *21*, 928–938. [CrossRef]

45. Rehman, K.; Imran, A.; Amin, I.; Afzal, M. Inoculation with bacteria in floating treatment wetlands positively modulates the phytoremediation of oil field wastewater. *J. Hazard. Mater.* **2018**, *349*, 242–251. [CrossRef]

46. Ye, C.; Chen, D.; Hall, S.J.; Pan, S.; Yan, X.; Bai, T.; Guo, H.; Zhang, Y.; Bai, Y.; Hu, S. Reconciling multiple impacts of nitrogen enrichment on soil carbon: Plant, microbial and geochemical controls. *Ecol. Lett.* **2018**, *21*, 1162–1173. [CrossRef]

47. Wang, X.; Tang, C.; Severi, J.; Butterly, C.R.; Baldock, J.A. Rhizosphere priming effect on soil organic carbon decomposition under plant species differing in soil acidification and root exudation. *New Phytol.* **2016**, *211*, 864–873. [CrossRef]

48. Saleem, H.; Rehman, K.; Arslan, M.; Afzal, M. Enhanced degradation of phenol in floating treatment wetlands by plant-bacterial synergism. *Int. J. Phytoremediat.* **2018**, *20*, 692–698. [CrossRef]

49. Ijaz, A.; Imran, A.; ul Haq, M.A.; Khan, Q.M.; Afzal, M. Phytoremediation: Recent advances in plant-endophytic synergistic interactions. *Plant Soil* **2016**, *405*, 179–195. [CrossRef]

50. Cui, L.; Ouyang, Y.; Lou, Q.; Yang, F.; Chen, Y.; Zhu, W.; Luo, S. Removal of nutrients from wastewater with *Canna indica* L. under different vertical-flow constructed wetland conditions. *Ecol. Eng.* **2016**, *36*, 1083–1088. [CrossRef]

51. Bordoloi, N.; Konwar, B. Bacterial biosurfactant in enhancing solubility and metabolism of petroleum hydrocarbons. *J. Hazard. Mater.* **2009**, *170*, 495–505. [CrossRef]

52. Ijaz, A.; Iqbal, Z.; Afzal, M. Remediation of sewage and industrial effluent using bacterially assisted floating treatment wetlands vegetated with Typha domingensis. *Water Sci. Technol.* **2016**, *74*, 2192–2201. [CrossRef]

53. Saxena, G.; Purchase, D.; Mulla, S.I.; Saratale, G.D.; Bharagava, R.N. Phytoremediation of heavy metal-contaminated sites: Eco-environmental concerns, field studies, sustainability issues, and future prospects. *Rev. Environ. Contam. Toxicol.* **2019**, *249*, 71–131.

54. Pal, S.; Kundu, A.; Banerjee, T.D.; Mohapatra, B.; Roy, A.; Manna, R.; Sar, P.; Kazy, S.K. Genome analysis of crude oil degrading Franconibacter pulveris strain DJ34 revealed its genetic basis for hydrocarbon degradation and survival in oil contaminated environment. *Genomics* **2017**, *109*, 374–382. [CrossRef]

55. Stottmeister, U.; Wießner, A.; Kuschk, P.; Kappelmeyer, U.; Kästner, M.; Bederski, O.; Müller, R.; Moormann, H. Effects of plants and microorganisms in constructed wetlands for wastewater treatment. *Biotechnol. Adv.* **2003**, *22*, 93–117. [CrossRef]

56. Hussain, Z.; Arslan, M.; Malik, M.H.; Mohsin, M.; Iqbal, S.; Afzal, M. Integrated perspectives on the use of bacterial endophytes in horizontal flow constructed wetlands for the treatment of liquid textile effluent: Phytoremediation advances in the field. *J. Environ. Manag.* **2018**, *224*, 387–395. [CrossRef]

57. Zhou, W.; Yang, J.; Lou, L.; Zhu, L. Solubilization properties of polycyclic aromatic hydrocarbons by saponin, a plant-derived biosurfactant. *Environ. Pollut.* **2011**, *159*, 1198–1204. [CrossRef] [PubMed]

58. Gogosz, A.; Bona, C.; Santos, G.; Botosso, P. Germination and initial growth of Campomanesia xanthocarpa O. Berg.(Myrtaceae), in petroleum-contaminated soil and bioremediated soil. *Braz. J. Biol* **2010**, *70*, 977–986. [CrossRef] [PubMed]

59. Afzal, M.; Yousaf, S.; Reichenauer, T.G.; Sessitsch, A. Ecology of Alkane-Degrading Bacteria and Their Interaction with the Plant. *Mol. Microb. Ecol. Rhizosphere* **2013**, 975–989. [CrossRef]

60. Weyens, N.; van der Lelie, D.; Taghavi, S.; Vangronsveld, J. Phytoremediation: Plant–endophyte partnerships take the challenge. *Curr. Opin. Biotechnol.* **2009**, *20*, 248–254. [CrossRef]

61. Mesa, V.; Navazas, A.; González-Gil, R.; González, A.; Weyens, N.; Lauga, B.; Gallego, J.L.R.; Sánchez, J.; Peláez, A.I. Use of endophytic and rhizosphere bacteria to improve phytoremediation of arsenic-contaminated industrial soils by autochthonous Betula celtiberica. *Appl. Environ. Microbiol.* **2017**. [CrossRef]

62. Ijaz, A.; Shabir, G.; Khan, Q.M.; Afzal, M. Enhanced remediation of sewage effluent by endophyte-assisted floating treatment wetlands. *Ecol. Eng.* **2015**, *84*, 58–66. [CrossRef]

63. Bai, H.; Wu, M.; Zhang, H.; Tang, G. Chronic polycyclic aromatic hydrocarbon exposure causes DNA damage and genomic instability in lung epithelial cells. *Oncotarget* **2017**, *8*, 79034. [CrossRef] [PubMed]

sustainability

Article

Implementation of Floating Treatment Wetlands for Textile Wastewater Management: A Review

Fan Wei [1], Munazzam Jawad Shahid [2,*], Ghalia S. H. Alnusairi [3,4], Muhammad Afzal [5], Aziz Khan [1], Mohamed A. El-Esawi [6], Zohaib Abbas [2], Kunhua Wei [1], Ihsan Elahi Zaheer [2], Muhammad Rizwan [2] and Shafaqat Ali [2,7,*]

[1] Guangxi Key Laboratory of Medicinal Resources Protection and Genetic Improvement, Guangxi Botanical Garden of Medicinal Plants, Nanning 530005, China; fanwei_gx@163.com (F.W.); aziz.hzau@gmail.com (A.K.); divinekh@163.com (K.W.)

[2] Department of Environmental Sciences and Engineering, Government College University, Faisalabad 38000, Pakistan; zohaib.abbas83@gmail.com (Z.A.); ihsankhanlashari@gmail.com (I.E.Z.); mrazi1532@yahoo.com (M.R.)

[3] Department of Biology, College of Science, Jouf University, Sakaka 2014, Saudi Arabia; gshalnusairi@ju.edu.sa

[4] Department of Biology, College of Science, Princess Nourah Bint Abdulrahman University, Riyadh 11451, Saudi Arabia

[5] Soil and Environmental Biotechnology Division, National Institute of Biotechnology and Genetic Engineering, Faisalabad 38000, Pakistan; manibge@yahoo.com

[6] Botany Department, Faculty of Science, Tanta University, Tanta 31527, Egypt; mohamed.elesawi@science.tanta.edu.eg

[7] Department of Biological Sciences and Technology, China Medical University, Taichung 40402, Taiwan

* Correspondence: munazzam01@gmail.com (M.J.S.); shafaqataligill@gcuf.edu.pk or shafaqataligill@yahoo.com (S.A.)

Received: 12 June 2020; Accepted: 14 July 2020; Published: 19 July 2020

Abstract: The textile industry is one of the most chemically intensive industries, and its wastewater is comprised of harmful dyes, pigments, dissolved/suspended solids, and heavy metals. The treatment of textile wastewater has become a necessary task before discharge into the environment. The textile effluent can be treated by conventional methods, however, the limitations of these techniques are high cost, incomplete removal, and production of concentrated sludge. This review illustrates recent knowledge about the application of floating treatment wetlands (FTWs) for remediation of textile wastewater. The FTWs system is a potential alternative technology for textile wastewater treatment. FTWs efficiently removed the dyes, pigments, organic matter, nutrients, heavy metals, and other pollutants from the textile effluent. Plants and bacteria are essential components of FTWs, which contribute to the pollutant removal process through their physical effects and metabolic process. Plants species with extensive roots structure and large biomass are recommended for vegetation on floating mats. The pollutant removal efficiency can be enhanced by the right selection of plants, managing plant coverage, improving aeration, and inoculation by specific bacterial strains. The proper installation and maintenance practices can further enhance the efficiency, sustainability, and aesthetic value of the FTWs. Further research is suggested to develop guidelines for the selection of right plants and bacterial strains for the efficient remediation of textile effluent by FTWs at large scales.

Keywords: bacteria; floating treatment wetlands; plants; textile effluent

1. Introduction

The major sources of water pollution are industries, domestic discharges, urbanization, pesticides, fertilizers, and poorly managed farm wastes [1,2]. The textile industry significantly contributes to the economy of a country. However, it consumes a large amount of water, and thus generates a larger quantity of wastewater [3]. Textile industry wastewater contains harmful dyes, different pigments, oil, surfactants, heavy metals, sulphates, and chlorides [4]. All these pollutants unfavorably affect the quality of water and aquatic life.

Dyes are key constituents of textile effluent. Textile dyes are considered as one of the worst polluters of our environment, including water bodies and soils [5]. These dyes also have adverse effects on human health. The dyes in textile wastewaters are carcinogenic, mutagenic, and genotoxic for all life forms [6]. Dyes in wastewater hinder the sunlight reaching to water, and thus decrease photosynthetic activity, reduce transparency, and disturb the ecosystem [3,4]. Additionally, different chemicals are used in the textile industry and cause problems for life forms, as well as the environment upon direct contact with them [7]. Existing wastewater treatment technologies are inefficient for the removal of dyes and associated pollutants from wastewater because of their persistent nature and resistance to degradation [8].

Incompletely treated or untreated water is harmful to the environment and other living creatures [9,10]. All types of wastewater should be treated before dumping into open water bodies in order to minimize the spread of water pollution [11]. Textile wastewater can be treated by various methods based upon physical, chemical, and biological approaches. However, the by-products of these treatment processes can be toxic and difficult to dispose of safely [6,12]. Consequently, it is essential to devise and adopt an environmentally friendly and sustainable technique to treat textile wastewater.

Phytoremediation, i.e., use of plants to remove pollutants, is one of the best economical and sustainable approaches for wastewater treatment [13,14]. Plants can take up contaminants from water, soil, and air [15]. Over the past years, different plants have been used to remediate dyes from textile wastewater. Different plants species have different nutrients/pollutants removal potential, and could exhibit great phytoremediation and stress tolerance [15–17]. Along with the applications of plants, different eco-friendly mechanisms are now being adopted to treat textile wastewater, and they include plant seeds [18], bacteria [8], fungi [19,20], yeast [21,22], and microalgae [23]. Recently, helminths have also been used to degrade dyes, for example, the nematode *Ascans lumbncoides* and the cestode *Momezia expansa* have been found to reduce azo dyes by anaerobic methods [24,25].

Although dyes are resistant to degradation, many microorganisms can completely decolorize and mineralize them [26]. The application of bacteria is an efficient way to treat dyes, as they are not harmful for the environment. Different bacteria have a high ability to degrade different dyes; for example, *Pseudomonas* sp. and *Sphingomonas* sp. have been found useful in the degradation of dyes [3]. The specifically adaptive bacteria can produce reductase enzymes that can reductively cut the dyes in the presence of molecular oxygen [27]. In the current scenario, we must seek efficient, eco-friendly, and economical technologies to treat textile wastewater with a minimum generation of waste materials [3]. Application of plants and bacteria has become a sustainable approach for wastewater treatment [18].

2. Potential Pollutants in Textile Wastewater

Textile wastewater contains highly variable dyes that have structural varieties including basic, acidic, reactive, azo, metal complex, and diazo dyes [28]. Typical characteristics of textile effluents include high temperature, the extensive range of pH, chemical oxygen demand (COD), biological oxygen demand (BOD), heavy metals, and a variety of contaminants such as dyes, salts, surfactants, dissolved solids, and suspended solids (Table 1) [28–30].

Table 1. Characteristics of textile wastewater.

Reference	[12]	[31]	[4]	[32]	[29]	[33]	[34]	[35]	[36]	[37]	[38]
Country	India	India	Iraq	Canada	Pakistan	Pakistan	India	Pakistan	Pakistan	India	Pakistan
Temp (°C)	35–45		33–45	35–45	25.4	42			38		38
pH	6.0–10.0	9.2–11	5.5–10.5	6–10	8.5	12.93			7.8	10.7	8.8
EC (µS/cm)						8.07		7.1	8.4		8.2
DO (mg/L)					0.84						
Color (Pt–Co)	50–2500			50–2500	456	53 (m^{-1})		35.5 (m^{-1})	68 (m^{-1})		66 (m^{-1})
COD (mg/L)	150–10,000	465–1400	150–10,000	150–12,000	433.7	813	1090	471	493	1734	513
BOD (mg/L)	100–4000	130–820	100–4000	80–6000	224.6	422	141	249	190	1478	201
Total solids (mg/L)		3600–6540				5125		4961	5420		5420
TSS (mg/L)	100–5000	360–370	100–5000	15–8000	244	391	1004	391	324	6438	324
TDS (mg/L)	1800–6000	3230–6180	1500–6000	2900–3100	2570	4834		4569	5164	9060	5251
Total Alkalinity (mg/L)	500–800	1250–3160									
Hardness (mg/L)						410			380		
Total settleable solids (mg/L)						24			38		
Total Organic Carbon (mg/L)						301		166	230		201
TN (mg/L)					55.8						
TP (mg/L)					13						
Phenol (mg/L)									0.86		0.85
Chlorine (mg/L)				1000–6000		600					
Chlorides (mg/L)	1000–6000		200–6000		846				1382		1383
Free chlorine (mg/L)				<10							
TA (mg/L) as CaCo3			500–800								
TH (mg/L) as CaCo3											
TKN (mg/L)			70–80	70–80							
Phosphate (mg/L)				<10							16.4
Sulphates (mg/L)			500–700	600–1000	412				310		311
Sulphides (mg/L)			5–20								
Oil and grease (mg/L)			10–50	10–30	28						

Table 1. *Cont.*

Reference	[12]	[31]	[4]	[32]	[29]	[33]	[34]	[35]	[36]	[37]	[38]
Nitrogen (mg/L)									28.6		28.7
Zink (mg/L)			3–6	<10							
Nickel (mg/L)				<10		2.0		0.125	7.6		7.57
Manganese (mg/L)				<10							
Iron (mg/L)				<10		3.3		1.171	14.3		14.4
Copper (mg/L)			2–6	<10				0.503			
Boron (mg/L)				<10							
Arsenic (mg/L)				<10					0.025	0.90	
Silica (mg/L)				<15							
Mercury (mg/L)				<10							
Fluorine (mg/L)				<10							
Chromium (mg/L)			2–5			0.21		0.812	9.7	3.7	9.67
Potassium (mg/L)			30–50			858			242		
Sodium (mg/L)	610–2175		400–2175	7000		1656			1560		
Cadmium (mg/L)						0.27			0.88	0.80	0.88
Calcium (mg/L)						80.16			110		
Magnesium (mg/L)						48.6			65		
Sulfate (mg/L)						412.54					
Phosphate (mg/L)						10.08					
Nitrate (mg/L)						24					
Lead (mg/L)								0.880		0.40	
Phosphorous (mg/L)									16.4		
Aluminum (mg/L)									2.5		

EC: Electrical Conductivity; DO: Dissolved Oxygen; COD: Chemical Oxygen Demand; BOD: Biological Oxygen Demand; TSS: Total Suspended Solids; TDS: Total Dissolved Solids; TN: Total Nitrogen; TP: Total Phosphorus; TA: Total Alkalinity; TH: Total Hardness; TKN: Total Kjeldahl Nitrogen.

2.1. Dyes

Discharge of wastewater from the finishing and dying process in the textile sector is a substantial cause of environmental pollution [39]. Discharge of dying effluents in the environment is the primary cause of a significant decline in freshwater bodies [40]. Dyes are the substances that, when applied, give color to the substrate by altering the crystal structure of the colored materials. Textile industries extensively use extensive dyes primarily due to their capacity to bind with the textile fibers via formation of covalent bonds [41]. Moreover, dyes are those contaminants that are not only toxic, but they also can change the color of the wastewater [42]. The main environmental risk associated with their use is their subsequent loss during the dying process. Consequently, significant quantities of unfixed dyes are regrettably discharged into the wastewater. The release of toxic textile wastewater causes adverse health risks to humans, plants, animals, and micro-organisms [43].

Colored textile dyes not only degrade the water bodies, but also hamper the penetration of sunlight via water, which causes a decrease in the rate of photosynthesis and level of dissolved oxygen, thereby affecting the whole aquatic ecosystem [44]. Textile dyes are composed of two key elements, auxochromes and chromophores. Chromophores are responsible for coloring the dyes, while auxochromes provide chromophores with additional assistance [45]. Azo dyes are most commonly used among all textile dyes in coloring multiple substrates. They have the large molecular structure, and their degradation products are sometimes more toxic [46]. When they get adsorbed by the soil from the wastewater, they can easily alter the chemical and physical characteristics of the soil. It may lead to a reduction of flora in the surrounding environment. The presence of azo dyes in the soil for a longer period dramatically disturbs the productivity of the crops and also kills the beneficial microbes [44]. Different studies reported that textile dyes also act as carcinogenic, mutagenic, and toxic agents [47,48]. An increase in textile industry means more use of dyes that may lead to severe toxicity disturbing the surrounding environment. Textile dyes pose a major risk to healthy living due to their xenobiotic effects [7]. The textile sector releases huge concentrations of colored effluents into the water bodies without prior treatment. Therefore, saving water from pollutants and prior treatment of textile effluents has indeed received emerging attention.

2.2. Dissolved Solids

Textile wastewater is contaminated heavily with dissolved and suspended solids [28]. Total dissolved solids (TDS) are consistently associated with conductivity and salinity of the water. Estimation of solids in water is a vital factor in making it safe for drinking purposes [49]. The World Health Organization (WHO) sets a minimum limit of 500 mg/L for TDS and 2000 mg/L as a maximum limit [50]. A higher value of TDS corresponds to the extensive use of several human-made dyes [51]. In TDS, soluble salts usually exist as cations and anions. Slight changes in the physiochemical characteristics of wastewater completely change the nature of deposit and ions concentration in the bottom. Higher values of TDS result in extreme salinity upon discharging into the water streams used for irrigation [52]. Much higher values of TDS can significantly produce harmful impacts on the biological, chemical, and physical characteristics of water bodies [51].

2.3. Suspended Solids

Suspended solids are considered as major pollutants in textile wastewater. They contain phosphate, chlorides, and nitrates of K, Ca, Mg, Na, organic matter, carbonates, and bio-carbonates [53]. A higher concentration of suspended solids hinders the prolific transfer setup of oxygen between air and water. Excess of a suspended solid released from the textile effluents can block the breathing organs of aquatic animals [54]. Suspended solid in aquatic medium leads to increasing turbidity, which subsequently results in depletion of oxygen. Likewise, suspended solids also can restrict the necessary penetration of light into the aquatic system, which decreased the capability of various algae and different flora to produce oxygen and food. Suspended solids directly absorb the sunlight, which enhances the temperature of the water and, at the same time, reduces the amount of dissolved oxygen [28].

Durotoye et al. (2018) conducted a study to examine the quality of effluents discharged from the textile industry [55]. It was found that the total suspended solids (TSS) exceeded the set limits specified by the national standards for textile effluents by 10 to 110% in all analyzed samples. Similarly, Ubale and Salkar [56] also reported a higher value of TSS (1910 mg/L) in cotton textile effluents [56]. Discharge of untreated textile wastewater with a higher concentration of TSS may potentially be very toxic for all living organisms.

2.4. Heavy Metals

Effluents from the textile industries comprise of several organic and inorganic chemical, organic salts, dyes, and heavy metals [42]. Heavy metals are more evident and non-biodegradable when released into the surrounding environment. Heavy metals can easily accumulate in the food chain as well [57,58]. High non-biodegradability, toxicity, and biological enrichment of heavy metals pollution has gravely threatened the sustainability of the ecological system and human health [59]. High risk of deterioration in water quality is prominent due to the heavy metal pollution [60].

The existence of heavy metals that greatly characterizes textile effluents. Heavy metals present in untreated textile wastewater can easily accumulate into the bio-system leading to various health repercussions [61]. Discharge of untreated textile wastewater is primarily associated with the concentration of several heavy metals such as Arsenic (Ar), Copper (Cu), Zinc (Zn), Cadmium (Cd), Lead (Pb), Mercury (Hg), Nickel (Ni), Chromium (Cr), and many others [62]. Because of the health hazards of heavy metals, numerous regulations and standards have been introduced to avoid any accumulation of heavy metals that would otherwise be lethal to human. Unfortunately, the discharged untreated textile effluents exceed the admissible limits set for heavy metals, especially in developing countries. As reported by Noreen et al. [63] and Mulugeta and Tibebe [64] the discharge of heavy metals from textile wastewater was high as compared to the permissible limits. Similarly, Wijeyaratne and Wickramasinghe [65] also reported that the concentration of Cu and Zn were higher than the permissible limits. Therefore, it is a matter of extreme importance to remediate these metals from the textile effluents before their discharge into the surrounding environment in order to prevent water pollution.

3. Available Technologies for Treatment of Textile Effluent

Textile effluent can be treated by several chemical, biological, and physical methods and reused for irrigation and industrial processes [66,67]. All methods work in some ways, but they all have some constraints. Textile wastewater remediation techniques include, but are not limited to, filtration, chemical oxidation, flocculation, Fenton's reagent oxidation, foam flotation, fixed-film bioreactors, anaerobic digestion, and electrolysis [68,69]. Among these coagulation-flocculation are the most commonly used methods [70]. Coagulation is the addition of a coagulant into wastewater to treat it, and is also a popular method of textile wastewater removal [71]. Electrodialysis, reverse osmosis, and ion exchange process are some of the tertiary treatment processes for textile wastewater treatment [72]. Adsorption is remarkably known as an equilibrium separation process and is widely used to remove contaminants [73,74]. Furthermore, advanced chemical oxidation processes are also commonly used for such purposes [66]. In many effluent treatment plants, first chemicals are added to make the wastewater constituents biodegradable. Then biological methods are applied, as biological methods alone cannot treat textile wastewaters up to the standard [67].

In certain cases, a combination of two or more techniques can be used to improve water quality, such as the aeration and filtration after coagulation. Filtration is applied as a tertiary treatment to improve the quality of treated wastewater. Carbon filter and sand filters are used to eliminate fine suspended solids and residual colors [75] Recently, a combination of coagulation and ultrafiltration has been applied for better results [76].

Biological treatments include aerobic treatments (activated sludge, trickling filtration, oxidation, ponds, lagoons, and aerobic digestion) and anaerobic treatment (anaerobic digestion, septic tanks and lagoons) [72]. It also includes the treatment by fungal biomass, such as *Aspergillus fumigates*, effectively used to remove reactive dyes from textile wastewater [77]. A large number of microbes can degrade dyes, and this approach is gaining momentum [78]. Some of the techniques used previously for textile wastewater treatment, along with their disadvantages, are shown in Table 2.

Table 2. Various techniques for the treatment of textile wastewater and their drawbacks.

Type	Technique	Drawbacks	References
Chemical	Combined Electrocoagulation	The pH should be maintained below 6 during the process	[79,80]
	Coagulation and Adsorption by Alum	Increase the concentration of sulfate and sulfide	[81]
	Ozonation	It has low COD reduction capacity	[82]
	Chemical coagulation	It is a slow technique and large amount of sludge is produced	[83,84]
	Electrochemical oxidation	Secondary salt contamination	[66]
	Coagulation	Coagulants can be associated with diseases like cancer or Alzheimer's	[85,86]
	Electrochemical technology	Produce undesirable by-products that can be harmful for environment	[87,88]
	Ion exchange method	Not effective for all dyes	[89]
	Photochemical Sonolysis	Requires a lot of dissolved oxygen, high cost, and produces undesirable by-products	[90]
	Coagulation-photocatalytic treatment by nanoparticles	Sludge production, difficulty of light penetration in dark and colored wastewaters, high costs of nanoparticles preparation, and limited cycles of nanoparticles usage	[91]
	Fenton and Photo-Fenton process	Sludge production, accumulation of unused ferrous ions, and difficult maintenance of pH	[92]
Physical	Adsorption/filtration (commercially activated carbon)	High cost of materials, costly operation, may not work with certain dyes and metals, performance depends upon the material types	[11]
	Adsorption	It is a costly process	[93]
	Membrane based treatment	Membrane failing may happen, and costly method	[67,94]
	Pilot-scale bio-filter	Bio-filter has low efficiency to metabolize hydrophobic volatile organic compounds because of the massive transfer limitations	[95]
	Pressure-driven membranes	Sensitivity to fouling and scaling	[96,97]
Biological	Constructed wetlands	High retention time and large area required for establishment	[66,98]
	Use of White-rot fungi along with bioreactor	It has long hydraulic retention time and requires large reactors	[99,100]
	Microalgae	Conditions hard to maintain, selection of suitable algae is critical	[101]
	Duckweed and algae ponds	Inefficient removal of heavy metals	[102,103]

Though many of these technologies have excellent performance, they have many limitations [67,104]. Many physicochemical treatment options are costly because of the equipment [67]. Conventional treatment methods achieve incomplete removal of dyes and produce concentrated sludge, which causes issue of secondary disposal [72]. In flocculation, the floc is difficult to control, and sludge underneath can re-suspend solids in water [54]. Trickling filters also have drawbacks such as high capital cost and a heavy odor [105].

Comparatively, biological methods have various advantages. They are cost-effective, produce a smaller amount of sludge, and are eco-friendly [106,107]. Ecological engineering has the advantage of being cheap. They are also able to treat non-point source wastewater effluents [107,108]. Though some biological processes also have many limitations, such as they are somewhat lengthy processes, some dyes can be a non-biodegradable, and a large amount of heavy metals in wastewater may hamper the microbial growth [54].

4. Floating Treatment Wetlands for Textile Effluent Treatment

Constructed wetlands are engineered systems composed of emergent plants and microbes with tremendous potential to remediate wastewater. Microbes proved great potential in enhancing phytoremediation potential and tolerance of plants to various environmental stresses [109,110]. Floating treatment wetlands (FTWs) are an innovative variant of constructed wetlands that make use of floating macrophytes and microbes for treatment of wastewater [111,112]. The application of FTWs (Figure 1) is a practical, eco-friendly, sustainable, and economical approach for the treatment of wastewater [112,113]. In addition to their high economic importance [114,115], plants have a key role in wastewater treatment. Mats float on the water surface, and plants are grown on these mats in such a way that the plant's roots are completely submerged in water and the plant's aerial parts are above the water [116]. Vegetation is supported on buoyant mats, which make these mats easy to retrofit in any water body where they need to be used [112]. Mostly halophytic grasses are explicitly selected for FTWs due to their rhizome, which can trap air [108,117]. FTWs share properties of both a pond and a wetland system. There is a hydraulic gradient between the bottom of the pond and the plant roots, so that the pollutants are degraded, trapped, and/or filtered by the plant roots and associated bacteria [118]. FTWs make use of plants and associated biofilms to reduce the nutrients load; that is why they are described as biofilm reactors with plants [117].

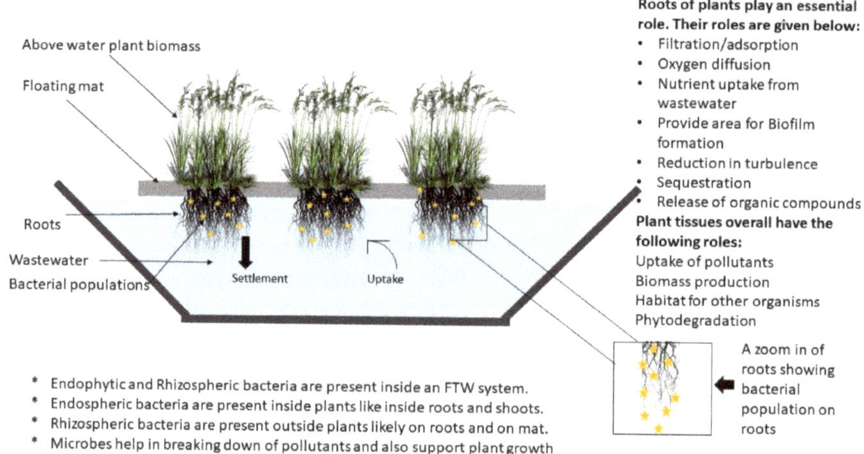

Figure 1. Schematic representation of floating treatment wetlands (FTWs) and pollutants removal process.

4.1. Role of Plants

The success of pollutant removal from water largely depends upon the selection of the plant species [119]. Plants play an essential role in the removal of pollutants from a water body. The roots of plants play a significant part in this process. The roots act as a physical filter in a FTWs system.

Roots filter the suspended particles in water and settle the filtrates at the bottom of the tank or water pond [111].

The key functions of plant in a FTW are:

1. Direct uptake of pollutants by the roots [120].
2. Extracellular enzyme production by roots [113].
3. Provide a surface area for the growth of biofilm [117].
4. Roots secrete root exudates that help in denitrification [121].
5. Suspended particles are entrapped in the roots [111].
6. Macrophytes also enhance flocculation of suspended matter [113].

Pollutant uptake by the roots of the plants is a significant process of pollutant removal from wastewater [122,123]. Dyes are phyto-transformed and then absorbed by the roots of plants [124]. The physical characteristics of the roots of plants and the nutrient uptake are interdependent/interlinked. The type of medium and nutrients in which the root exists specify the root's physical characteristics. In FTWs, the roots of the plants remain hanging in water and obtain their nutrition directly from the water. It leads to faster movement of nutrients and pollutants in the water towards the roots, thus leading to their accumulation in plant biomass. In a comparison between original plants and only plant roots in FTWs, the plants exhibited an excellent percentage of pollutant removal than artificial roots [111,125]. It suggests that the roots of the plants release some bioactive compounds in the water, and there is also a change in physicochemical processes in water. These bioactive compounds help in the change of metal species to an insoluble form, and it also enhances sorption characteristics of the biofilm, which help in pollutant removal from water [108]. Plants in FTWs support the activities of microbes already present in the wastewater as plant-microbe interactions play a prominent role in the treatment of water [113,126]. Plant roots provide spaces for the microbial growth that are necessary for water treatment [127].

Nitrogen and phosphorus are important pollutants of wastewater discharged by the textile industry. Several plants in FTWs have found efficient in removing total nitrogen (TN) and total phosphorus (TP) from wastewater [128]. Nitrogen is extracted from water by denitrification and sedimentation, while phosphorus is removed by plant uptake [129,130]. FTWs can also remove particle bind metals easily [130]. The ammonia-oxidizing bacteria and archaea on the rhizoplane have a major role in the nitrification and denitrification process [130,131].

Heavy metals are also present in textile wastewater. Macrophytes can take up these metals from the contaminated water effectively. *Phragmites australis* has excellent capacity for heavy metals removal from water, which is also a significant constituent of textile effluents [132]. FTWs vegetated with *P. australis* achieved 87–99% removal of heavy metals from textile wastewater [121]. The vegetation and floating mats minimize the penetration of sunlight in the water and stop the production of algal blooms in the water [133].

4.2. Role of Microorganism

Microbes are a key component of the biogeochemical cycle and energy flow in the aquatic ecosystem [134]. Microbes can decompose and demineralize the organic/inorganic pollutants and play a crucial role in pollutant removal from textile wastewater (Table 3) [135]. Microorganisms possess a different mechanism for the remediation of contaminated water, likely bio-sorption, bio-accumulation, bio-transformation, and bio-mineralization of organic and inorganic pollutants [127,135]. The presence of bacteria and their survival in FTWs, along with their activities, is mainly dependent on the type of plants [35]. In addition to the roots, mats also serve as a growth point for microbes [103]. These bacteria in FTWs can be rhizospheric and endophytic [136]. Rhizospheric bacteria reside outside the plant, and are sometimes attached to plant roots or on floating mats. Whereas endophytic bacteria reside inside the roots and shoots of plants [137]. The microorganisms present on roots and inside the plant tissues aid in the pollutant removal process of plants [138]. Microbes also promote plant growth by stimulating plant growth promoting activities like the release of indole-3-acetic acid, siderophore, and 1-amino-cyclopropane-1-carboxylic acid deaminase. They also solubilize inorganic phosphorous [138,139].

Bacteria have a unique ability to adhere and grow to almost every surface, and form complex communities termed biofilms [117]. The rhizoplane of FTWs release roots exudate to attract microbial cells to form biofilms and maintain large microbial biomass. In biofilms, bacterial cells grow in multicellular aggregates that are contained in an extracellular matrix, such as polysaccharide biopolymers together with protein and DNA produced by the bacteria [140]. This biofilm formation is very beneficial for bacteria themselves, such as resistance to many antimicrobial, protozoan, and environmental stresses [141].

In FTWs, different groups of bacteria have been identified; however, the nature and abundance of the bacterial community may vary depending upon the growth conditions, substrate, growth medium, and plant species [142]. In a study on floating wetland's plants with *Eichhorina crassipes*, 40 phyla of bacteria were identified, among these most common was Proteobacteria, followed by Actinobacteria, Bacteroidetes, and Cyanobacteria [143]. In another study, Actinobacteria were found dominant in water samples, while proteobacteria were the largest group of bacteria in roots and biofilms samples of a floating wetland planted with *Canna* and *Juncus* [142]. The second-largest group of bacteria found in water and roots samples was Cyanobacteria, but was not found in biofilm. The roots of floating macrophytes also harbored the sulfate-reducing and sulfur-oxidizing bacteria [144]. In FTWs, the production of reduced sulfide acts as potential phytotoxin and sulfur oxidizing bacteria contribute in the detoxification of plants [142,145]. The presence of nitrosamines on the plants roost also confirms the abundance of nitrifiers in the aquatic system that contribute to the ammonia-oxidation process [142]. The anoxic and anaerobic microbes ubiquitous on floating mats, soil, and roots contribute to denitrification and retain metals, and thus remove pollutants from contaminated water [146,147]. The metals acquired by bacteria can be sequestered through bioaccumulation and adsorption by binding to different functional groups such as carboxylate, hydroxyl, amino, and phosphate offered by cell walls [148].

Table 3. Application of bacteria for dye removal from textile wastewater.

Bacteria	Dye	Reference
Bacillus firmus	Reactive Blue 160	[149]
Oerskovia paurometabola	Acid Red 14	[150]
Pseudomonas aeruginosa and Thiosphaera pantotropha	Reactive Yellow 14	[151]
Enterobacter sp. CV–S1	Crystal Violet	[152]
Serratia sp. RN34	Reactive Yellow 2	[153]
Paracoccus sp. GSM2	Reactive Violet 5	[154]
Staphylococcus hominis RMLRT03	Acid Orange	[155]
Bacillus cereus RMLAU1	Orange II (Acid Orange 7)	[156]
Enterococcus faecalis strain ZL	Acid Orange 7	[157]
Pseudomonas aeruginosa strain BCH	Orange 3R (RO3R)	[158]
Anoxybacillus pushchinoensis, Anoxybacillus kamchatkensis and Anoxybacillus flavithermus	Reactive Black 5	[159]
Citrobacter sp. CK3	Reactive Red 180	[160]
Bacillus Fusiformis kmk 5	Disperse Blue 79 (DB79) and Acid Orange 10 (AO10)	[161]
Pseudomonas sp. SUK1	Red BLI	[162]
Brevibacillus sp.	Toluidine Blue dye (TB)	[163]
Bacterial strains 1CX and SAD4i	Acid Orange 7	[164]
Pseudomonas luteola	Azo Dye RP2B	[165]

5. Removal of Pollutants

5.1. Removal of Dissolved and Suspended Solids

Textile effluents usually contain a high concentration of total dissolved solids (TDS) as compared to the other industrial discharge mainly due to dying, bleaching, and fixing agent. TDS is correspondingly related to conductivity and salinity of the water [49,166]. Similarly, total suspended solids (TSS) consist of nitrates, phosphates, carbonates, and bicarbonates of K, Na, Mg, Ca, salt, organic matters, and other particles. The maximum concentration of TSS in textile effluents is due to the increased concentration of suspended particles, which increases the turbidity of the water [111,124]. It also erases the level of oxygen from the aqueous medium, resulting in the disturbance of principal food chain balance in the aquatic ecosystem [166]. The much higher value of TDS was observed in textile wastewater from an extended range of 1000–10,000 mg/L [167].

In general, total dissolved solids (TDS) and total suspended solids (TSS) are removed via the filtration and physical settling in FTWs. Plant roots have a crucial role in extracellular trapping of suspended solids and the pollutants in order to neutralize the risk and avoid cell injury [168]. In FTWs, plant roots provide a living, high surface area for the effective development of successive biofilms that hold several communities of micro-organisms responsible for entrapping and filtering of suspended particles [111,169]. The root-related network of biofilms has proven active in physical trapping of fine particulates [170]. The presence of disturbance-free atmosphere and unrestricted water layers among the floating roots provides idyllic conditions for sedimentation of particles [171]. Floating treatment wetlands demonstrated productive potential to remediate TDS, TS, TSS, and other suspended pollutants from various types of wastewater [139,172]. Tara et al. (2019) reported an effectual decline of TSS from 391 to 141 mg/L, TDS from 4569 to 1632, and TS from 4961 to 1733 by using FTWs for textile wastewater treatment [35]. Another report showed the achievement of FTWs applied for textile wastewater remediation, showing a significant decline in TDS and TSS after the end of the experiment [173]. The presence of microbial community directly affects the treatment performance of wetland treatment systems [174]. Key role of microbial communities in the effective removal of suspended solid particles is evident by different studies of FTWs [132,146,175].

5.2. Removal of Organic Matter

In textile wastewater substantial organic matter load is present in terms of biological oxygen demand (BOD) and chemical oxygen demand (COD). Wastewater effluents from the dyeing and printing systems are distinguished by significant BOD and COD fluctuations. Dye wastewater with high concentrations of COD and BOD will lead to eutrophication in the receiving water bodies, and raise environmental concerns about possible toxicity [176]. Various recent studies reported a high concentration of BOD and COD in textile wastewater effluents [177–179].

In different forms of wastewater, effective removal of organic matter by application of FTWs has been achieved. Darajeh et al. 2016 reported 96% and 94% reduction in BOD and COD from palm oil mill effluents treated with FTWs [180]. Queiroz et al. (2019) treated dairy wastewater by employing eleven different species of floating aquatic plants and observed a considerable reduction in both BOD and COD [71]. Recently, plant-bacteria partnership in FTWs proved to be very promising in the successful removal of organic matter. The maximum reduction in BOD is attributed to the microbial degradation of organic components coupled with the ample oxygen supply in the root zone [181]. Adsorption, sedimentation, and microbial degradation are the primary mechanisms for the effective removal of BOD [182,183]. Meanwhile, reduction of COD is credited to microbial degradation of substrate through plant roots [146,184]. Microbial activities are usually more vigorous in the root zone [158]. Plants roots provide an active settling medium and surface area for essential attachment and food for microbial population [185].

5.3. Removal of Heavy Metals

In FTWs, different processes such as adsorption, the formation of metal sulfide, direct accumulation by plants, algae, bacteria, and entrapments by biofilms in the roots zone play a promising role in successful remediation of heavy metals [123,130]. Various potentially toxic heavy metals settle down in the system bottom once they bind with minute clay particles in the roots zone [186,187]. Endophytic and rhizospheric microbes performed a variety of important chemical reactions, including adsorption, chelation, complexation, sulfide formation, and micro-precipitation, reduction-oxidation, and ion exchange [188]. Root exudates in the root zone speed up these reactions for the subsequent formation of metals hydroxide and sulfide, which in turn improve the sorption of trace heavy metals [189,190].

Recent studies have shown that successful inoculation of various degrading bacteria improves the efficiency of aquatic wetland plants in removing metal ions/metalloids from textile wastewater, resulting in safe disposal or reuse of treated wastewater [172,191]. When bacteria enter into plant tissues, they offer more productive effects for plants as compared to those bacteria present outside the plant body. Endophyte bacteria increase contaminant accumulation and reduce their phytotoxicity in the host plant by mineralizing recalcitrant elements that would be otherwise not degradable by plants [136]. Combining use of plant and endophyte bacteria is a promising approach in the remediation of heavy metals [126,192]. In line with the prospect mentioned above, Tara et al. (2018) determined the positive impacts of bacterial augmentation on two FTWs plants, *Phragmites australis* and *Typha domingensis*. Bacteria partnership with *T. domingensis* reduced copper to 0.009 mg/L, nickel to 0.034 mg/L, chromium to 0.101 mg/L, lead to 0.147 mg/L, and iron to 0.054 mg/L, while *P. australis* decreased copper to 0.007 mg/L, nickel to 0.027 mg/L, chromium to 0.032 mg/L, lead to 0.079 mg/L, and iron to 0.016 mg/L from their initial concentrations [35].

6. Factors Affecting the Performance of FTWs

6.1. Plant Selection

The selection of the right plants at floating mats is essential to achieve optimal remediation of pollutants. The plants for the vegetation of FTWs should be a non-invasive, native species, perennial, with a quick growth rate, extensive root system, high biomass yield, high tolerance to pollutants, and high ability to uptake and accumulate pollutants in above-ground parts, and which can grow in a hydroponic environment [118,193,194]. The roots' morphology, plant tolerance to pollutants, and root exudate profile play a major role in determining the plant's potential for phytoremediation [119]. Many kinds of grass are selected for phytoremediation due to their dense root structure that can harbor a vibrant microbial community. The production of root exudate and its quality also vary significantly even in closely related genotypes. It results in substantial differences in associated microbial community and their stimulation in the rhizoplane [195,196]. Thus, the selection of the right plant in FTWs increases the remediation performance, such as cattails (*Typha* spp.) are specially used for the treatment of acid mine drainage [197,198]. However, FTWs are planted with several species, and there is no precise pattern of using specific species for certain types of wastewater or pollutants [199,200]. In the past, various plant species have been used effectively in FTWs (Table 4).

Each plant species has a different phytoremediation potential and different metals uptake mechanisms such as accumulation, exclusion, translocation, osmoregulation, distribution, and concentration [201]. Different types of vegetation can be used in FTWs such as terrestrial, aquatic emergent, sub-emergent, and free-floating species. However, emergent plants are most widely used in FTWs due to their extensive root structure [201,202].

The terrestrial and emergent plant species have mostly long and extensive root structures as compared to free-floating aquatic plants, and provide ample surface area for the pollutant removal process [203,204]. The dense root structure and ability of plants to grow hydroponically are important to obtain maximum pollutant removal by FTWs [116].

Table 4. Use of various species of macrophytes in floating treatment wetlands.

Country	Plant Name	Wastewater	Removal Efficiency	Reference
Argentina	*Typha domingensis*	Synthetic runoff effluent	Achieved 95% removal of total phosphorus, soluble reactive phosphorus, NH_4^+ and NO_3^-	[205]
Australia	*Carex appressa*	Runoff from low density residential area	The pollutants removal performance was 80% for TSS, 53% for total phosphorus, 17% for total nitrogen	[206]
China	*Iris pseudacorus*	Synthetic secondary effluent	Achieved 89.4% removal of TN in one day retention time	[207]
China	*Cyperus ustulatus*	Domestic wastewater	The average removal efficiency for total microcystin-RR and microcystin-LR were 63.0% and 66.7%, respectively	[208]
Indonesia	*Chrysopogon zizanioides*	Textile wastewater	The average removal rate for chromium was 40%, BOD was 98.47%, and COD was 89.05%	[209]
Italy	*Phragmites australis, Carex elata, Juncus effusus, Typha latifolia, Chrysopogon zizanioides, Sparganium erectum,* and *Dactylis glomerata*	Resurgent water	The COD, BOD, and TP were reduced by 66%, 52%, and 65%, respectively	[210]
New Zealand	*Carex virgate*	Storm water	The pond with FTWs achieved 41% TSS, 40% particulate zinc, 39% copper, and 16% dissolved copper removal more than pond without FTWs	[111]
New Zealand	*Carex virgate*	Domestic wastewater	The removal rate for both TSS and BOD was more than 93%, TP and dissolve reactive phosphorus removal rate were 44.9% and 29.7%	[211]
Pakistan	*Phragmites australis*	Synthetic diesel oil contaminated water	The hydrocarbons concentration was reduced to 95.8%, COD to 98.6%, BOD to 97.7%, and phenol to 98.9%	[212]
Pakistan	*Phragmites australis, T. domingensis, Leptochloa fusca* and *Brachia mutica*	Oil contaminated stabilization pit	Reduced COD 97.4%, BOD 98.9%, TDS 82.4%, hydrocarbons 99.1%, and heavy metals 80%.	[108]
Pakistan	*Brachia mutica and Phragmites australis*	Oil field-produced wastewater	The COD, BOD, and oil contents reduced by 93%, 97%, and 97%, respectively	[139]
Pakistan	*Phragmites australis and Typha domingensis*	Textile wastewater	The color, COD, and BOD were reduced by 97%, 87%, and 92%, respectively	[35]
Pakistan	*Brachiaria mutica*	Sewage effluent	The COD, BOD, and oil contents were approximately reduced by 80%, 95%, and 50%	[112]
Pakistan	*Typha domingensis, Pistia stratiotes* and *Eichhornia crassipes*	Textile effluent	The average reduction rate for color, COD, and BOD was 57%, 72%, and 78%, respectively	[138]
Pakistan	*Phragmites australis, T. domingensis, Leptochloa fusca* and *Brachia mutica*	Oil contaminated stabilization pit	The COD, BOD, and TDS contents were reduced by 79%, 88%, and 65%	[213]
Sri Lanka	*Eichhornia crassipes*	Sewage water	The removal rate was 74.8% for TP and 55.8% for TN	[214]

Table 4. *Cont.*

Country	Plant Name	Wastewater	Removal Efficiency	Reference
Sri Lanka	*Typha angustifolia* and *Canna iridiflora*	Sewage wastewater	Achieved 80% reduction in BOD and NH_4^+-N, and 40% reduction in NO_3^--N	[215]
USA	*Spartina patens*	Synthetic marine aquaculture effluent	The TP concentration was dropped to ranging from 17–40%	[216]
USA	*Pontederia cordata* and *Schoenoplectus tabernaemontani*	Urban runoff	The TP and TN concentration were dropped to 60% and 40% in treated wastewater	[217]

It is well reported that plants with small root structure and slow growth rates are not suitable for phytofiltration [218]. The dense root structure also favors the bio-adsorption and biochemical mechanism essential for the pollutant removal process [219]. Although terrestrial plants demonstrated good potential for phytoremediation in the hydroponic system, they were not commonly used in FTWs [201]. Species with good potential for rhizo-filtration, such as *Brassica juncea* and *Helianthus annus*, can be used in FTWs. The most commonly used emergent plants genera/species are *Phragmites* (*Phragmites australis*), *Typha* (*Typha angustifolia, Typha latifolia*), *Scripus* (*Scripus lacustris, Scripus californicus*), *Juncus, Eleocharis, Cyperus*, and *Elode* [33,146,220]. Among all these *Phragmites australis* is the most frequently used species in free water surface wetlands followed by *Typha* (*T. angustifolia* and *T. latifolia*) [221]. The features such as perennial, flood-tolerant, toxic pollutants tolerant, extensive rhizome system, and rigid stems make it the best contestant for wetlands [204].

A combination of more than one species of plant has also been used many times to see the effect of using multiple species instead of one. Moreover, different plants have different pollutant capacities that vary from species to species. Under same conditions *Typha angustifolia* removes more nutrients from wastewater as compared to *Polygonum barbatum* [133]. *P. australis* produced the highest amount of biomass, followed by *T. domingensis, B. mutica, L. fusca, C. indica*, and *R. indica*, whereas *L. fusca* showed the highest plant density followed by *B. mutica, P. australis, T. domingensis, C. indica*, and *R. indica* [108,213]. Plants can uptake dyes, which are the principal constituent of textile wastewater. Previously, *Myriophyllum spicatum* and *Ceratophyllum demersum* species of plants efficiently removed dyes from synthetic textile wastewater [10].

6.2. Plant Coverage

Plant coverage on a floating mat has a prominent role in the wastewater remediation process. An increase or decrease in plant density may also increase or decrease the decontamination process. However, an increase in plant density does not equate with an increase in pollutant removal [118]. The increasing plant density will ultimately decrease the dissolved oxygen level of water under the floating mats. In a constructed wetland dominated by cattails and reeds, results indicated that microbial community and nitrate removal rates were high in wetlands with 50% plant coverage than 100% plant coverage [222]. Chance and White [223] reported that non-aerated floating wetlands with 100% planting coverage had a low dissolved oxygen level as compared to floating wetlands with 50% planting coverage. The dense plant coverage limits the gaseous exchange, which mostly occurs through the uncovered portion of the system [223]. There is little information in the literature on plant density, but plant coverage is suggested to be less than 80 percent for most FTWs [224].

6.3. Aeration and Dissolve Oxygen

In constructed wetlands, the dissolved oxygen level is an essential factor that can influence the pollutant removal process. In traditional wetlands, often the problem of insufficient oxygen supply and inappropriate oxygen distribution are found [225]. The atmospheric reaeration is one of the most important sources of oxygen supply in wetlands. Plants produce oxygen during the photosynthesis process, which can be released from plant leaves and roots into their surrounding environment [226]. The microbial degradation of organic matter can be achieved under both aerobic and anaerobic conditions. The aerobic degradation is mostly applied for less polluted wastewater to achieve high removal efficiency, and anaerobic conditions are favorable for the treatment of highly polluted wastewater [227]. It is well reported that higher oxygen contents in wetlands enhance the organic pollutant degradation process [228]. In wetlands, mostly oxygen is consumed by the organic matter degradation and left insufficient oxygen for the nitrification process and total nitrogen removal process [229]. The phosphorus removal bacteria in constructed wetlands can uptake more phosphorus in aerobic conditions as compared to anoxic conditions [230].

The leakage of oxygen from roots facilitates oxygen in FTWs. The extensive roots system, attached microbial communities, organic growth media, and organic pollutants under floating mats develop a substantial requirement of oxygen [223]. In addition, the photosynthesis process in water, gaseous exchange, and aeration may be reduced due to limited sunlight and air circulation, depending on the coverage area of the floating mat. It may lead to a low oxygen level under the floating mats [118]. It is widely reported that water under planted floating mats had a low dissolved oxygen level as compared to floating mats without plants or with artificial roots [111,116,217].

However, an increase in oxygen level does not mean an equal increase in the pollutant removal process. In some cases, the increasing level of oxygen in wetlands did not result in increased removal of total nitrogen and total phosphorus [231,232]. Similarly, in FTWs augmented with biofilm, the increased aeration improved the ammonium and phosphorus removal from polluted river water. In contrast, this increasing dissolved oxygen level decreased the denitrification process and overall total nitrogen removal [233]. In another study, while treating the nutrients enriched agricultural runoff, the aerated water column achieved less nitrogen and phosphorus removal as compared to the non-aerated water column [223]. Although aerated and non-aerated systems removed an almost similar amount of ammonia and nitrate, the aerated system showed higher uptake of nitrogen by plants than the non-aerated system. Park et al. (2019) treated the domestic wastewater through aerated and non-aerated FTWs coupled with biofilms and concluded that aerated FTWs with biofilms enhanced the organic matter, nitrogen, and *E.coli* removal [211]. Furthermore, it showed that FTWs can effectively perform under aerobic as well as anaerobic conditions.

6.4. Bacterial Inoculation

Plants-microbes interaction in FTWs has been widely studied [211,234,235], and signified the crucial role of plants-microbes interaction in mitigating the pollutants from wastewater. The plants and microbe interaction in wetlands largely depend upon plant species, availability of nitrogen, phosphorus, and various nutrients and minerals. Many plant species could cope with the adverse impacts of heavy metals and other abiotic stresses via regulating their antioxidants and nutrient uptake [236–238]. In FTWs, the hanging roots of the plants provide surface area for microbes' attachment and biofilm formation. Where these bacteria contribute to pollutant removal process and, in return, get organic carbon and oxygen from plants for their growth and survival [169]. Often, these symbiotic bacteria are not competent enough to remediate the diverse and potentially toxic pollutants from the wastewater [231]. The remediation potential of FTWs can be enhanced by inoculating the plants with purposefully isolated bacterial strains [232]. These inoculated bacteria not only enhance the pollutants remediation process, but also reduce the pollutants induced toxicity in plants and favor the plant growth by secreting multiple plant growth promoting hormones.

Rehman et al. [175] reported that the inoculation of FTWs with hydrocarbon-degrading bacteria enhanced the remediation of oil field contaminated water. Further, this plant-bacteria synergism improved the plant growth by reducing level of hydrocarbon induced toxicity in plants by producing siderophores and indole acetic acid and some other enzymes. Similarly, the inoculation of plant roots with rhizospheric and endophytic bacteria enhanced the removal of potentially toxic metals from the polluted river water and metals uptake and accumulation by plants [123]. Tara et al., 2019 applied FTWs vegetated with *P. australis* in combination with three dye degrading and plant growth promoting bacteria to treat textile effluent. The combined application of *P. australis* and bacteria enhanced the organic and inorganic pollutant removal and showed a reduction of 92% in COD, 91% in BOD, 86% in color, and 87% in trace metals [35]. Dyes degrading bacteria with the ability to degrade dyes can be isolated from the effluent of textile mills. Bacteria were isolated from textile effluent to degrade reactive dyes and it was found that three bacterial species, *Alcaligenes faecalis*, *Bacillus cereus*, and *Bacillus* sp., exhibited the potential to achieve more than 25% decolorization [239]. The bacteria can also be isolated from the plant parts to use as inoculum in FTWs for the degradation of textile effluent [38]. Some examples of successful application of bacteria in FTWs are given in Table 5.

Table 5. Inoculation of bacteria in floating treatment wetlands to enhance remediation potential.

Wastewater	Plant Specie	Inoculated Bacteria	Pollutant Removal	Retention Period	Reference
River water	*Phragmites australis, Typha domingensis, Brachia mutica, Leptochloa fusca*	*Aeromonas salmonicida, Pseudomonas indoloxydans, Bacillus cerus, Pseudomonas gessardii,* and *Rhodococcus* sp.	Significant reduction in trace metals contents (Fe, Mn, Ni, Pb, and Cr)	5 weeks	[123]
Diesel contaminated water (1%, w/v)	*Phragmites australis*	*Acinetobacter* sp. BRRH61, *Bacillus megaterium* RGR14 32, and *Acinetobacter iwoffii* AKR1	95.8% hydrocarbon, 98.6% chemical oxygen demand (COD), 97.7% biological oxygen demand (BOD), 95.2%, total organic carbon (TOC), 98.9% Phenol removal	3 months	[212]
Textile effluent	*Phragmites australis*	*Acinetobacter junii, Pseudomonas indoloxydans,* and *Rhodococcus* sp.	97% color, 87% COD, and 92% BOD removal	8 days	[38]
Oil contaminated water	*Phragmites australis T. domingensis Leptochloa fusca Brachiaria mutica* inoculated with bacteria	*Ochrobactrum intermedium* R2, *Microbacterium oryzae* R4, *Pseudomonas aeruginosa* R25, *P. aeruginosa* R21, *Acinetobacter* sp. LCRH81, *Klebsiella* sp. LCRI-87, *Acinetobacter* sp. BRSI56, *P. aeruginosa* BRRI54, *Bacillus subtilus* LORI66, and *Acinetobacter junii* TYRH47.	97.43% COD, 98.83% BOD, 82.4% TDS, 99.1% hydrocarbon content, and 80% heavy metal removal	18 months	[108]
Phenol contaminated water	*Typha domingensis*	*Acinetobacter lwofii* ACRH76, *Bacillus cereus* LORH97, and *Pseudomonas* sp. LCRH90	COD was reduced from 1057 to 97 mg/L; BOD5 from 423 to 64 mg/L, and TOC from 359 to 37 mg/L Phenol removal of 0.166 g/m2/day	15 days	[235]
River water	*Phragmites australis, Brachia mutica*	*Aeromonas Salmonicida, Bacillus cerus Pseudomonas indoloxydans, Pseudomonas gessardii,* and *Rhodococcus* sp.	85.9% COD, 83.3% BOD, and 86.6% TOC reduction, respectively	96 h	[146]
Oil field wastewater	*Brachiara mutica* and *Phragmites australis*	*Bacillus subtilis* LORI66, *Klebsiella* sp. LCRI87, *Acinetobacter Junii* TYRH47, *Acinetobacter* sp. LCRH81	97% COD 93%, and 97% BOD reduction, respectively	42 days	[139]
Oil field produced wastewater	*Typha domingensis*	*Bacillus subtilis* LORI66, *Klebsiella* sp. LCRI87, *Acinetobacter Junii* TYRH47, and *Acinetobacter* sp. BRSI56	95% Hydrocarbon, 90% COD, and 93% BOD content removal	42 days	[175]
Sewage effluent	*Brachiaria mutica*	*Acinetobacter* sp. strain BRSI56, *Bacillus cereus* strain BRSI57, and *Bacillus licheniformis* strain BRSI58	Reduction in COD, BOD, Total nitrogen (TN), and phosphate (PO_4)	8 days	[112]

7. Care and Maintenance of FTWs

FTWs can be constructed by using different types of materials including polyvinyl chloride (PVC) pipes, bamboo, polystyrene foam, wire mesh, fibrous material, and many more [116,127]. The most critical factors that should be considered while selecting appropriate material are buoyancy, durability, performance, eco-friendly, local availability, and cost [116,206]. In general, buoyancy is provided by floating mats/rafts, which also provide the base for plantation of vegetation. Sometimes plants can be grown in other structures such as wire mesh structure, and buoyancy can be provided by different materials such as PVC pipes [116,240]. The floating mats should be strong enough to support the load of plants, growth media, and be resistant to damage by sun, water, and heavy wind for long term sustainability. Floating mats should be designed to extend over the width of the retention pond, making a closed area between the inlets and FTWs for better flow distribution and to prevent short-circuiting [118,223].

The plants can be established on floating mats by direct seeding, planting cuttings and seedlings of the plants. The choice of method depends upon plant species, the structure of the floating mat, and environmental conditions, and availability of plants [118,171]. Direct seeding may be a cost-effective and rapid method for the vegetation of large-scale FTWs. The species, such as *Typha* and *Phragmites*, are commonly vegetated on floating mats through their cuttings [146]. The planting of seedlings may be an expensive approach in the sort-term, however, it results in rapid establishment of plants and a high growth rate. For plants establishment, selection of appropriate growth media is very important, especially during the initial stage of the plant vegetation. The most commonly used growth media are coconut fiber, peat, and soil [118,127]. The organic and inorganic fertilizers are also often applied to ensure better growth and development of plants on floating mats [136]. Care should be taken that growth media must have the ability to hold enough water for plant uptake and air circulation to maintain aerobic conditions, be resistant to waterlogging, and have ideal pH for plant growth [118,223].

It is suggested to avoid tall plants for FTWs, as during windy periods, these plants may cause the floating mat to drift, and laying of these plants in one direction may cause the salting or turnover of the floating mats [118,171]. Further, the plants with lose and large above-ground biomass should be avoided to limit the accumulation of dead plant biomass and release of accumulated pollutants in the water column [118]. Care should be taken while planting that plant roots should be able to reach the water column to ensure the availability of water during the initial stage of development [119,127]. After initial days of plantation, some plant may die due to unfavorable environmental conditions or toxicity of polluted water. Additionally, plants may die back during regular weather changes and by severe deoxygenation conditions below the floating mats. This issue can be solved by replanting the new plants at the free area of the floating mats [130]. Periodic trimming and harvesting of the plants may boost plant growth and prevent the accumulation of plant detritus and biomass on the floating mats [168].

Floating mats should be secured appropriately in the aquatic ponds to prevent drifting due to wind and waves [116]. The floating mats can be supported by fastening the floating mat's corners to the side of the ponds and anchoring them. Care must be taken to ensure that there should be some flexibility in the anchored ropes to adjust floating mats with changing water levels to the prevent sinking or submerging of floating mats during rising water level. In the windy area, the chances of over-turn of floating mats can be minimized by installing small floating mats rather than a large one with low height vegetation [118]. In a warmer climate, vegetation on floating mats may become a habitat for mosquito and other similar insects. This problem can be controlled by maintaining aerobic conditions in the pond, water spray on plants, frequent harvesting of the plants, and by use of approved chemical and biological control agents for these insects [241,242]. The periodic harvesting of the plants also improves the ability of plants to uptake nutrients and phosphorus from the polluted water.

The growth of invasive species on the FTWs may pose a potential issue for specific vegetation. The predominance of selected species can be maintained by regularly checking and pulling the weeds from the floating mats [243,244]. Regular monitoring of the FTWs is also vital to maintain the aesthetic value and prevent the clogging of inlet and outlet by the accumulation of plastic bottles, plant branches, and other non-biodegradable materials [118].

8. Conclusions and Recommendations

FTWs can be a viable option for remediation of textile wastewater as an alternative to costly and partially effective conventional wastewater treatment methods. The combined action of plants and associated biofilm in FTWs can efficiently remove the solid particles, organic matter, dyes, pigments, and heavy metals. The *P. australis* and *T. domingensis* have been widely used for FTWs and found efficient for remediation of textile effluent. FTWs are cost-effective, but need proper care and maintenance for long term performance. The harvesting of plants on floating mats can further boost the pollutant removal process and reduce the addition of litter and plant material in water.

The manipulation of characteristics such as plant selection, biofilms, plant coverage, and oxidation/aeration can be used further to enhance the remediation potential of FTWs. The development of guidelines for the right selections of plants for specific types of textile effluents can increase the success rate of FTWs. Further research is required to isolate and characterize the specific bacterial strains capable of colonizing the plants for remediation of textile effluent according to pollutant load. One of the main hindrances in the application of FTWs is the availability of land, which can be solved by the installation of FTWs on already existing water ponds.

Most of the studies conducted on the application of FTWs for treatment of textile effluent were on lab or pilot scale for a short duration. Therefore, it is suggested to research large scale application of FTWs for remediation of textile wastewater under natural environmental conditions. Further, the effect of weather should be deeply observed to analyze the performance of FTWs under changing temperature, precipitation, and other environmental conditions. The proper disposal of harvested plant biomass and litter also needs extensive research for safe disposal of extracted pollutants from the treated wastewater. The use of harvested grasses from the floating mat as the fodder of livestock needs careful investigation of nutritional values of plants, accumulated pollutants in plant parts, and the ultimate effect on animal and animal products and transportation in the food chain.

Author Contributions: This review article was written by F.W., M.J.S., S.A. and Z.A. The data was collected and coordinated by A.K., M.A.E.-E., K.W., and I.E.Z. The manuscript was reviewed, edited and revised by M.R., M.A., M.A.E.-E., and G.S.H.A. All authors have read and agreed to the published version of the manuscript.

Funding: This work was funded by Guangxi Natural Science Foundation (2018GXNSFBA294016), Guangxi Innovation-Driven Development Project (GuiKe AA18242040), "Guangxi Bagui Scholars" and Research Innovation Team Project (GuiYaoChuang2019005).

Acknowledgments: The authors are also grateful to the Higher Education Commission (HEC) Islamabad, Pakistan, for its support.

Conflicts of Interest: The authors declare no conflict of interest.

References

1. Owa, F. Water pollution: Sources, effects, control and management. *Mediterr. J. Soc. Sci.* **2013**, *4*, 65. [CrossRef]
2. Keiser, D.A. The missing benefits of clean water and the role of mismeasured pollution. *J. Assoc. Environ. Resour. Econ.* **2019**, *6*, 669–707. [CrossRef]
3. Júnior, M.A.C. Advances in the Treatment of Textile Effluents: A Review. *OALib. J.* **2019**, *6*, 1–13. [CrossRef]
4. Hussein, F.H. Chemical Properties of Treated Textile Dyeing Wastewater. *Asian J. Chem.* **2013**, *25*, 9393–9400. [CrossRef]
5. Khandare, R.V.; Govindwar, S.P. Phytoremediation of textile dyes and effluents: Current scenario and future prospects. *Biotechnol. Adv.* **2015**, *33*, 1697–1714. [CrossRef]
6. Ayadi, I.; Souissi, Y.; Jlassi, I.; Peixoto, F.; Mnif, W. Chemical synonyms, molecular structure and toxicological risk assessment of synthetic textile dyes: A critical review. *J. Dev. Drugs* **2016**, *5*, 2. [CrossRef]
7. Rovira, J.; Domingo, J.L. Human health risks due to exposure to inorganic and organic chemicals from textiles: A review. *Environ. Res.* **2019**, *168*, 62–69. [CrossRef] [PubMed]

8. Saratale, R.G.; Saratale, G.D.; Chang, J.-S.; Govindwar, S.P. Bacterial decolorization and degradation of azo dyes: A review. *J. Taiwan Inst. Chem. Eng.* **2011**, *42*, 138–157. [CrossRef]

9. Khan, S.; Malik, A. Environmental and health effects of textile industry wastewater. In *Environmental Deterioration and Human Health*; Springer: Berlin/Heidelberg, Germany, 2014; pp. 55–71.

10. Yaseen, D.A.; Scholz, M. Textile dye removal using experimental wetland ponds planted with common duckweed under semi-natural conditions. *Environ. Prot. Eng.* **2017**, *43*, 39–60. [CrossRef]

11. Crini, G.; Lichtfouse, E. Advantages and disadvantages of techniques used for wastewater treatment. *Environ. Chem. Lett.* **2019**, *17*, 145–155. [CrossRef]

12. Kalra, S.S.; Mohan, S.; Sinha, A.; Singh, G. *Advanced Oxidation Processes for Treatment of Textile And Dye Wastewater: A Review*; 2nd International Conference on Environmental Science and Development, 2011; IACSIT Press: Singapore, 2011; pp. 271–275.

13. Ashraf, S.; Ali, Q.; Zahir, Z.A.; Ashraf, S.; Asghar, H.N. Phytoremediation: Environmentally sustainable way for reclamation of heavy metal polluted soils. *Ecotoxicol. Environ. Saf.* **2019**, *174*, 714–727. [CrossRef] [PubMed]

14. Anning, A.K.; Akoto, R. Assisted phytoremediation of heavy metal contaminated soil from a mined site with *Typha latifolia* and *Chrysopogon zizanioides*. *Ecotoxicol. Environ. Saf.* **2018**, *148*, 97–104. [CrossRef] [PubMed]

15. Jeevanantham, S.; Saravanan, A.; Hemavathy, R.; Kumar, P.S.; Yaashikaa, P.; Yuvaraj, D. Removal of toxic pollutants from water environment by phytoremediation: A survey on application and future prospects. *Environ. Technol. Inno.* **2019**, *13*, 264–276. [CrossRef]

16. El-Esawi, M.A.; Al-Ghamdi, A.A.; Ali, H.M.; Ahmad, M. Overexpression of *AtWRKY30* Transcription Factor Enhances Heat and Drought Stress Tolerance in Wheat (*Triticum aestivum* L.). *Genes* **2019**, *10*, 163. [CrossRef] [PubMed]

17. El-Esawi, M.A.; Alayafi, A.A. Overexpression of rice *Rab7* gene improves drought and heat tolerance and increases grain yield in rice (*Oryza sativa* L.). *Genes* **2019**, *10*, 56. [CrossRef]

18. Khamparia, S.; Jaspal, D.; Malviya, A. Optimization of adsorption process for removal of sulphonated di azo textile dye. *OPTIMIZATION* **2015**, *1*, 61–66. [CrossRef]

19. Ali, N.; El-Mohamedy, R. Microbial decolourization of textile waste water. *J. Saudi Chem. Soc.* **2012**, *16*, 117–123. [CrossRef]

20. Wesenberg, D.; Kyriakides, I.; Agathos, S.N. White-rot fungi and their enzymes for the treatment of industrial dye effluents. *Biotechnol. Adv.* **2003**, *22*, 161–187. [CrossRef]

21. Feng, C.; Fang-yan, C.; Yu-bin, T. Isolation, identification of a halotolerant acid red B degrading strain and its decolorization performance. *Apcbee Procedia* **2014**, *9*, 131–139. [CrossRef]

22. Mahmoud, M. Decolorization of certain reactive dye from aqueous solution using Baker's Yeast (Saccharomyces cerevisiae) strain. *HBRC J.* **2016**, *12*, 88–98. [CrossRef]

23. El-Kassas, H.Y.; Mohamed, L.A. Bioremediation of the textile waste effluent by Chlorella vulgaris. *Egypt. J. Aqua. Res.* **2014**, *40*, 301–308. [CrossRef]

24. Rather, L.J.; Akhter, S.; Hassan, Q.P. Bioremediation: Green and sustainable technology for textile effluent treatment. In *Sustainable Innovations in Textile Chemistry and Dyes*; Springer: Berlin/Heidelberg, Germany, 2018; pp. 75–91.

25. Chung, K.T.; Stevens, S.E., Jr. Degradation azo dyes by environmental microorganisms and helminths. *Environ. Toxicol. Chem. Int. J.* **1993**, *12*, 2121–2132.

26. Sandesh, K.; Kumar, G.; Chidananda, B.; Ujwal, P. Optimization of direct blue-14 dye degradation by Bacillus fermus (KX898362) an alkaliphilic plant endophyte and assessment of degraded metabolite toxicity. *J. Hazard. Mater.* **2019**, *364*, 742–751.

27. Stolz, A. Basic and applied aspects in the microbial degradation of azo dyes. *Appl. Microbiol. Biotechnol.* **2001**, *56*, 69–80. [CrossRef]

28. Kumar, P.S.; Saravanan, A. Sustainable wastewater treatments in textile sector. In *Sustainable Fibres and Textiles*; Elsevier: Amsterdam, The Netherlands, 2017; pp. 323–346.

29. Nawaz, M.S.; Ahsan, M. Comparison of physico-chemical, advanced oxidation and biological techniques for the textile wastewater treatment. *Alex. Eng. J.* **2014**, *53*, 717–722. [CrossRef]

30. Dey, S.; Islam, A. A review on textile wastewater characterization in Bangladesh. *Resour. Environ.* **2015**, *5*, 15–44.

31. Kant, R. Textile dyeing industry an environmental hazard. *Nat. Sci.* **2011**, *4*, 22–26. [CrossRef]

32. Ghaly, A.; Ananthashankar, R.; Alhattab, M.; Ramakrishnan, V. Production, characterization and treatment of textile effluents: A critical review. *J. Chem. Eng. Process Technol.* **2014**, *5*, 1–19.

33. Shehzadi, M.; Afzal, M.; Khan, M.U.; Islam, E.; Mobin, A.; Anwar, S.; Khan, Q.M. Enhanced degradation of textile effluent in constructed wetland system using *Typha domingensis* and textile effluent-degrading endophytic bacteria. *Water Res.* **2014**, *58*, 152–159. [CrossRef]

34. Watharkar, A.D.; Khandare, R.V.; Waghmare, P.R.; Jagadale, A.D.; Govindwar, S.P.; Jadhav, J.P. Treatment of textile effluent in a developed phytoreactor with immobilized bacterial augmentation and subsequent toxicity studies on *Etheostoma olmstedi* fish. *J. Hazard. Mater.* **2015**, *283*, 698–704. [CrossRef]

35. Tara, N.; Iqbal, M.; Mahmood Khan, Q.; Afzal, M. Bioaugmentation of floating treatment wetlands for the remediation of textile effluent. *Water Environ. J.* **2019**, *33*, 124–134. [CrossRef]

36. Hussain, Z.; Arslan, M.; Malik, M.H.; Mohsin, M.; Iqbal, S.; Afzal, M. Integrated perspectives on the use of bacterial endophytes in horizontal flow constructed wetlands for the treatment of liquid textile effluent: Phytoremediation advances in the field. *J. Environ. Manag.* **2018**, *224*, 387–395. [CrossRef] [PubMed]

37. Kadam, S.K.; Watharkar, A.D.; Chandanshive, V.V.; Khandare, R.V.; Jeon, B.-H.; Jadhav, J.P.; Govindwar, S.P. Co-planted floating phyto-bed along with microbial fuel cell for enhanced textile effluent treatment. *J. Clean. Prod.* **2018**, *203*, 788–798. [CrossRef]

38. Tara, N.; Arslan, M.; Hussain, Z.; Iqbal, M.; Khan, Q.M.; Afzal, M. On-site performance of floating treatment wetland macrocosms augmented with dye-degrading bacteria for the remediation of textile industry wastewater. *J. Clean. Prod.* **2019**, *217*, 541–548. [CrossRef]

39. Bafana, A.; Devi, S.S.; Chakrabarti, T. Azo dyes: Past, present and the future. *Environ. Rev.* **2011**, *19*, 350–371. [CrossRef]

40. Essandoh, M.; Garcia, R.A. Efficient removal of dyes from aqueous solutions using a novel hemoglobin/iron oxide composite. *Chemosphere* **2018**, *206*, 502–512. [CrossRef]

41. Uddin, M.G.; Islam, M.M.; Islam, M.R. Effects of reductive stripping of reactive dyes on the quality of cotton fabric. *Fash. Text.* **2015**, *2*, 8. [CrossRef]

42. El Harfi, S.; El Harfi, A. Classifications, properties and applications of textile dyes: A review. *Appl. J. Environ. Eng. Sci.* **2017**, *3*, 311–320.

43. Benkhaya, S.; El Harfi, A. A critical review of surface water contaminated with dyes from textile industry effluent: Possible approaches. *Appl. J. Environ. Eng. Sci.* **2018**, *4*, 1–12.

44. Imran, M.; Shaharoona, B.; Crowley, D.E.; Khalid, A.; Hussain, S.; Arshad, M. The stability of textile azo dyes in soil and their impact on microbial phospholipid fatty acid profiles. *Ecotoxicol. Environ. Saf.* **2015**, *120*, 163–168. [CrossRef]

45. Uday, U.S.P.; Mahata, N.; Sasmal, S.; Bandyopadhyay, T.K.; Mondal, A.; Bhunia, B. Dyes Contamination in the Environment: Ecotoxicological Effects, Health Hazards, and Biodegradation and Bioremediation Mechanisms for Environmental Cleanup. In *Environmental Pollutants and Their Bioremediation Approaches*; CRC Press: Boca Raton, FL, USA, 2017; pp. 127–176.

46. Bokare, A.D.; Chikate, R.C.; Rode, C.V.; Paknikar, K.M. Iron-nickel bimetallic nanoparticles for reductive degradation of azo dye Orange G in aqueous solution. *Appl. Catal. B-Environ.* **2008**, *79*, 270–278. [CrossRef]

47. Ismail, A.; Toriman, M.E.; Juahir, H.; Zain, S.M.; Habir, N.L.A.; Retnam, A.; Kamaruddin, M.K.A.; Umar, R.; Azid, A. Spatial assessment and source identification of heavy metals pollution in surface water using several chemometric techniques. *Mar. Pollut. Bull.* **2016**, *106*, 292–300. [CrossRef]

48. Khatri, J.; Nidheesh, P.; Singh, T.A.; Kumar, M.S. Advanced oxidation processes based on zero-valent aluminium for treating textile wastewater. *Chem. Eng. J.* **2018**, *348*, 67–73. [CrossRef]

49. Bhatia, D.; Sharma, N.R.; Kanwar, R.; Singh, J. Physicochemical assessment of industrial textile effluents of Punjab (India). *Appl. Water Sci.* **2018**, *8*, 83. [CrossRef]

50. Maruthi, Y.; Rao, S.; Kiran, D. Evaluation of ground water pollution potential in Chandranagar, Visakhapatnam: A case study. *J. Ecobiol.* **2004**, *16*, 423–430.

51. Mohabansi, N.P.; Tekade, P.; Bawankar, S. Physico-chemical and microbiological analysis of textile industry effluent of Wardha region. *Water Res. Dev.* **2011**, *1*, 40–44.

52. Kolhe, A.; Pawar, V. Physico-chemical analysis of effluents from dairy industry. *Rec. Res. Sci. Technol.* **2011**, *3*, 29–32.

53. Elango, G.; Govindasamy, R. Removal of Colour from Textile Dyeing Effluent Using Temple Waste Flowers as Ecofriendly Adsorbent. *IOSR J. Appl. Chem.* **2018**, *11*, 19–28.

54. Bilotta, G.; Brazier, R. Understanding the influence of suspended solids on water quality and aquatic biota. *Water Res.* **2008**, *42*, 2849–2861. [CrossRef]

55. Durotoye, T.O.; Adeyemi, A.A.; Omole, D.O.; Onakunle, O. Impact assessment of wastewater discharge from a textile industry in Lagos, Nigeria. *Cogent Eng.* **2018**, *5*, 1531687. [CrossRef]

56. Ubale, M.A.; Salkar, V.D. Experimental study on electrocoagulation of textile wastewater by continuous horizontal flow through aluminum baffles. *Korean J. Chem. Eng.* **2017**, *34*, 1044–1050. [CrossRef]

57. Ismail, M.; Akhtar, K.; Khan, M.; Kamal, T.; Khan, M.A.; M Asiri, A.; Seo, J.; Khan, S.B. Pollution, toxicity and carcinogenicity of organic dyes and their catalytic bio-remediation. *Curr. Pharm. Des.* **2019**, *25*, 3645–3663. [CrossRef] [PubMed]

58. Sultana, T.; Arooj, F.; Nawaz, M.; Alam, S. Removal of Heavy Metals from Contaminated Soil using Plants: A Mini-Review. *PSM Biol. Res.* **2019**, *4*, 113–117.

59. Yin, K.; Wang, Q.; Lv, M.; Chen, L. Microorganism remediation strategies towards heavy metals. *Chem. Eng. J.* **2019**, *360*, 1553–1563. [CrossRef]

60. Rezaei, A.; Hassani, H.; Hassani, S.; Jabbari, N.; Mousavi, S.B.F.; Rezaei, S. Evaluation of groundwater quality and heavy metal pollution indices in Bazman basin, southeastern Iran. *Groundw. Sustain. Dev.* **2019**, *9*, 100245. [CrossRef]

61. Roque, F.; Diaz, K.; Ancco, M.; Delgado, D.; Tejada, K. Biodepuration of domestic sewage, textile effluents and acid mine drainage using starch-based xerogel from recycled potato peels. *Water Sci. Technol.* **2018**, *77*, 1250–1261. [CrossRef]

62. Järup, L. Hazards of heavy metal contamination. *Br. Med. Bull.* **2003**, *68*, 167–182. [CrossRef]

63. Noreen, M.; Shahid, M.; Iqbal, M.; Nisar, J. Measurement of cytotoxicity and heavy metal load in drains water receiving textile effluents and drinking water in vicinity of drains. *Measurement* **2017**, *109*, 88–99. [CrossRef]

64. Mulugeta, M.; Tibebe, D. Assessment of some selected metals from textile effluents in amhara region using AAS and ICPOES. *Assessment* **2019**, *7*, 27–31. [CrossRef]

65. Wijeyaratne, W.D.N.; Wickramasinghe, P.M.U. Treated textile effluents: Cytotoxic and genotoxic effects in the natural aquatic environment. *Bull. Environ. Contam. Toxicol.* **2020**, *104*, 245–252. [CrossRef]

66. Paździor, K.; Bilińska, L.; Ledakowicz, S. A review of the existing and emerging technologies in the combination of AOPs and biological processes in industrial textile wastewater treatment. *Chem. Eng. J.* **2019**, *376*, 120597. [CrossRef]

67. Holkar, C.R.; Jadhav, A.J.; Pinjari, D.V.; Mahamuni, N.M.; Pandit, A.B. A critical review on textile wastewater treatments: Possible approaches. *J. Environ. Manag.* **2016**, *182*, 351–366. [CrossRef] [PubMed]

68. Deng, D.; Aryal, N.; Ofori-Boadu, A.; Jha, M.K. Textiles wastewater treatment. *Water Environ. Res* **2018**, *90*, 1648–1662. [CrossRef] [PubMed]

69. Vandevivere, P.C.; Bianchi, R.; Verstraete, W. Treatment and reuse of wastewater from the textile wet-processing industry: Review of emerging technologies. *J. Chem. Technol. Biotechnol. Int. Res. Process Environ. Clean Technol.* **1998**, *72*, 289–302. [CrossRef]

70. Rodrigues, C.S.; Madeira, L.M.; Boaventura, R.A. Synthetic textile wastewaters treatment by coagulation/flocculation using ferric salt as coagulant. *Environ. Eng. Manag. J.* **2017**, *16*, 1881–1889.

71. Queiroz, R.D.C.S.D.; Lôbo, I.P.; de Ribeiro, V.S.; Rodrigues, L.B.; de Almeida Neto, J.A. Assessment of autochthonous aquatic macrophytes with phytoremediation potential for dairy wastewater treatment in floating constructed wetlands. *Int. J. Phytorem.* **2020**, *22*, 518–528. [CrossRef]

72. Chandran, D. A review of the textile industries waste water treatment methodologies. *Int. J. Sci. Eng. Res.* **2016**, *7*, 2229–5518.

73. Dąbrowski, A. Adsorption—From theory to practice. *Adv. Colloid Interface Sci.* **2001**, *93*, 135–224. [CrossRef]

74. Amrial, M.; Kusumandari, K.; Saraswati, T.; Suselo, Y. *Textile Wastewater Treatment by Using Plasma Corona Discharge in a Continuous Flow System*; IOP Conference Series: Materials Science and Engineering, 2019; IOP Publishing: Bristol, UK, 2019; p. 012016.

75. Paprowicz, J.; Słodczyk, S. Application of biologically activated sorptive columns for textile waste water treatment. *Environ. Technol.* **1988**, *9*, 271–280. [CrossRef]

76. Choo, K.-H.; Choi, S.-J.; Hwang, E.-D. Effect of coagulant types on textile wastewater reclamation in a combined coagulation/ultrafiltration system. *Desalination* **2007**, *202*, 262–270. [CrossRef]

77. Kalaiarasi, K.; Lavanya, A.; Amsamani, S.; Bagyalakshmi, G. Decolourization of textile dye effluent by non-viable biomass of Aspergillus fumigatus. *Braz. Arc. Biol. Technol.* **2012**, *55*, 471–476. [CrossRef]

78. Forss, J.; Lindh, M.V.; Pinhassi, J.; Welander, U. Microbial biotreatment of actual textile wastewater in a continuous sequential rice husk biofilter and the microbial community involved. *PLoS ONE* **2017**, *12*, e0170562. [CrossRef] [PubMed]

79. Can, O.; Kobya, M.; Demirbas, E.; Bayramoglu, M. Treatment of the textile wastewater by combined electrocoagulation. *Chemosphere* **2006**, *62*, 181–187. [CrossRef]

80. GilPavas, E.; Dobrosz-Gómez, I.; Gómez-García, M.-Á. Optimization and toxicity assessment of a combined electrocoagulation, H2O2/Fe2+/UV and activated carbon adsorption for textile wastewater treatment. *Sci. Total Environ.* **2019**, *651*, 551–560. [CrossRef] [PubMed]

81. Aleem, M.; Cao, J.; Li, C.; Rashid, H.; Wu, Y.; Nawaz, M.I.; Abbas, M.; Akram, M.W. Coagulation-and Adsorption-Based Environmental Impact Assessment and Textile Effluent Treatment. *Water Air Soil Pollut.* **2020**, *231*, 45. [CrossRef]

82. Bilińska, L.; Blus, K.; Gmurek, M.; Ledakowicz, S. Coupling of electrocoagulation and ozone treatment for textile wastewater reuse. *Chem. Eng. J.* **2019**, *358*, 992–1001. [CrossRef]

83. Dotto, J.; Fagundes-Klen, M.R.; Veit, M.T.; Palácio, S.M.; Bergamasco, R. Performance of different coagulants in the coagulation/flocculation process of textile wastewater. *J. Clean. Prod.* **2019**, *208*, 656–665. [CrossRef]

84. Khandegar, V.; Saroha, A.K. Electrocoagulation for the treatment of textile industry effluent–a review. *J. Environ. Manag.* **2013**, *128*, 949–963. [CrossRef]

85. Beltrán-Heredia, J.; Sánchez-Martín, J.; Rodríguez-Sánchez, M. Textile wastewater purification through natural coagulants. *Appl. Water Sci.* **2011**, *1*, 25–33. [CrossRef]

86. Abouri, M.; Souabi, S.; Jada, A. Optimization of Coagulation Flocculation Process for the Removal of Heavy Metals from Real Textile Wastewater. *Adv. Intell. Syst. Sustain. Dev. (AI2SD'2018) Vol 3 Adv. Intell. Syst. Appl. Environ.* **2019**, *913*, 257.

87. Solano, A.M.S.; de Araújo, C.K.C.; de Melo, J.V.; Peralta-Hernandez, J.M.; da Silva, D.R.; Martínez-Huitle, C.A. Decontamination of real textile industrial effluent by strong oxidant species electrogenerated on diamond electrode: Viability and disadvantages of this electrochemical technology. *Appl. Catal. B Environ.* **2013**, *130*, 112–120. [CrossRef]

88. Martínez-Huitle, C.A.; Panizza, M. Electrochemical oxidation of organic pollutants for wastewater treatment. *Curr. Opin. Electrochem.* **2018**, *11*, 62–71. [CrossRef]

89. Marin, N.M.; Pascu, L.F.; Demba, A.; Nita-Lazar, M.; Badea, I.A.; Aboul-Enein, H. Removal of the Acid Orange 10 by ion exchange and microbiological methods. *Int. J. Environ. Sci. Technol.* **2019**, *16*, 6357–6366. [CrossRef]

90. Yetim, T.; Tekin, T. A kinetic study on photocatalytic and sonophotocatalytic degradation of textile dyes. *Period. Polytech. Chem. Eng.* **2017**, *61*, 102–108. [CrossRef]

91. Jorfi, S.; Barzegar, G.; Ahmadi, M.; Soltani, R.D.C.; Takdastan, A.; Saeedi, R.; Abtahi, M. Enhanced coagulation-photocatalytic treatment of Acid red 73 dye and real textile wastewater using UVA/synthesized MgO nanoparticles. *J. Environ. Manag.* **2016**, *177*, 111–118. [CrossRef]

92. Naseem, Z.; Bhatti, H.N.; Iqbal, M.; Noreen, S.; Zahid, M. Fenton and photo-fenton oxidation for the remediation of textile effluents: An experimental study. *Text. Cloth.* **2019**, 235–251. [CrossRef]

93. Sharma, A.; Syed, Z.; Brighu, U.; Gupta, A.B.; Ram, C. Adsorption of textile wastewater on alkali-activated sand. *J. Clean. Prod.* **2019**, *220*, 23–32. [CrossRef]

94. Dasgupta, J.; Sikder, J.; Chakraborty, S.; Curcio, S.; Drioli, E. Remediation of textile effluents by membrane based treatment techniques: A state of the art review. *J. Environ. Manag.* **2015**, *147*, 55–72. [CrossRef] [PubMed]

95. Liang, Y.; Zhu, H.; Bañuelos, G.; Yan, B.; Zhou, Q.; Yu, X.; Cheng, X. Constructed wetlands for saline wastewater treatment: A review. *Ecol. Eng.* **2017**, *98*, 275–285. [CrossRef]

96. Shi, L.; Huang, J.; Zeng, G.; Zhu, L.; Gu, Y.; Shi, Y.; Yi, K.; Li, X. Roles of surfactants in pressure-driven membrane separation processes: A review. *Environ. Sci. Pollut. Res.* **2019**, *26*, 30731–30754. [CrossRef]

97. Van der Bruggen, B.; Canbolat, Ç.B.; Lin, J.; Luis, P. The potential of membrane technology for treatment of textile wastewater. In *Sustainable Membrane Technology for Water and Wastewater Treatment*; Springer: Berlin/Heidelberg, Germany, 2017; pp. 349–380.

98. Saeed, T.; Khan, T. Constructed wetlands for industrial wastewater treatment: Alternative media, input biodegradation ratio and unstable loading. *J. Environ. Chem. Eng.* **2019**, *7*, 103042. [CrossRef]

99. Periasamy, D.; Mani, S.; Ambikapathi, R. White Rot Fungi and Their Enzymes for the Treatment of Industrial Dye Effluents. In *Recent Advancement in White Biotechnology Through Fungi*; Springer: Berlin/Heidelberg, Germany, 2019; pp. 73–100.

100. Anastasi, A.; Spina, F.; Prigione, V.; Tigini, V.; Giansanti, P.; Varese, G.C. Scale-up of a bioprocess for textile wastewater treatment using *Bjerkandera adusta*. *Bioresour. Technol.* **2010**, *101*, 3067–3075. [CrossRef] [PubMed]

101. Fazal, T.; Mushtaq, A.; Rehman, F.; Khan, A.U.; Rashid, N.; Farooq, W.; Rehman, M.S.U.; Xu, J. Bioremediation of textile wastewater and successive biodiesel production using microalgae. *Renew. Sust. Energ. Rev.* **2018**, *82*, 3107–3126. [CrossRef]

102. Sekomo, C.B.; Kagisha, V.; Rousseau, D.; Lens, P. Heavy metal removal by combining anaerobic upflow packed bed reactors with water hyacinth ponds. *Environ. Technol.* **2012**, *33*, 1455–1464. [CrossRef]

103. Yaseen, D.A.; Scholz, M. Treatment of synthetic textile wastewater containing dye mixtures with microcosms. *Environ. Sci. Pollut. Rese.* **2018**, *25*, 1980–1997. [CrossRef] [PubMed]

104. Siddique, K.; Rizwan, M.; Shahid, M.J.; Ali, S.; Ahmad, R.; Rizvi, H. Textile wastewater treatment options: A critical review. In *Enhancing Cleanup of Environmental Pollutants*; Springer: Berlin/Heidelberg, Germany, 2017; pp. 183–207.

105. Simsek, H.; Kasi, M.; Ohm, J.-B.; Blonigen, M.; Khan, E. Bioavailable and biodegradable dissolved organic nitrogen in activated sludge and trickling filter wastewater treatment plants. *Water Res.* **2013**, *47*, 3201–3210. [CrossRef] [PubMed]

106. Singh, R.P.; Singh, P.K.; Gupta, R.; Singh, R.L. Treatment and recycling of wastewater from textile industry. In *Advances in Biological Treatment of Industrial Waste Water and their Recycling for a Sustainable Future*; Springer: Berlin/Heidelberg, Germany, 2019; pp. 225–266.

107. Winanti, E.; Rahmadyanti, E.; Fajarwati, I. *Ecological Approach of Campus Wastewater Treatment Using Constructed Wetland*; IOP Conference Series: Materials Science and Engineering, 2018; IOP Publishing: Bristol, UK, 2018; p. 012062.

108. Afzal, M.; Rehman, K.; Shabir, G.; Tahseen, R.; Ijaz, A.; Hashmat, A.J.; Brix, H. Large-scale remediation of oil-contaminated water using floating treatment wetlands. *NPJ Clean Water* **2019**, *2*, 1–10. [CrossRef]

109. El-Esawi, M.A.; Alaraidh, I.A.; Alsahli, A.A.; Alzahrani, S.M.; Ali, H.M.; Alayafi, A.A.; Ahmad, M. *Serratia liquefaciens* KM4 Improves Salt Stress Tolerance in Maize by Regulating Redox Potential, Ion Homeostasis, Leaf Gas Exchange and Stress-Related Gene Expression. *Int. J. Mol. Sci.* **2018**, *19*, 3310. [CrossRef]

110. El-Esawi, M.A.; Al-Ghamdi, A.A.; Ali, H.M.; Alayafi, A.A. *Azospirillum lipoferum* FK1 confers improved salt tolerance in chickpea (*Cicer arietinum* L.) by modulating osmolytes, antioxidant machinery and stress-related genes expression. *Environ. Exp. Bot.* **2019**, *159*, 55–65. [CrossRef]

111. Borne, K.E.; Fassman, E.A.; Tanner, C.C. Floating treatment wetland retrofit to improve stormwater pond performance for suspended solids, copper and zinc. *Ecol. Eng.* **2013**, *54*, 173–182. [CrossRef]

112. Ijaz, A.; Shabir, G.; Khan, Q.M.; Afzal, M. Enhanced remediation of sewage effluent by endophyte-assisted floating treatment wetlands. *Ecol. Eng.* **2015**, *84*, 58–66. [CrossRef]

113. Yeh, N.; Yeh, P.; Chang, Y.-H. Artificial floating islands for environmental improvement. *Renew Sustain Energy Rev.* **2015**, *47*, 616–622. [CrossRef]

114. El-Esawi, M.A. Genetic diversity and evolution of *Brassica* genetic resources: From morphology to novel genomic technologies—A review. *Plant Genet. Resour.* **2017**, *15*, 388–399. [CrossRef]

115. El-Esawi, M.A.; Germaine, K.; Bourke, P.; Malone, R. AFLP analysis of genetic diversity and phylogenetic relationships of *Brassica oleracea* in Ireland. *Comptes Rendus Biol.* **2016**, *339*, 163–170. [CrossRef]

116. Tanner, C.C.; Headley, T.R. Components of floating emergent macrophyte treatment wetlands influencing removal of stormwater pollutants. *Ecol. Eng.* **2011**, *37*, 474–486. [CrossRef]

117. Zhang, L.; Zhao, J.; Cui, N.; Dai, Y.; Kong, L.; Wu, J.; Cheng, S. Enhancing the water purification efficiency of a floating treatment wetland using a biofilm carrier. *Environ. Sci. Pollut. Rese.* **2016**, *23*, 7437–7443. [CrossRef] [PubMed]

118. Borne, K.E.; Fassman-Beck, E.A.; Winston, R.J.; Hunt, W.F.; Tanner, C.C. Implementation and maintenance of floating treatment wetlands for urban stormwater management. *J. Environ. Eng.* **2015**, *141*, 04015030. [CrossRef]

119. Ali, S.; Abbas, Z.; Rizwan, M.; Zaheer, I.E.; Yavaş, İ.; Ünay, A.; Abdel-Daim, M.M.; Bin-Jumah, M.; Hasanuzzaman, M.; Kalderis, D. Application of Floating Aquatic Plants in Phytoremediation of Heavy Metals Polluted Water: A Review. *Sustainability* **2020**, *12*, 1927. [CrossRef]

120. Chang, N.-B.; Xuan, Z.; Marimon, Z.; Islam, K.; Wanielista, M.P. Exploring hydrobiogeochemical processes of floating treatment wetlands in a subtropical stormwater wet detention pond. *Ecol. Eng.* **2013**, *54*, 66–76. [CrossRef]

121. Wand, H.; Vacca, G.; Kuschk, P.; Krüger, M.; Kästner, M. Removal of bacteria by filtration in planted and non-planted sand columns. *Water Res.* **2007**, *41*, 159–167. [CrossRef] [PubMed]

122. Vymazal, J. Constructed wetlands for treatment of industrial wastewaters: A review. *Ecol. Eng.* **2014**, *73*, 724–751. [CrossRef]

123. Shahid, M.J.; Ali, S.; Shabir, G.; Siddique, M.; Rizwan, M.; Seleiman, M.F.; Afzal, M. Comparing the performance of four macrophytes in bacterial assisted floating treatment wetlands for the removal of trace metals (Fe, Mn, Ni, Pb, and Cr) from polluted river water. *Chemosphere* **2020**, *243*, 125353. [CrossRef] [PubMed]

124. Nawaz, N.; Ali, S.; Shabir, G.; Rizwan, M.; Shakoor, M.B.; Shahid, M.J.; Afzal, M.; Arslan, M.; Hashem, A.; Abd_Allah, E.F. Bacterial Augmented Floating Treatment Wetlands for Efficient Treatment of Synthetic Textile Dye Wastewater. *Sustainability* **2020**, *12*, 3731. [CrossRef]

125. Borne, K.E. Floating treatment wetland influences on the fate and removal performance of phosphorus in stormwater retention ponds. *Ecol. Eng.* **2014**, *69*, 76–82. [CrossRef]

126. Ashraf, S.; Afzal, M.; Naveed, M.; Shahid, M.; Ahmad Zahir, Z. Endophytic bacteria enhance remediation of tannery effluent in constructed wetlands vegetated with Leptochloa fusca. *Int. J. Phytorem.* **2018**, *20*, 121–128. [CrossRef]

127. Shahid, M.; Arslan, M.; Ali, S.; Siddique, M.; Afzal, M. Floating wetlands: A sustainable tool for wastewater treatment. *CLEAN Soil Air Water* **2018**, *46*, 1–13. [CrossRef]

128. Zhu, L.; Li, Z.; Ketola, T. Biomass accumulations and nutrient uptake of plants cultivated on artificial floating beds in China's rural area. *Ecol. Eng.* **2011**, *37*, 1460–1466. [CrossRef]

129. Khin, T.; Annachhatre, A.P. Novel microbial nitrogen removal processes. *Biotechnol. Adv.* **2004**, *22*, 519–532. [CrossRef]

130. Borne, K.E.; Fassman-Beck, E.A.; Tanner, C.C. Floating treatment wetland influences on the fate of metals in road runoff retention ponds. *Water Res.* **2014**, *48*, 430–442. [CrossRef] [PubMed]

131. Zhang, M.; Qiao, S.; Shao, D.; Jin, R.; Zhou, J. Simultaneous nitrogen and phosphorus removal by combined anammox and denitrifying phosphorus removal process. *J. Chem. Technol. Biotechnol.* **2018**, *93*, 94–104. [CrossRef]

132. Shahid, M.J.; Tahseen, R.; Siddique, M.; Ali, S.; Iqbal, S.; Afzal, M. Remediation of polluted river water by floating treatment wetlands. *Water Supply* **2019**, *19*, 967–977. [CrossRef]

133. Chua, L.H.; Tan, S.B.; Sim, C.; Goyal, M.K. Treatment of baseflow from an urban catchment by a floating wetland system. *Ecol. Eng.* **2012**, *49*, 170–180. [CrossRef]

134. Nakphet, S.; Ritchie, R.J.; Kiriratnikom, S. Aquatic plants for bioremediation in red hybrid tilapia (Oreochromis niloticus× Oreochromis mossambicus) recirculating aquaculture. *Aquacult. Int.* **2017**, *25*, 619–633. [CrossRef]

135. Chaturvedi, H.; Singh, V.; Gupta, G. Potential of bacterial endophytes as plant growth promoting factors. *J. Plant Pathol. Microbiol.* **2016**, *7*, 1–6. [CrossRef]

136. Arslan, M.; Imran, A.; Khan, Q.M.; Afzal, M. Plant–bacteria partnerships for the remediation of persistent organic pollutants. *Environ. Sci. Pollut. Res.* **2017**, *24*, 4322–4336. [CrossRef] [PubMed]

137. Fatima, K.; Afzal, M.; Imran, A.; Khan, Q.M. Bacterial rhizosphere and endosphere populations associated with grasses and trees to be used for phytoremediation of crude oil contaminated soil. *Bull. Environ. Contam. Toxicol.* **2015**, *94*, 314–320. [CrossRef]

138. Shehzadi, M.; Fatima, K.; Imran, A.; Mirza, M.; Khan, Q.; Afzal, M. Ecology of bacterial endophytes associated with wetland plants growing in textile effluent for pollutant-degradation and plant growth-promotion potentials. *Plant Biosyst.* **2016**, *150*, 1261–1270. [CrossRef]

139. Rehman, K.; Imran, A.; Amin, I.; Afzal, M. Inoculation with bacteria in floating treatment wetlands positively modulates the phytoremediation of oil field wastewater. *J. Hazard. Mater.* **2018**, *349*, 242–251. [CrossRef]

140. Branda, S.S.; Vik, Å.; Friedman, L.; Kolter, R. Biofilms: The matrix revisited. *Trends Microbiol.* **2005**, *13*, 20–26. [CrossRef] [PubMed]

141. Singh, S.; Singh, S.K.; Chowdhury, I.; Singh, R. Understanding the mechanism of bacterial biofilms resistance to antimicrobial agents. *Open Microbiol. J.* **2017**, *11*, 53–62. [CrossRef] [PubMed]

142. Urakawa, H.; Dettmar, D.L.; Thomas, S. The uniqueness and biogeochemical cycling of plant root microbial communities in a floating treatment wetland. *Ecol. Eng.* **2017**, *108*, 573–580. [CrossRef]

143. Sun, Z.; Xie, D.; Jiang, X.; Fu, G.; Xiao, D.; Zheng, L. Effect of eco-remediation and microbial community using multilayer solar planted floating island (MS-PFI) in the drainage channel. *BioRxiv* **2018**, 327965. [CrossRef]

144. Achá, D.; Iniguez, V.; Roulet, M.; Guimaraes, J.R.D.; Luna, R.; Alanoca, L.; Sanchez, S. Sulfate-reducing bacteria in floating macrophyte rhizospheres from an Amazonian floodplain lake in Bolivia and their association with Hg methylation. *Appl. Environ. Microbiol.* **2005**, *71*, 7531–7535. [CrossRef]

145. Lamers, L.P.; Van Diggelen, J.M.; Op Den Camp, H.J.; Visser, E.J.; Lucassen, E.C.; Vile, M.A.; Jetten, M.S.; Smolders, A.J.; Roelofs, J.G. Microbial transformations of nitrogen, sulfur, and iron dictate vegetation composition in wetlands: A review. *Front. Microbiol.* **2012**, *3*, 156. [CrossRef]

146. Shahid, M.J.; Arslan, M.; Siddique, M.; Ali, S.; Tahseen, R.; Afzal, M. Potentialities of floating wetlands for the treatment of polluted water of river Ravi, Pakistan. *Ecol. Eng.* **2019**, *133*, 167–176. [CrossRef]

147. Zhao, T.; Fan, P.; Yao, L.; Yan, G.; Li, D.; Zhang, W. Ammonifying bacteria in plant floating island of constructed wetland for strengthening decomposition of organic nitrogen. *Trans. Chin. Soc. Agric. Eng.* **2011**, *27*, 223–226.

148. Govarthanan, M.; Mythili, R.; Selvankumar, T.; Kamala-Kannan, S.; Rajasekar, A.; Chang, Y.-C. Bioremediation of heavy metals using an endophytic bacterium Paenibacillus sp. RM isolated from the roots of Tridax procumbens. *3 Biotech* **2016**, *6*, 242. [CrossRef] [PubMed]

149. Barathi, S.; Karthik, C.; Nadanasabapathi, S.; Padikasan, I.A. Biodegradation of textile dye Reactive Blue 160 by Bacillus firmus (Bacillaceae: Bacillales) and non-target toxicity screening of their degraded products. *Toxicol. Rep.* **2020**, *7*, 16–22. [CrossRef]

150. Franca, R.D.G.; Vieira, A.; Carvalho, G.; Oehmen, A.; Pinheiro, H.M.; Crespo, M.T.B.; Lourenço, N.D. Oerskovia paurometabola can efficiently decolorize azo dye Acid Red 14 and remove its recalcitrant metabolite. *Ecotoxicol. Environ. Saf.* **2020**, *191*, 110007. [CrossRef] [PubMed]

151. Garg, N.; Garg, A.; Mukherji, S. Eco-friendly decolorization and degradation of reactive yellow 145 textile dye by Pseudomonas aeruginosa and Thiosphaera pantotropha. *J. Environ. Manag.* **2020**, *263*, 110383. [CrossRef] [PubMed]

152. Roy, D.C.; Biswas, S.K.; Saha, A.K.; Sikdar, B.; Rahman, M.; Roy, A.K.; Prodhan, Z.H.; Tang, S.-S. Biodegradation of Crystal Violet dye by bacteria isolated from textile industry effluents. *PeerJ* **2018**, *6*, e5015. [CrossRef]

153. Najme, R.; Hussain, S.; Maqbool, Z.; Imran, M.; Mahmood, F.; Manzoor, H.; Yasmeen, T.; Shehzad, T. Biodecolorization of Reactive Yellow-2 by Serratia sp. RN34 Isolated from textile wastewater. *Water Environ. Res.* **2015**, *87*, 2065–2075. [CrossRef]

154. Bheemaraddi, M.C.; Patil, S.; Shivannavar, C.T.; Gaddad, S.M. Isolation and characterization of Paracoccus sp. GSM2 capable of degrading textile azo dye reactive violet 5. *Sci. World J.* **2014**, *2014*, 410704. [CrossRef]

155. Singh, R.P.; Singh, P.K.; Singh, R.L. Bacterial decolorization of textile azo dye acid orange by Staphylococcus hominis RMLRT03. *Toxicol. Int.* **2014**, *21*, 160. [CrossRef] [PubMed]

156. Garg, S.K.; Tripathi, M. Process parameters for decolorization and biodegradation of orange II (Acid Orange 7) in dye-simulated minimal salt medium and subsequent textile effluent treatment by Bacillus cereus (MTCC 9777) RMLAU1. *Environ. Monit. Assess.* **2013**, *185*, 8909–8923. [CrossRef]

157. Lim, C.K.; Bay, H.H.; Aris, A.; Majid, Z.A.; Ibrahim, Z. Biosorption and biodegradation of Acid Orange 7 by Enterococcus faecalis strain ZL: Optimization by response surface methodological approach. *Environ. Sci. Pollut. Res.* **2013**, *20*, 5056–5066. [CrossRef]

158. Surwase, S.V.; Deshpande, K.K.; Phugare, S.S.; Jadhav, J.P. Biotransformation studies of textile dye Remazol Orange 3R. *3 Biotech* **2013**, *3*, 267–275. [CrossRef]

159. Deive, F.; Domínguez, A.; Barrio, T.; Moscoso, F.; Morán, P.; Longo, M.; Sanromán, M. Decolorization of dye Reactive Black 5 by newly isolated thermophilic microorganisms from geothermal sites in Galicia (Spain). *J. Hazard. Mater.* **2010**, *182*, 735–742. [CrossRef]

160. Wang, H.; Su, J.Q.; Zheng, X.W.; Tian, Y.; Xiong, X.J.; Zheng, T.L. Bacterial decolorization and degradation of the reactive dye Reactive Red 180 by Citrobacter sp. CK3. *Int. Biodeterior. Biodegrad.* **2009**, *63*, 395–399. [CrossRef]

161. Kolekar, Y.M.; Pawar, S.P.; Gawai, K.R.; Lokhande, P.D.; Shouche, Y.S.; Kodam, K.M. Decolorization and degradation of Disperse Blue 79 and Acid Orange 10, by *Bacillus fusiformis* KMK5 isolated from the textile dye contaminated soil. *Bioresour. Technol.* **2008**, *99*, 8999–9003. [CrossRef] [PubMed]

162. Kalyani, D.C.; Patil, P.S.; Jadhav, J.P.; Govindwar, S.P. Biodegradation of reactive textile dye Red BLI by an isolated bacterium Pseudomonas sp. SUK1. *Bioresour. Technol.* **2008**, *99*, 4635–4641. [CrossRef] [PubMed]

163. Alhassani, H.A.; Rauf, M.A.; Ashraf, S.S. Efficient microbial degradation of Toluidine Blue dye by Brevibacillus sp. *Dyes Pigments* **2007**, *75*, 395–400. [CrossRef]

164. Coughlin, M.F.; Kinkle, B.K.; Bishop, P.L. High performance degradation of azo dye Acid Orange 7 and sulfanilic acid in a laboratory scale reactor after seeding with cultured bacterial strains. *Water Res.* **2003**, *37*, 2757–2763. [CrossRef]

165. Hu, T. Degradation of azo dye RP2B by Pseudomonas luteola. *Water Sci. Technol.* **1998**, *38*, 299–306. [CrossRef]

166. Elango, G.; Rathika, G.; Elango, S. Physico-chemical parameters of textile dyeing effluent and its impacts with case study. *Int. J. Res Chem. Environ.* **2017**, *7*, 17–24.

167. Mirbolooki, H.; Amirnezhad, R.; Pendashteh, A.R. Treatment of high saline textile wastewater by activated sludge microorganisms. *J. Appl Res. Technol.* **2017**, *15*, 167–172. [CrossRef]

168. Hawes, M.C.; McLain, J.; Ramirez-Andreotta, M.; Curlango-Rivera, G.; Flores-Lara, Y.; Brigham, L.A. Extracellular trapping of soil contaminants by root border cells: New insights into plant defense. *Agronomy* **2016**, *6*, 5. [CrossRef]

169. Cao, W.; Zhang, H.; Wang, Y.; Pan, J. Bioremediation of polluted surface water by using biofilms on filamentous bamboo. *Ecol. Eng.* **2012**, *42*, 146–149. [CrossRef]

170. Prajapati, M.; van Bruggen, J.J.; Dalu, T.; Malla, R. Assessing the effectiveness of pollutant removal by macrophytes in a floating wetland for wastewater treatment. *Appl Water Sci.* **2017**, *7*, 4801–4809. [CrossRef]

171. Chen, Z.; Cuervo, D.P.; Müller, J.A.; Wiessner, A.; Köser, H.; Vymazal, J.; Kästner, M.; Kuschk, P. Hydroponic root mats for wastewater treatment—A review. *Environ. Sci. Pollut. Res.* **2016**, *23*, 15911–15928. [CrossRef]

172. Ijaz, A.; Iqbal, Z.; Afzal, M. Remediation of sewage and industrial effluent using bacterially assisted floating treatment wetlands vegetated with Typha domingensis. *Water Sci. Technol.* **2016**, *74*, 2192–2201. [CrossRef]

173. Qamar, M.T.; Mumtaz, H.M.; Mohsin, M.; Asghar, H.N.; Iqbal, M.; Nasir, M. Development of floating treatment wetlands with plant-bacteria partnership to clean textile bleaching effluent. *Ind. Text.* **2019**, *70*, 502–511. [CrossRef]

174. Kumar, S.; Pratap, B.; Dubey, D.; Dutta, V. Microbial Communities in Constructed Wetland Microcosms and Their Role in Treatment of Domestic Wastewater. In *Emerging Eco-Friendly Green Technologies for Wastewater Treatment*; Springer: Berlin/Heidelberg, Germany, 2020; pp. 311–327.

175. Rehman, K.; Imran, A.; Amin, I.; Afzal, M. Enhancement of oil field-produced wastewater remediation by bacterially-augmented floating treatment wetlands. *Chemosphere* **2019**, *217*, 576–583. [CrossRef]

176. Suryawan, I.; Helmy, Q.; Notodarmojo, S. *Textile Wastewater Treatment: Colour and COD Removal of Reactive Black-5 by Ozonation*; IOP Conference Series: Earth and Environmental Science, 2018; IOP Publishing: Bristol, UK, 2018; p. 012102.

177. Mohan, S.; Vidhya, K.; Sivakumar, C.; Sugnathi, M.; Shanmugavadivu, V.; Devi, M. Textile Waste Water Treatment by Using Natural Coagulant (Neem-Azadirachta India). *Int. Res. J. Multidis. Technovation* **2019**, *1*, 636–642.

178. Sathya, U.; Nithya, M.; Balasubramanian, N. Evaluation of advanced oxidation processes (AOPs) integrated membrane bioreactor (MBR) for the real textile wastewater treatment. *J. Environ. Manag.* **2019**, *246*, 768–775. [CrossRef]

179. Tusief, M.Q.; Malik, M.H.; Asghar, H.N.; Mohsin, M.; Mahmood, N. Bioremediation of textile wastewater through floating treatment wetland system. *Int. J. Agric. Biol.* **2019**, *22*, 821–826.

180. Darajeh, N.; Idris, A.; Masoumi, H.R.F.; Nourani, A.; Truong, P.; Sairi, N.A. Modeling BOD and COD removal from Palm Oil Mill Secondary Effluent in floating wetland by *Chrysopogon zizanioides* (L.) using response surface methodology. *J. Environ. Manag.* **2016**, *181*, 343–352. [CrossRef] [PubMed]

181. Klomjek, P.; Nitisoravut, S. Constructed treatment wetland: A study of eight plant species under saline conditions. *Chemosphere* **2005**, *58*, 585–593. [CrossRef] [PubMed]

182. Yulistyorini, A.; Puspasari, A.K.; Sari, A.A. *Removal of BOD and TSS of Student Dormitory Greywater Using Vertical Sub-Surface Flow Constructed Wetland of Ipomoea Aquatica*; IOP Conference Series: Materials Science and Engineering, 2019; IOP Publishing: Bristol, UK, 2019; p. 012056.

183. Saeed, T.; Haque, I.; Khan, T. Organic matter and nutrients removal in hybrid constructed wetlands: Influence of saturation. *Chem. Eng. J.* **2019**, *371*, 154–165. [CrossRef]

184. Vymazal, J. Removal of nutrients in various types of constructed wetlands. *Sci. Total Environ.* **2007**, *380*, 48–65. [CrossRef] [PubMed]
185. Arivoli, A.; Sathiamoorthi, T.; Satheeshkumar, M. Treatment of Textile Effluent by Phytoremediation with the Aquatic Plants: Alternanthera sessilis. In *Bioremediation and Sustainable Technologies for Cleaner Environment*; Springer: Berlin/Heidelberg, Germany, 2017; pp. 185–197.
186. Ladislas, S.; Gerente, C.; Chazarenc, F.; Brisson, J.; Andres, Y. Floating treatment wetlands for heavy metal removal in highway stormwater ponds. *Ecol. Eng.* **2015**, *80*, 85–91. [CrossRef]
187. Ladislas, S.; El-Mufleh, A.; Gérente, C.; Chazarenc, F.; Andrès, Y.; Béchet, B. Potential of aquatic macrophytes as bioindicators of heavy metal pollution in urban stormwater runoff. *Water Air Soil Pollut.* **2012**, *223*, 877–888. [CrossRef]
188. Gadd, G.M. Metals, minerals and microbes: Geomicrobiology and bioremediation. *Microbiol.* **2010**, *156*, 609–643. [CrossRef] [PubMed]
189. Haq, S.; Bhatti, A.A.; Dar, Z.A.; Bhat, S.A. Phytoremediation of Heavy Metals: An Eco-Friendly and Sustainable Approach. In *Bioremediation and Biotechnology*; Springer: Berlin/Heidelberg, Germany, 2020; pp. 215–231.
190. Ullah, A.; Heng, S.; Munis, M.F.H.; Fahad, S.; Yang, X. Phytoremediation of heavy metals assisted by plant growth promoting (PGP) bacteria: A review. *Environ. Exp. Bot.* **2015**, *117*, 28–40. [CrossRef]
191. Hussain, I.; Aleti, G.; Naidu, R.; Puschenreiter, M.; Mahmood, Q.; Rahman, M.M.; Wang, F.; Shaheen, S.; Syed, J.H.; Reichenauer, T.G. Microbe and plant assisted-remediation of organic xenobiotics and its enhancement by genetically modified organisms and recombinant technology: A review. *Sci. Total Environ.* **2018**, *628*, 1582–1599. [CrossRef]
192. Doty, S.L. Enhancing phytoremediation through the use of transgenics and endophytes. *New Phytol.* **2008**, *179*, 318–333. [CrossRef]
193. Wang, C.-Y.; Sample, D.J.; Bell, C. Vegetation effects on floating treatment wetland nutrient removal and harvesting strategies in urban stormwater ponds. *Sci. Total Environ.* **2014**, *499*, 384–393. [CrossRef]
194. Headley, T.; Tanner, C. Constructed wetlands with floating emergent macrophytes: An innovative stormwater treatment technology. *Crit. Rev. Environ. Sci. Technol.* **2012**, *42*, 2261–2310. [CrossRef]
195. Correa-García, S.; Pande, P.; Séguin, A.; St-Arnaud, M.; Yergeau, E. Rhizoremediation of petroleum hydrocarbons: A model system for plant microbiome manipulation. *Microb. Biotechnol.* **2018**, *11*, 819–832. [CrossRef] [PubMed]
196. Yergeau, E.; Tremblay, J.; Joly, S.; Labrecque, M.; Maynard, C.; Pitre, F.E.; St-Arnaud, M.; Greer, C.W. Soil contamination alters the willow root and rhizosphere metatranscriptome and the root–rhizosphere interactome. *ISME J.* **2018**, *12*, 869–884. [CrossRef]
197. Sheoran, A. A laboratory treatment study of acid mine water of wetlands with emergent macrophyte (Typha angustata). *Int. J. Min. Reclam. Environ.* **2006**, *20*, 209–222. [CrossRef]
198. Kumar, K.; Kumar, D.; Teja, V.; Venkateswarlu, V.; Kumar, M.; Nadendla, R. A review on Typha angustata. *Int. J. Phytopharm.* **2013**, *4*, 277–281.
199. Vymazal, J. Plants used in constructed wetlands with horizontal subsurface flow: A review. *Hydrobiologia* **2011**, *674*, 133–156. [CrossRef]
200. Vymazal, J. The use of hybrid constructed wetlands for wastewater treatment with special attention to nitrogen removal: A review of a recent development. *Water Res.* **2013**, *47*, 4795–4811. [CrossRef] [PubMed]
201. Rezania, S.; Taib, S.M.; Din, M.F.M.; Dahalan, F.A.; Kamyab, H. Comprehensive review on phytotechnology: Heavy metals removal by diverse aquatic plants species from wastewater. *J. Hazard. Mater.* **2016**, *318*, 587–599. [CrossRef] [PubMed]
202. Valipour, A.; Ahn, Y.-H. Constructed wetlands as sustainable ecotechnologies in decentralization practices: A review. *Environ. Sci. Pollut. Res.* **2016**, *23*, 180–197. [CrossRef]
203. Fritioff, Å.; Greger, M. Aquatic and terrestrial plant species with potential to remove heavy metals from stormwater. *Int. J. Phytorem.* **2003**, *5*, 211–224. [CrossRef]
204. Vymazal, J. Emergent plants used in free water surface constructed wetlands: A review. *Ecol. Eng.* **2013**, *61*, 582–592. [CrossRef]
205. Di Luca, G.A.; Mufarrege, M.; Hadad, H.R.; Maine, M.A. Nitrogen and phosphorus removal and Typha domingensis tolerance in a floating treatment wetland. *Sci. Total Environ.* **2019**, *650*, 233–240. [CrossRef] [PubMed]

206. Nichols, P.; Lucke, T.; Drapper, D.; Walker, C. Performance evaluation of a floating treatment wetland in an urban catchment. *Water* **2016**, *8*, 244. [CrossRef]

207. Gao, L.; Zhou, W.; Huang, J.; He, S.; Yan, Y.; Zhu, W.; Wu, S.; Zhang, X. Nitrogen removal by the enhanced floating treatment wetlands from the secondary effluent. *Bioresour. Technol.* **2017**, *234*, 243–252. [CrossRef]

208. Song, H.-L.; Li, X.-N.; Lu, X.-W.; Inamori, Y. Investigation of microcystin removal from eutrophic surface water by aquatic vegetable bed. *Ecol. Eng.* **2009**, *35*, 1589–1598. [CrossRef]

209. Tambunan, J.A.M.; Effendi, H.; Krisanti, M. Phytoremediating Batik Wastewater Using Vetiver *Chrysopogon zizanioides* (L). *Pol. J. Environ. Stud.* **2018**, *27*, 1281–1288. [CrossRef]

210. De Stefani, G.; Tocchetto, D.; Salvato, M.; Borin, M. Performance of a floating treatment wetland for in-stream water amelioration in NE Italy. *Hydrobiologia* **2011**, *674*, 157–167. [CrossRef]

211. Park, J.B.; Sukias, J.P.; Tanner, C.C. Floating treatment wetlands supplemented with aeration and biofilm attachment surfaces for efficient domestic wastewater treatment. *Ecol. Eng.* **2019**, *139*, 105582. [CrossRef]

212. Fahid, M.; Arslan, M.; Shabir, G.; Younus, S.; Yasmeen, T.; Rizwan, M.; Siddique, K.; Ahmad, S.R.; Tahseen, R.; Iqbal, S. Phragmites australis in combination with hydrocarbons degrading bacteria is a suitable option for remediation of diesel-contaminated water in floating wetlands. *Chemosphere* **2020**, *240*, 124890. [CrossRef]

213. Afzal, M.; Arslan, M.; Müller, J.A.; Shabir, G.; Islam, E.; Tahseen, R.; Anwar-ul-Haq, M.; Hashmat, A.J.; Iqbal, S.; Khan, Q.M. Floating treatment wetlands as a suitable option for large-scale wastewater treatment. *Nat. Sustain.* **2019**, *2*, 863–871. [CrossRef]

214. Nanayakkara, C. Floating wetlands for management of algal washout from waste stabilization pond effluent: Case study at hikkaduwa waste stabilization ponds. *Engineer* **2013**, *46*, 63–74.

215. Weragoda, S.; Jinadasa, K.; Zhang, D.Q.; Gersberg, R.M.; Tan, S.K.; Tanaka, N.; Jern, N.W. Tropical application of floating treatment wetlands. *Wetlands* **2012**, *32*, 955–961. [CrossRef]

216. Lopardo, C.R.; Zhang, L.; Mitsch, W.J.; Urakawa, H. Comparison of nutrient retention efficiency between vertical-flow and floating treatment wetland mesocosms with and without biodegradable plastic. *Ecol. Eng.* **2019**, *131*, 120–130. [CrossRef]

217. Wang, C.-Y.; Sample, D.J. Assessment of the nutrient removal effectiveness of floating treatment wetlands applied to urban retention ponds. *J. Environ. Manag.* **2014**, *137*, 23–35. [CrossRef] [PubMed]

218. Abhilash, P.; Jamil, S.; Singh, N. Transgenic plants for enhanced biodegradation and phytoremediation of organic xenobiotics. *Biotechnol. Adv.* **2009**, *27*, 474–488. [CrossRef]

219. Xie, W.-Y.; Huang, Q.; Li, G.; Rensing, C.; Zhu, Y.-G. Cadmium accumulation in the rootless macrophyte Wolffia globosa and its potential for phytoremediation. *Int. J. Phytorem.* **2013**, *15*, 385–397. [CrossRef] [PubMed]

220. Rezania, S.; Din, M.F.M.; Taib, S.M.; Dahalan, F.A.; Songip, A.R.; Singh, L.; Kamyab, H. The efficient role of aquatic plant (water hyacinth) in treating domestic wastewater in continuous system. *Int. J. Phytorem.* **2016**, *18*, 679–685. [CrossRef]

221. Colares, G.S.; Dell'Osbel, N.; Wiesel, P.G.; Oliveira, G.A.; Lemos, P.H.Z.; da Silva, F.P.; Lutterbeck, C.A.; Kist, L.T.; Machado, Ê.L. Floating treatment wetlands: A review and bibliometric analysis. *Sci. Total Environ.* **2020**, *714*, 136776. [CrossRef]

222. Ibekwe, A.M.; Lyon, S.; Leddy, M.; Jacobson-Meyers, M. Impact of plant density and microbial composition on water quality from a free water surface constructed wetland. *J. Appl. Microbiol.* **2007**, *102*, 921–936. [CrossRef]

223. Chance, L.M.G.; White, S.A. Aeration and plant coverage influence floating treatment wetland remediation efficacy. *Ecol. Eng.* **2018**, *122*, 62–68. [CrossRef]

224. Wu, H.; Zhang, J.; Ngo, H.H.; Guo, W.; Hu, Z.; Liang, S.; Fan, J.; Liu, H. A review on the sustainability of constructed wetlands for wastewater treatment: Design and operation. *Bioresour. Technol.* **2015**, *175*, 594–601. [CrossRef] [PubMed]

225. Liu, H.; Hu, Z.; Zhang, J.; Ngo, H.H.; Guo, W.; Liang, S.; Fan, J.; Lu, S.; Wu, H. Optimizations on supply and distribution of dissolved oxygen in constructed wetlands: A review. *Bioresour. Technol.* **2016**, *214*, 797–805. [CrossRef] [PubMed]

226. Austin, D.; Nivala, J. Energy requirements for nitrification and biological nitrogen removal in engineered wetlands. *Ecol. Eng.* **2009**, *35*, 184–192. [CrossRef]

227. Chen, M.; Wu, X.; Chen, Y.; Dong, M. Mechanism of nitrogen removal by adsorption and bio-transformation in constructed wetland systems. *Chin. J. Environ. Eng.* **2009**, *3*, 223–228.

228. Pan, J.; Fei, H.; Song, S.; Yuan, F.; Yu, L. Effects of intermittent aeration on pollutants removal in subsurface wastewater infiltration system. *Bioresour. Technol.* **2015**, *191*, 327–331. [CrossRef]

229. Hu, Y.; Zhao, Y.; Zhao, X.; Kumar, J.L. High rate nitrogen removal in an alum sludge-based intermittent aeration constructed wetland. *Environ. Sci. Technol.* **2012**, *46*, 4583–4590. [CrossRef]

230. Ahn, K.-H.; Song, K.-G.; Choa, E.; Cho, J.; Yun, H.; Lee, S.; Me, J. Enhanced biological phosphorus and nitrogen removal using a sequencing anoxic/anaerobic membrane bioreactor (SAM) process. *Desalination* **2003**, *157*, 345–352. [CrossRef]

231. Wang, C.-Y.; Sample, D.J.; Day, S.D.; Grizzard, T.J. Floating treatment wetland nutrient removal through vegetation harvest and observations from a field study. *Ecol. Eng.* **2015**, *78*, 15–26. [CrossRef]

232. Ong, S.-A.; Uchiyama, K.; Inadama, D.; Ishida, Y.; Yamagiwa, K. Performance evaluation of laboratory scale up-flow constructed wetlands with different designs and emergent plants. *Bioresour. Technol.* **2010**, *101*, 7239–7244. [CrossRef]

233. Dunqiu, W.; Shaoyuan, B.; Mingyu, W.; Qinglin, X.; Yinian, Z.; Hua, Z. Effect of artificial aeration, temperature, and structure on nutrient removal in constructed floating islands. *Water Environ. Res.* **2012**, *84*, 405–410. [CrossRef]

234. Fahid, M.; Ali, S.; Shabir, G.; Rashid Ahmad, S.; Yasmeen, T.; Afzal, M.; Arslan, M.; Hussain, A.; Hashem, A.; Abd Allah, E.F. Cyperus laevigatus L. Enhances Diesel Oil Remediation in Synergism with Bacterial Inoculation in Floating Treatment Wetlands. *Sustainability* **2020**, *12*, 2353. [CrossRef]

235. Saleem, H.; Rehman, K.; Arslan, M.; Afzal, M. Enhanced degradation of phenol in floating treatment wetlands by plant-bacterial synergism. *Int. J. Phytorem.* **2018**, *20*, 692–698. [CrossRef] [PubMed]

236. El-Esawi, M.A.; Alaraidh, I.A.; Alsahli, A.A.; Ali, H.M.; Alayafi, A.A.; Witczak, J.; Ahmad, M. Genetic variation and alleviation of salinity stress in barley. *Molecules* **2018**, *23*, 2488. [CrossRef] [PubMed]

237. Vwioko, E.; Adinkwu, O.; El-Esawi, M.A. Comparative Physiological, Biochemical and Genetic Responses to Prolonged Waterlogging Stress in Okra and Maize Given Exogenous Ethylene Priming. *Front. Physiol.* **2017**, *8*, 632. [CrossRef]

238. El-Esawi, M.A.; Al-Ghamdi, A.A.; Ali, H.M.; Alayafi, A.A.; Witczak, J.; Ahmad, M. Analysis of Genetic Variation and Enhancement of Salt Tolerance in French Pea (*Pisum Sativum* L.). *Int. J. Mol. Sci.* **2018**, *19*, 2433. [CrossRef]

239. Hossen, M.Z.; Hussain, M.E.; Hakim, A.; Islam, K.; Uddin, M.N.; Azad, A.K. Biodegradation of reactive textile dye Novacron Super Black G by free cells of newly isolated Alcaligenes faecalis AZ26 and Bacillus spp obtained from textile effluents. *Heliyon* **2019**, *5*, e02068. [CrossRef]

240. Pavlineri, N.; Skoulikidis, N.T.; Tsihrintzis, V.A. Constructed floating wetlands: A review of research, design, operation and management aspects, and data meta-analysis. *Chem. Eng. J.* **2017**, *308*, 1120–1132. [CrossRef]

241. Thullen, J.S.; Sartoris, J.J.; Walton, W.E. Effects of vegetation management in constructed wetland treatment cells on water quality and mosquito production. *Ecol. Eng.* **2002**, *18*, 441–457. [CrossRef]

242. Walton, W.E.; Jiannino, J.A. Vegetation management to stimulate denitrification increases mosquito abundance in multipurpose constructed treatment wetlands. *J. Am. Mosq. Control Assoc.* **2005**, *21*, 22–27. [CrossRef]

243. Francesco, D. Biomass recovery from invasive species management in wetlands. *Biomass Bioenergy* **2017**, *105*, 259–265.

244. Lishawa, S.C.; Lawrence, B.A.; Albert, D.A.; Larkin, D.J.; Tuchman, N.C. Invasive species removal increases species and phylogenetic diversity of wetland plant communities. *Ecol. Evol.* **2019**, *9*, 6231–6244. [CrossRef] [PubMed]

sustainability

Article

Plant-Microbe Synergism in Floating Treatment Wetlands for the Enhanced Removal of Sodium Dodecyl Sulphate from Water

Momina Yasin [1,2], Muhammad Tauseef [3], Zaniab Zafar [1,2], Moazur Rahman [4], Ejazul Islam [1,2], Samina Iqbal [1,2] and Muhammad Afzal [1,2,*]

[1] Soil and Environmental Biotechnology Division, National Institute for Biotechnology and Genetic Engineering (NIBGE), Faisalabad 38000, Pakistan; mominayasin4@gmail.com (M.Y.); zainabzafar39@gmail.com (Z.Z.); ejazulislam75@yahoo.com (E.I.); siqbaleb@gmail.com (S.I.)
[2] Pakistan Institute of Engineering and Applied Sciences (PIEAS), Islamabad 44000, Pakistan
[3] Department of Physics, Kohsar University, Murree 47118, Pakistan; tauseefqau@gmail.com
[4] School of Biological Sciences, University of the Punjab, Quaid-i-Azam Campus, Lahore 53720, Pakistan; moaz.sbs@pu.edu.pk
* Correspondence: manibge@yahoo.com

Abstract: Excessive use of detergents in wide industrial processes results in unwanted surfactant pollution. Among them, sodium dodecyl sulphate (SDS) has well-known history to be used in pharmaceutical and industrial applications. However, if discharged without treatment, it can cause toxic effects on living organisms especially to the aquatic life. Floating treatment wetlands (FTWs) could be a cost-effective and eco-friendly options for the treatment of wastewater containing SDS. In this study, FTWs mesocosms were established in the presence of hydrocarbons-degrading bacteria. Two plant species (*Brachiaria mutica* and *Leptochloa fusca*) were vegetated and a consortium of bacteria (*Acinetobacter* sp. strain BRSI56, *Acinetobacter junii* strain TYRH47, and *Acinetobacter* sp. strain CYRH21) was applied to enhance degradation in a short-time. Results illustrated that FTWs vegetated with both plants successfully removed SDS from water, however, bacterial augmentation further enhanced the removal efficiency. Maximum reduction in SDS concentration (97.5%), chemical oxygen demand (92.0%), biological oxygen demand (94.2%), and turbidity (99.4%) was observed in the water having FTWs vegetated with *B. mutica* and inoculated with the bacteria. The inoculated bacteria showed more survival in the roots and shoots of B. mutica as compared to *L. fusca*. This study concludes that FTWs have the potential for the removal of SDS from contaminated water and their remediation efficiency can be enhanced by bacterial augmentation.

Keywords: hydroponic root mats; plant-bacteria partnership; detergents; phytoremediation; wastewater

Citation: Yasin, M.; Tauseef, M.; Zafar, Z.; Rahman, M.; Islam, E.; Iqbal, S.; Afzal, M. Plant-Microbe Synergism in Floating Treatment Wetlands for the Enhanced Removal of Sodium Dodecyl Sulphate from Water. *Sustainability* **2021**, *13*, 2883. https://doi.org/10.3390/su13052883

Academic Editor: Helvi Heinonen-Tanski

Received: 20 January 2021
Accepted: 2 March 2021
Published: 7 March 2021

Publisher's Note: MDPI stays neutral with regard to jurisdictional claims in published maps and institutional affiliations.

1. Introduction

The consumption of detergents is increasing due to industrialization and urbanization, which results in the discharge of a higher concentration of these pollutants in the environment. Detergents are synthetic organic compounds that are used extensively in different cleansing activities such as car washing facilities, laundries, household as well as in many industries such as cosmetics, textile, paper, etc. [1–3]. These compounds can cause complications in sewage treatment due to their high foaming ability, lower oxygenation potential, and subsequently kill aquatic organisms including fish [4,5]. A typical detergent contains surfactants (10 to 20%), bleach (7%), phosphate builders (50%), and additives (23–33%). Among these, surfactants are the components that are responsible for the cleaning action of detergents [6,7]. Surfactant molecules are composed of a polar head group which may either be charged or uncharged and a non-polar hydrocarbon tail [8–10]. The hydrophilic and hydrophobic properties of these molecules make them suitable for a cleansing purpose [11]. One of the main surfactants in detergents is sodium dodecyl sulfate (SDS) or sodium lauryl sulphate (SLS), which has extensive applications in various sectors [12–15]. Further, they are used to modify many adsorbents to increase the efficiency of removing many pollutants. Traditionally, surfactants have been categorized into four types based on the hydrophilic heads, namely, anionic, cationic, amphoteric, and non ionic [16]. Among them, ionic surfactants have received tremendous attention because they are very popular, strongest, and inexpensive agents [17,18]. They have a negative charge on their hydrophilic end that helps the surfactant molecules lift and suspend particles in the micelles. In the micelle, the surfactants are oriented with their charged head groups toward the solid surface while the hydrophobic hydrocarbon chains protrude into aqueous phase followed by their effective removal from the contaminated environment [18,19]. SDS is an anionic surfactant that exhibits these properties. However, different researchers have also reported the toxic effects of SDS on living organisms especially fish and microbes like yeast and bacteria. It is also toxic to mammals such as human beings though to a lesser extent [20].

Many traditional methods, such as coagulation, filtration with coagulation, distillation, precipitation, ozonation, adsorption, ion exchange, sedimentation, filtration, reverse osmosis, and advanced oxidation have been reported for the removal of SDS from the wastewater [21–23]. However, these methods are less sustainable because of their high operational, capital, and maintenance costs [24–26]. Moreover, one of the drawbacks of these methods is the generation of toxic sludge which may produce secondary pollution [27–29]. Therefore, remediation of the wastewater by a technology that has low capital and operational costs, self-sustaining, and environmentally friendly is required [30,31]. Recently, the use of floating treatment wetlands (FTWs) is considered a promising method for the treatment of contaminated water [32–36]. These FTWs are small artificial self-buoyant mats or hydroponics platforms [37,38] that permit the aquatic plants to grow in water that is typically too deep for them [6]. These have been considered applicable and proved to be very effective and suitable in the restoration of contaminated water due to their multi-pollutant treatment capability, low cost, easy operation and do not require any technical skills for operation and maintenance [24,39–41]. In FTWs, plants' roots provide a wide surface area for the growth and proliferation of microorganisms which results in the formation of biofilm on the roots [42–44]. The biofilm is the place where the majority of nutrients uptake and organic pollutants degradation takes place in FTWs [45,46]. The microorganisms contain certain enzymes such as alkyl sulfatases which initiate the degradation of detergents by catalyzing the hydrolytic cleavage of ester bonds to release inorganic sulfates [47–50]. The resulting parental alcohol upon β-oxidation is degraded and transformed into water and carbon dioxide [1].

Brachiaria mutica (Para grass) and *Leptochloa fusca* (Kallar grass) are the two different plant species used in the remediation of contaminated soil and water [42,43]. These are common salt-tolerant grasses with an extensive root system and biomass that allow them to withstand stress conditions, such as wastewater [42,43]. The present work aimed to evaluate the effect of inoculation of the bacteria in FTWs, vegetated with *B. mutica* and *L. fusca*, on the remediation of SDS contaminated water. We selected SDS as a model compound because it is abundantly found in the domestic wastewater. It has extensive application in cleaning and hygiene products such as household detergents, car wash shampoos, soaps, and toothpaste [15]. The adsorption potential of SDS on various substrates is already investigated [16–18]; nevertheless, removal efficiency by CWs from the contaminated wastewater is still unknown. In this study, we investigated the treatment performance of the system by evaluating temporal decrease in SDS concentration, COD, and BOD reduction in the water. Moreover, the persistence of the inoculated bacteria was monitored in the water, and root and shoot of the plants.

2. Materials and Methods

2.1. Bacterial Strains

Seven bacterial strains (*Bacillus pumilus* strain RT1, *Acinetobacter* sp. strain BRSI56, *Acinetobacter junii* strain TYRH47, *Acinetobacter* sp. strain CYRH21, *Pseudomonas aeruginosa* strain BRRI54, *Burkholderia phytofirmans* PsJN, and *Klebsiella* sp. strain LCRI-87) were tested for their ability to degrade SDS. These strains were already isolated and characterized for oil degradation and plant growth promoting activities [51,52]. The strains were cultivated in an M9 medium containing 50 mg L^{-1} SDS at 37 °C in a shaking incubator. Their SDS degradation potential and growth were monitored. Among these, three (*Acinetobacter* sp. strain BRSI56, *Acinetobacter* sp. strain CYRH21, and *B. phytofirmans* PsJN) were selected based on their maximum growth and SDS degradation potential. The selected strains were grown in LB broth overnight, and their cells were harvested by centrifugation. The bacterial consortium was prepared by mixing equal numbers of cells of each strain. One hundred mL of this consortium was used for inoculation in each FTWs mesocosm.

2.2. Development of FTWs

Eighteen FTWs mesocosms were established in 50 L plastic drums in the vicinity of NIBGE, Faisalabad (Figure 1). For the development of FTWs, a polyethylene sheet (Jumbolon Roll) was used for preparing hydroponic mats. The octagon-shaped sheet having 3 inches thickness was drilled from the middle to make a hole for vegetation. The cuttings (20) of *B. mutica* and *L. fusca* were taken from the nursery (developed in the vicinity of NIBGE, Faisalabad), and placed in each hole, and then soil and coconut shavings were used to support the cuttings in these holes. The cuttings were allowed to grow for two months in tap water to develop fresh roots and leaves, after that the water of the pots was replaced with SDS contaminated water (50 mg L^{-1}) and bacterial consortium (100 mL) was added to the required treatments. Different treatments having floating mats were: SDS contaminated water without vegetation and bacteria (C), SDS contaminated water with *L. fusca* (T1), SDS contaminated water with *L. fusca* and bacteria (T2), SDS contaminated water with *B. mutica* (T3), SDS contaminated water with *B. mutica* and *bacteria* (T4), and SDS contaminated water with bacteria (T5). Each treatment was in triplicate and the whole experimental setup was placed at a place having ambient conditions of temperature and light from June to August 2020.

2.3. Water Analysis

The water samples were collected every 24 h from each treatment and analyzed for pH, turbidity, BOD, and COD as described earlier (APHA, 2005) [53]. Turbidity was determined by using Spectro Quadrant Nova 60. The benchtop digital AccumetModel-25 pH meter (Denver Instrument, Denver, CO, USA) was used to determine the pH. The COD was analyzed by colorimetric method using a Spectrophotometer. The BOD was determined by a 5-Day BOD test. The concentration of SDS in the water was determined spectrophotometrically as described earlier [42,43]. At first, 1 drop of 1% phenolphthalein solution was added to the solution as an indicator. Then, 1 M NaOH was added until the color was changed to pink, which was followed by the addition of 1 M H_2SO_4 until the solution became colorless. The chloroform and methylene blue reagents were then added in the solution. All of the procedure was done in a separatory funnel. The flasks were shaken for about 30 s and for the phase separation, these flasks were left for 30 min. The chloroform layer was extracted in a 100 mL of Erlenmeyer flask. The procedure was repeated thrice by adding 5 mL of chloroform. Three layers of chloroform were obtained. The chloroform layer was extracted in a volumetric flask. The absorbance of chloroform was measured by a spectrophotometer at 652 nm against the blank chloroform. The blank was prepared by adding 5 mL phosphate buffer solution, 2 mL cationic dye (methylene blue) solution, and 5 mL of extracting solvent (chloroform) in 100 mL distilled water. Standard solution was prepared by adding 5 mL phosphate buffer solution, 2 mL the dye solution, and 5 mL of extracting solvent in 100 mL of 10 ppm standard SDS solution. The calculations were performed as follows:

Standard Factor = 10/OD of standard solution

Detergents (mg/L) in wastewater sample = OD of sample × SF

Figure 1. (**A**) Development of floating treatment wetlands. Buoyant mat, (**B**) vegetation of the plant in plastic pot, (**C**) fixing of the pot in the mat, (**D**) placement of the vegetated mat in the tank, (**E**) and different treatments for the remediation of SDS contaminated water.

2.4. Determination of Persistence of Inoculated Bacteria

The number of inoculated bacteria in the water was determined by the plate count method as described earlier [42,43]. Briefly, the water samples were plated on M9 agar plates containing SDS (50 mg L^{-1}). At the end of the experiment, the roots, and shoots of T3 treatments were collected and surface sterilized by 70% ethanol followed by washing with bleach (1%, *v/v*) and rinse with autoclaved distilled water three times. The surface-sterilized roots and shoots were then homogenized in 0.9% (*w/v*) NaCl solution and plated on the M9 media containing SDS (50 mg L^{-1}). The plates were incubated at 37 °C for 48 h. The number of colonies forming units (CFUs) was counted and the identity of the isolates was compared with the inoculated strains by restriction fragment polymorphism analysis as described earlier [32–36].

2.5. Data Analysis

The data were analyzed using the SPSS software package. Comparisons among treatments were carried out by one-way analysis of variance (ANOVA). Duncan's test was applied for ANOVA after testing homogeneity of variance.

3. Results and Discussion

3.1. Performance Evaluation of FTWs

The performance of FTWs with and without bacterial consortium was evaluated by analyzing the water quality parameters such as SDS, COD, BOD, pH, and turbidity of the water samples. FTWs, vegetated with both plants, efficiently reduced the level of all the tested water quality parameters. The reduction of SDS, COD, and BOD was more in the FTWs vegetated with *B. mutica* and *L. fusca* (T1 and T3) (Figures 2–4) as compared to unvegetated treatment (C). Usually, plants can take up the organic contaminants from the environment if the water octanol partition coefficient (log k_{ow}) ranges between 0.5 to 3.5. The log k_{ow} values of SDS is 1.6 that makes it an easy compound to be taken up by the plants. This might be the reason that even vegetation alone significantly reduced the SDS concentration and other pollution parameters in the contaminated water. Nevertheless, performance of *B. mutica* was better than *L. fusca* which could be due to the better adaptability of *B. mutica* in this kind of wastewater. It is previously reported that *B. mutica* outperforms in the wetlands even under harsh environmental conditions for the removal of variety of organic and inorganic contaminants [41]. On the other hand, a significantly better reduction in SDS, COD, and BOD concentration (90–97.5%) was observed in the FTWs having both vegetation and bacterial inoculation (T2 and T4) than in the FTWs having only vegetation (T1 and T3) or bacterial inoculation (T5). This could be due to the effective plant-microbe interplay in the FTWs: (1) inoculated bacteria were previously isolated from the shoot and root interior of plants so they could have already developed mechanisms of proliferation in the plant rhizo- and endosphere that allow the bacteria have helped degrade the SDS and supported the health of host plant in a synergistic manner, and (2) the bacteria possessed genes involved in pollutant degradation and plant growth promoting activities, i.e., 1-aminocyclopropane-1-carboxylate (ACC) deaminase, siderophores production, phosphorus solubilization [54–57]. Once again, maximum reduction in SDS (97.5%), COD (92%), and BOD (94%) were observed in the treatments vegetated with *B. mutica* and bacterial inoculation. This might be attributed to the better plant-bacteria synergism in FTWs having *B. mutica* as compared to *L. fusca*. In our earlier investigations, we found that *B. mutica* allows proliferation of diverse and rich bacteria in the rhizo- and endosphere that overall improves the health of the host plant and degradation potential of the wetland system [42]. The reduction in COD and BOD might be related to the bacterial enzymatic activities that cause the degradation and transformation of SDS into simpler metabolites which are then taken up by the plants in the form of nutrients [41,57]. Both COD and BOD are important water quality parameters, and their attenuation indicates the cleaning of contaminated water [42]. Alongside, high oxygen concentration is fundamental to such environment and successful interactions among plant roots and associated bacteria rely on the availability of oxygen diffusion. In a well growing FTW system, vegetation could have provided oxygen in the rhizosphere through the plant roots thus allowing the microbes to nurture and ultimately leads to the degradation of contaminants [42,54,55]. The synergistic interactions between plants and microorganisms intensified the oxidation and reduction processes which are responsible for the removal and degradation of a wide range of contaminants [56]. These results are following the previously published findings which reported that bacterial inoculation enhanced the efficiency of FTWs [42,57].

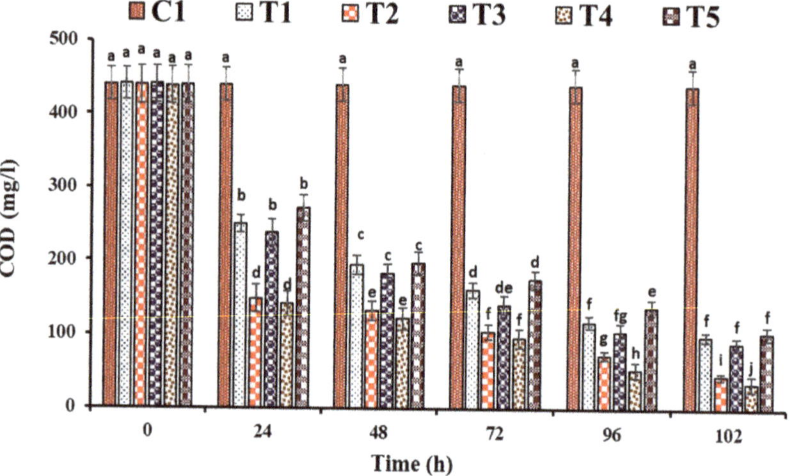

Figure 2. COD of water treated by floating treatment wetlands vegetated with *L. fusca* and *B. mutica* at different times. SDS contaminated water without vegetation and bacteria (C), SDS contaminated water with *L. fusca* (T1), SDS contaminated water with *L. fusca* and bacteria (T2), SDS contaminated water with *B. mutica* (T3), SDS contaminated water with *B. mutica* and bacteria (T4), and SDS contaminated water with bacteria (T5). Error bars indicate the standard error among the three replicates. Labels (a)–(j) indicate statistically significant differences between treatments at a 5% level of significance.

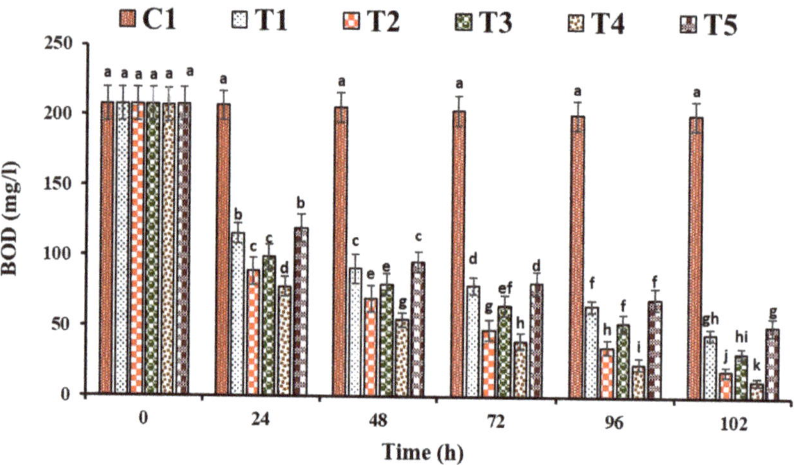

Figure 3. BOD of water treated by floating treatment wetlands vegetated with *L. fusca* and *B. mutica* at different times. SDS contaminated water without vegetation and bacteria (C), SDS contaminated water with *L. fusca* (T1), SDS contaminated water with *L. fusca* and bacteria (T2), SDS contaminated water with *B. mutica* (T3), SDS contaminated water with *B. mutica* and bacteria (T4), and SDS contaminated water with bacteria (T5). Error bars indicate the standard error among the three replicates. Labels (a)–(j) indicate statistically significant differences between treatments at a 5% level of significance.

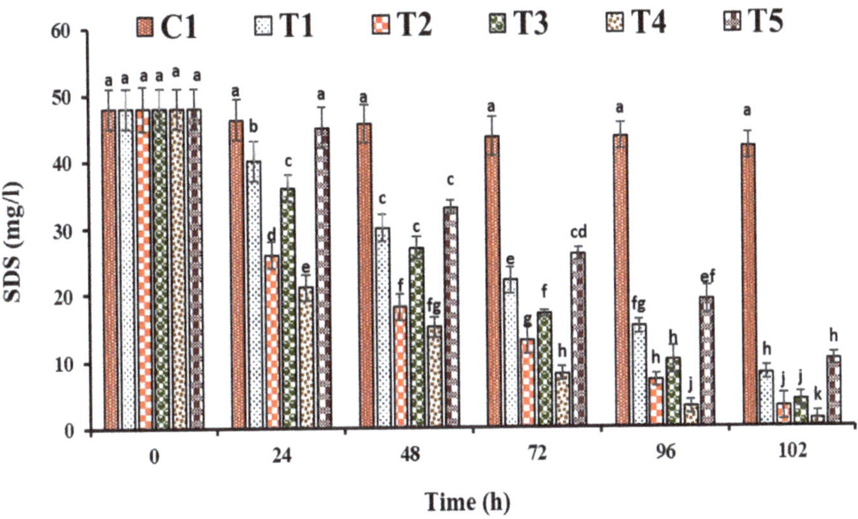

Figure 4. SDS concentrations after treatment with floating treatment wetlands vegetated with *L. fusca* and *B. mutica* at different times. SDS contaminated water without vegetation and bacteria (C), SDS contaminated water with *L. fusca* (T1), SDS contaminated water with *L. fusca* and bacteria (T2), SDS contaminated water with *B. mutica* (T3), SDS contaminated water with *B. mutica* and bacteria (T4), and SDS contaminated water with bacteria (T5). Error bars indicate the standard error among the three replicates. Labels (a)–(j) indicate statistically significant differences between treatments at a 5% level of significance.

In this study, 13% removal of SDS was also observed in the control when no vegetation and bacterial consortium was present. This could be attributed to natural factors such as photooxidation, adsorption, and indigenous role of microbial communities. However, the degradation was further enhanced by vegetation and bacterial inoculation. Specifically, SDS removal was enhanced to >90% when plants and bacterial consortium was applied together. Earlier studies revealed that some bacterial species could degrade the SDS and reduced its concentration to 0.1% within 11 days [11,58–60]. On the contrary, in this study, the plant-bacterial partnership in FTWs reduced the concentration of SDS to 0.1% only in five days.

Apart from the reduction in the above-mentioned parameters, the pH of the SDS contaminated water was slightly decreased, i.e., from 8.6 to 7.9 in FTWs vegetated with *L. fusca*, and from 8.6 to 7.8 in FTWs vegetated with *B. mutica* (Table 1). More reduction in the pH was observed in vegetated wetland units as compared to unvegetated units which might be related to the production of carbon dioxide (CO_2) during roots respiration [45]. Maximum pH reduction was observed in the treatment having *B. mutica* and bacterial inoculation. This decrease in pH might be related to the degradation of SDS by microorganisms that yield CO_2 that reacts with oxygen in the water and produces carbonic acid. Alongside, this reduction might have been related to the acidic root exudates as well which are released by the plants under standard conditions. Previous studies also reported pH reduction in water treated by FTWs vegetated with different plants [45,46]. A similar trend was observed for turbidity removal in all the treatments. However, the FTWs containing both B. mutica and bacterial inoculation showed maximum reduction (99.4%) of turbidity (Table 1).

Table 1. Effect of bacterial augmentation on pH and turbidity of SDS contaminated water treated by floating treatment wetlands vegetated with *Leptochloa fusca* and *Brachiaria mutica*.

Time (h)	Control (C)		L. fusca				B. mutica				T5	
			T1		T2		T3		T4			
	pH	Turbidity	pH	Turbidity	pH	Turbidity	pH	Turbidity	pH	Turbidity	pH	Turbidity
0	8.62 (0.13)	33.62 (2.07)	8.60 (0.22)	33.02 (1.52)	8.57 (0.21)	33.05 (2.78)	8.62 (0.31)	33.04 (1.65)	8.52 (0.22)	33.28 (1.72)	8.62 (0.31)	33.04 (1.15)
24	8.57 (0.28)	32.43 (2.19)	8.43 (0.12)	31.08 (1.23)	8.3 (0.22)	29.28 (1.55)	8.25 (0.12)	30.47 (1.53)	8.15 (0.14)	28.08 (1.56)	8.52 (0.70)	32.05 (1.23)
48	8.4 (0.12)	31.26 (2.54)	8.32 (0.11)	27.25 (1.05)	8.2 (0.14)	23.25 (1.05)	8.14 (0.17)	26.47 (2.05)	8.05 (0.11)	19.45 (1.03)	8.47 (0.20)	30.82 (1.52)
72	8.3 (0.14)	31.08 (3.04)	8.27 (0.15)	22.08 (3.35)	8.1 (0.17)	15.72 (2.08)	8.02 (0.13)	19.48 (1.08)	7.92 (0.20)	12.18 (1.64)	8.32 (0.10)	23.15 (1.58)
96	8.2 (0.21)	30.27 (2.45)	8.13 (0.23)	16.48 (2.95)	8.0 (0.11)	7.08 (1.52)	8.03 (0.15)	12.78 (2.05)	7.83 (0.10)	4.05 (1.17)	8.25 (0.21)	16.24 (1.39)
102	8.2 (0.22)	29.72 (2.06)	8.05 (0.16)	7.05 (1.85)	7.9 (0.23)	2.08 (1.62)	7.92 (0.22)	3.08 (1.04)	7.84 (0.25)	0.18 (0.05)	8.23 (0.22)	11.75 (1.05)

Each is the mean of three replicates, and values in parenthesis indicate standard deviation. SDS contaminated water without vegetation and bacteria (C), SDS contaminated water with *L. fusca* (T1), SDS contaminated water with *L. fusca* and bacteria (T2), SDS contaminated water with *B. mutica* (T3), SDS contaminated water with vegetation (*B. mutica*) and bacteria (T4), and SDS contaminated water with bacteria (T5).

3.2. Persistence of Inoculated Bacteria in FTWs

In phytoremediation, plant-bacteria synergism plays a key role in the degradation of organic contaminants. It has been proposed that the ability of a plant to remediate water is directly related to a number of the contaminants-degrading bacterial population in its different compartments [42,57]. In this study, the persistence of inoculated bacteria was determined in the water as well as the roots and shoots of the plants. In the water from unvegetated treatment (T5), a relatively lower number of inoculated bacteria were found as compared to the water collected from the treatment having vegetation (T2 and T4). This might be due to the lack of mutualistic partnership between the plant and bacteria. The plant provides nutrients, oxygen, and shelter to the residing microorganisms thus allow them to grow, proliferate, and nurture [42,57].

The survival and colonization of inoculated bacteria in the FTWs is highly crucial for efficient degradation of the contaminants [42]. In this study, inoculated bacteria showed survival in the root and shoot interiors of *B. mutica* and *L. fusca*. This could be due to the fact that all of the inoculated bacteria were previously isolated from the rhizosphere, roots, and shoots of the wetland plants. Therefore, they might have developed necessary mechanisms to colonize the plant interior for *B. mutica* and *L. fusca* as well [35,36,41,42,57]. Further, we observed that the bacterial population in the root interior was significantly higher than that of the shoot interior of both plants. The higher population in the roots could be attributed to the fact the inoculated strains were often observed in the rhizospheric and root interior of the wetland plants in earlier studies, which suggest their better colonization potential in the root environment compared to the shoot [52]. Also, in this study, the observations were made after a few days only and the time might not have been sufficient for the active migration of the bacterial communities to the aboveground tissues. Many earlier studies also demonstrated that a higher number of the inoculated bacteria was found in the root interior than the shoot interior [40–42]. Between two plants, the more bacterial population was observed in the root, shoot, and water of the FTWs vegetated with *B. mutica* than *L. fusca* (Table 2). This indicates that *B. mutica* is a more suitable host for the inoculated bacterial community than the *L. fusca*. Many earlier studies also reported that different plant species hosted different numbers of bacteria in their different compartments [42,43].

Table 2. Persistence of the inoculated bacteria in the water, root interior, and shoot interior of *Leptochloa fusca* and *Brachiaria mutica* vegetated in FTWs.

Treatment	Time			
	0 h	48 h	96 h	102 h
Water (CFU ml^{-1})				
L. fusca and bacteria	8.2×10^5 (5.2×10^3)	5.0×10^3 (2×10^2)	4.0×10^2 (2.2×10^2)	1.5×10^2 (0.9×10^2)
B. mutica and bacteria	8.2×10^5 (5.2×10^3)	7.3×10^3 (2.5×10^2)	9.1×10^2 (1.5×10^2)	2.1×10^2 (1.0×10^2)
Bacteria	8.2×10^5 (5.2×10^3)	4.5×10^2 (1.2×10^2)	2.6×10^2 (1.0×10^2)	1.2×10^2 (0.8×10^2)
Root interior (CFU g^{-1})				
L. fusca and bacteria	–	6.6×10^3 (2.3×10^2)	4.8×10^4 (2.2×10^2)	6.4×10^4 (2.7×10^2)
B. mutica and bacteria	–	3.0×10^4 (1.8×10^2)	6.0×10^5 (4.3×10^2)	8.7×10^5 (6.2×10^2)
Shoot interior (CFU g^{-1})				
L. fusca and bacteria	–	1.7×10^2 (0.9×10^2)	5.1×10^3 (2.6×10^2)	6.2×10^3 (3.5×10^2)
B. mutica and bacteria	–	2.7×10^2 (1.1×10^2)	5.8×10^3 (3.0×10^2)	7.0×10^3 (1.1×10^2)

4. Conclusions

This study establishes the usefulness of exploiting rhizospheric and endophytic bacteria in FTW in a partnership with two wetland plants namely *B. mutica* and *L. fusca* for reclamation of water contaminated with SDS. We argue that a traditional FTW can be an effective choice for enhanced SDS removal from the wastewater if well-screened bacterial communities are inoculated in the system. In this way, a successful attenuation in COD, BOD, and pollutant of interest (SDS) could be achieved in a very short time. This study also argues that, if inoculated bacteria are compatible with the host and do not compete for resources with each other, they can survive well *in planta*, support the host health, and improve pollutant degradation. The better performance of *B. mutica* nevertheless indicates that different plants have different capacity of effective plant-microbe interplay which should be investigated carefully before designing an experiment. In the end, this study strengthens the application of pollutant degrading bacteria in FTW for the remediation of water contaminated with organic compounds. Nevertheless, further studies on the activity of enzymes alkyl sulfatases for the degradation of SDS are suggested.

Author Contributions: Writing—original draft preparation, M.Y., Z.Z., E.I., M.T. and S.I.; writing—review and editing, M.A., E.I., and M.R.; supervision, M.A. and M.R. All authors have read and agreed to the published version of the manuscript.

Funding: This research was funded by Higher Education Commission (HEC), Pakistan, grant number 20-3854/R&D/HEC/14/918).

Conflicts of Interest: The authors declare no conflict of interest.

References

1. Shahbazi, R.; Kasra-Kermanshahi, R.; Gharavi, S.; Moosavi-Nejad, Z.; Borzooee, F. Screening of SDS-degrading bacteria from car wash wastewater and study of the alkylsulfatase enzyme activity. *Iran. J. Microbiol.* **2013**, *5*, 153.
2. Ambily, P.; Jisha, M. Metabolic profile of sodium dodecyl sulphate (SDS) biodegradation by *Pseudomonas aeruginosa* (MTCC 10311). *J. Environ. Biol.* **2014**, *35*, 827. [PubMed]
3. Hosseini, F.; Malekzadeh, F.; Amirmozafari, N.; Ghaemi, N. Biodegradation of anionic surfactants by isolated bacteria from activated sludge. *Int. J. Environ. Sci. Technol.* **2007**, *4*, 127–132. [CrossRef]
4. Dereszewska, A.; Cytawa, S.; Tomczak-Wandzel, R.; Medrzycka, K. The effect of anionic surfactant concentration on activated sludge condition and phosphate release in biological treatment plant. *Pol. J. Environ. Stud.* **2015**, *24*, 83–91. [CrossRef]
5. Mazumder, D.; Mukherjee, S. Treatment of automobile service station wastewater by coagulation and activated sludge process. *Int. J. Environ. Sci. Develop.* **2011**, *2*, 64–69. [CrossRef]
6. Jakovljević, V.D.; Vrvić, M.M. The Potential Application of Selected Fungi Strains in Removal of Commercial Detergents and Biotechnology. In *Application and Characterization of Surfactants*; IntechOpen: London, UK, 2017; p. 233. [CrossRef]

7.	Guerrini, A.; Romano, G.; Ferretti, S.; Fibbi, D.; Daddi, D. A performance measurement tool leading wastewater treatment plants toward economic efficiency and sustainability. *Sustainability* **2016**, *8*, 1250. [CrossRef]

8.	Muga, H.E.; Mihelcic, J.R. Sustainability of wastewater treatment technologies. *J. Environ. Manag.* **2008**, *88*, 437–447. [CrossRef] [PubMed]

9.	Balkema, A.J.; Preisig, H.A.; Otterpohl, R.; Lambert, F.J. Indicators for the sustainability assessment of wastewater treatment systems. *Urban Water* **2002**, *4*, 153–161. [CrossRef]

10.	Chen, K.-H.; Wang, H.-C.; Han, J.-L.; Liu, W.-Z.; Cheng, H.-Y.; Liang, B.; Wang, A.-J. The application of footprints for assessing the sustainability of wastewater treatment plants. *J. Clean. Prod.* **2020**, *277*, 124053. [CrossRef]

11.	Paulo, A.; Plugge, C.; García-Encina, P.; Stams, A. Anaerobic degradation of sodium dodecyl sulfate (SDS) by denitrifying bacteria. *Int. Biodeterior. Biodegrad.* **2013**, *84*, 14–20. [CrossRef]

12.	Headley, T.; Tanner, C. In Floating treatment wetlands: An innovative option for stormwater quality applications. In Proceedings of the 11th International Conference on Wetland Systems for Water Pollution Control, Indore, India, 1–7 November 2008; pp. 1101–1106.

13.	Weragoda, S.; Jinadasa, K.; Zhang, D.Q.; Gersberg, R.M.; Tan, S.K.; Tanaka, N.; Jern, N.W. Tropical application of floating treatment wetlands. *Wetlands* **2012**, *32*, 955–961. [CrossRef]

14.	Ladislas, S.; Gerente, C.; Chazarenc, F.; Brisson, J.; Andres, Y. Floating treatment wetlands for heavy metal removal in highway stormwater ponds. *Ecol. Eng.* **2015**, *80*, 85–91. [CrossRef]

15.	Yousaf, S.; Andria, V.; Reichenauer, T.G.; Smalla, K.; Sessitsch, A. Phylogenetic and functional diversity of alkane degrading bacteria associated with Italian ryegrass (*Lolium multiflorum*) and Birdsfoot trefoil (*Lotus corniculatus*) in a petroleum oil-contaminated environment. *J. Hazard. Mater.* **2010**, *184*, 523–532. [CrossRef]

16.	Olkowska, E.; Polkowska, Z.; Namiesnik, J. Analytics of surfactants in the environment: Problems and challenges. *Chem. Rev.* **2011**, *111*, 5667–5700. [CrossRef]

17.	Yen Doan, T.H.; Minh Chu, T.P.; Dinh, T.D.; Nguyen, T.H.; Tu Vo, T.C.; Nguyen, N.M.; Nguyen, B.H.; Pham, T.D. Adsorptive removal of Rhodamine B using novel adsorbent-based surfactant-nodified alpha alumina nanoparticles. *J. Anal. Methods Chem.* **2020**, *2020*. [CrossRef] [PubMed]

18.	Pham, H.D.; Dang, T.H.M.; Nguyen, T.T.N.; Nguyen, T.A.H.; Pham, T.N.M.; Pham, T.D. Separation and determination of alkyl sulfate surfactants in wastewater by capillary electrophoresis coupled with contactless conductivity detection after preconcentration by simultaneous adsorption using alumina beads. *Electrophoresis* **2021**, *42*, 191–199. [CrossRef] [PubMed]

19.	Li, P.; Ishiguro, M. Adsorption of anionic surfactant (sodium dodecyl sulfate) on silica. *Soil Sci. Plant Uutr.* **2016**, *62*, 223–229. [CrossRef]

20.	Kuppusamy, S.; Thavamani, P.; Venkateswarlu, K.; Lee, Y.B.; Naidu, R.; Megharaj, M. Remediation approaches for polycyclic aromatic hydrocarbons (PAHs) contaminated soils: Technological constraints, emerging trends and future directions. *Chemosphere* **2017**, *168*, 944–968. [CrossRef] [PubMed]

21.	Rashed, M.N. *Adsorption Technique for the Removal of Organic Pollutants from Water and Wastewater, Organic Pollutants—Monitoring, Risk and Treatment*; IntechOpen: London, UK, 2013; pp. 167–194. [CrossRef]

22.	Rasheed, T.; Bilal, M.; Nabeel, F.; Adeel, M.; Iqbal, H.M. Environmentally-related contaminants of high concern: Potential sources and analytical modalities for detection, quantification, and treatment. *Environ. Int.* **2019**, *122*, 52–66. [CrossRef]

23.	Yaseen, Z.M.; Zigale, T.T.; Kumar, R.; Salih, S.Q.; Awasthi, S.; Tung, T.M.; Al-Ansari, N.; Bhagat, S.K. Laundry wastewater treatment using a combination of sand filter, bio-char and teff straw media. *Sci. Rep.* **2019**, *9*, 1–11. [CrossRef]

24.	Ijaz, A.; Imran, A.; ul Haq, M.A.; Khan, Q.M.; Afzal, M. Phytoremediation: Recent advances in plant-endophytic synergistic interactions. *Plant Soil* **2016**, *405*, 179–195. [CrossRef]

25.	Ojo, O.A.; Oso, B.A. Biodegradation of synthetic detergents in wastewater. *Afr. J. Biotechnol.* **2009**, *8*, 229–249. [CrossRef]

26.	Sharman, S.H. Extensive biodegradation of synthetic detergents. *Nature* **1964**, *201*, 704–705. [CrossRef]

27.	Khan, M.S.; Zaidi, A.; Wani, P.A.; Oves, M. Role of plant growth promoting rhizobacteria in the remediation of metal contaminated soils. *Environ.Chem. Lett.* **2009**, *7*, 1–19. [CrossRef]

28.	Scott, M.J.; Jones, M.N. The biodegradation of surfactants in the environment. *Biochimica et Biophysica Acta (BBA)-Biomembranes* **2000**, *1508*, 235–251. [CrossRef]

29.	Khan, S.; Afzal, M.; Iqbal, S.; Khan, Q.M. Plant-bacteria partnerships for the remediation of hydrocarbon contaminated soils. *Chemosphere* **2013**, *90*, 1317–1332. [CrossRef]

30.	Abhilash, P.; Powell, J.R.; Singh, H.B.; Singh, B.K. Plant–microbe interactions: Novel applications for exploitation in multipurpose remediation technologies. *Trends Biotechnol.* **2012**, *30*, 416–420. [CrossRef] [PubMed]

31.	Hodson, M.E. The need for sustainable soil remediation. *Elements* **2010**, *6*, 363–368. [CrossRef]

32.	Verhoeven, J.T.; Meuleman, A.F. Wetlands for wastewater treatment: Opportunities and limitations. *Ecol. Eng.* **1999**, *12*, 5–12. [CrossRef]

33.	Song, J.; Li, Q.; Dzakpasu, M.; Wang, X.C.; Chang, N. Integrating stereo-elastic packing into ecological floating bed for enhanced denitrification in landscape water. *Bioresour. Technol.* **2020**, *299*, 122601. [CrossRef]

34.	Park, J.B.; Sukias, J.P.; Tanner, C.C. Floating treatment wetlands supplemented with aeration and biofilm attachment surfaces for efficient domestic wastewater treatment. *Ecol. Eng.* **2019**, *139*, 105582. [CrossRef]

35. Rehman, K.; Imran, A.; Amin, I.; Afzal, M. Inoculation with bacteria in floating treatment wetlands positively modulates the phytoremediation of oil field wastewater. *J. Hazard. Mater.* **2018**, *349*, 242–251. [CrossRef] [PubMed]
36. Afzal, M.; Arslan, M.; Müller, J.A.; Shabir, G.; Islam, E.; Tahseen, R.; Anwar-ul-Haq, M.; Hashmat, A.J.; Iqbal, S.; Khan, Q.M. Floating treatment wetlands as a suitable option for large-scale wastewater treatment. *Nat. Sustain.* **2019**, *2*, 863–871. [CrossRef]
37. Colares, G.S.; Dell'Osbel, N.; Wiesel, P.G.; Oliveira, G.A.; Lemos, P.H.Z.; da Silva, F.P.; Lutterbeck, C.A.; Kist, L.T.; Machado, Ê.L. Floating treatment wetlands: A review and bibliometric analysis. *Sci. Total Environ.* **2020**, *714*, 136776. [CrossRef] [PubMed]
38. Yeh, N.; Yeh, P.; Chang, Y.-H. Artificial floating islands for environmental improvement. *Renew. Sustain. Energy Rev.* **2015**, *47*, 616–622. [CrossRef]
39. Hu, Z.; Li, D.; Guan, D. Water quality retrieval and algae inhibition from eutrophic freshwaters with iron-rich substrate based ecological floating beds treatment. *Sci. Total Environ.* **2020**, *712*, 135584. [CrossRef]
40. Afzal, M.; Rehman, K.; Shabir, G.; Tahseen, R.; Ijaz, A.; Hashmat, A.J.; Brix, H. Large-scale remediation of oil-contaminated water using floating treatment wetlands. *NPJ Clean Water* **2019**, *2*, 3. [CrossRef]
41. Hussain, Z.; Arslan, M.; Malik, M.H.; Mohsin, M.; Iqbal, S.; Afzal, M. Treatment of the textile industry effluent in a pilot-scale vertical flow constructed wetland system augmented with bacterial endophytes. *Sci. Total Environ.* **2018**, *645*, 966–973. [CrossRef]
42. Ijaz, A.; Shabir, G.; Khan, Q.M.; Afzal, M. Enhanced remediation of sewage effluent by endophyte-assisted floating treatment wetlands. *Ecol. Eng.* **2015**, *84*, 58–66. [CrossRef]
43. Karigar, C.; Rao, S. Role of microbial enzymes in bioremediation of pollutans. *Enzyme Res.* **2011**, 2011805187. [CrossRef]
44. Hussain, Z.; Arslan, M.; Shabir, G.; Malik, M.H.; Mohsin, M.; Iqbal, S.; Afzal, M. Remediation of textile bleaching effluent by bacterial augmented horizontal flow and vertical flow constructed wetlands: A comparison at pilot scale. *Sci. Total Environ.* **2019**, *685*, 370–379. [CrossRef]
45. Nawaz, N.; Ali, S.; Shabir, G.; Rizwan, M.; Shakoor, M.B.; Shahid, M.J.; Afzal, M.; Arslan, M.; Hashem, A.; Abd-Allah, E.F. Bacterial augmented floating treatment wetlands for efficient treatment of synthetic textile dye wastewater. *Sustainability* **2020**, *12*, 3731. [CrossRef]
46. Chen, Z.; Cuervo, D.P.; Müller, J.A.; Wiessner, A.; Köser, H.; Vymazal, J.; Kästner, M.; Kuschk, P. Hydroponic root mats for wastewater treatment. *Environ. Sci. Pollut. Res.* **2016**, *23*, 15911–15928. [CrossRef] [PubMed]
47. Arslan, M.; Imran, A.; Khan, Q.M.; Afzal, M. Plant–bacteria partnerships for the remediation of persistent organic pollutants. *Environ. Sci. Pollut. Res.* **2017**, *24*, 4322–4336. [CrossRef] [PubMed]
48. Shahid, M.J.; Arslan, M.; Ali, S.; Siddique, M.; Afzal, M. Floating wetlands: A sustainable tool for wastewater treatment. *Clean Soil Air Water.* **2018**, *46*, 1800120. [CrossRef]
49. Lyu, J.; Lin, G.; Fan, Z.; Lin, W.; Dai, Z. Suitable plant combinations for ecological floating beds in eutrophic subtropical coastal wetlands under different salinities: Experimental evidences. *Int. J. Environ. Sci. Technol.* **2020**, *17*, 4505–4516. [CrossRef]
50. Mustafa, H.M.; Hayder, G. Recent studies on applications of aquatic weed plants in phytoremediation of wastewater. *Ain Shams Eng. J.* **2020**. [CrossRef]
51. Sessitsch, A.; Coenye, T.; Sturz, A.; Vandamme, P.; Barka, E.A.; Salles, J.; Van Elsas, J.; Faure, D.; Reiter, B.; Glick, B. *Burkholderia phytofirmans* sp. nov., a novel plant-associated bacterium with plant-beneficial properties. *Int. J. Syst. Evol. Microbiol.* **2005**, *55*, 1187–1192. [CrossRef] [PubMed]
52. Fatima, K.; Afzal, M.; Imran, A.; Khan, Q.M. Bacterial rhizosphere and endosphere populations associated with grasses and trees to be used for phytoremediation of crude oil contaminated soil. *Bull. Environ. Contam. Toxicol.* **2015**, *94*, 314–320. [CrossRef]
53. Federation, Water Environmental, and APH Association. *Standard Methods for the Examination of Water and Wastewater*; American Public Health Association (APHA): Washington, DC, USA, 2005.
54. Rehman, K.; Imran, A.; Amin, I.; Afzal, M. Enhancement of oil field-produced wastewater remediation by bacterially-augmented floating treatment wetlands. *Chemosphere* **2019**, *217*, 576–583. [CrossRef]
55. Shehzadi, M.; Afzal, M.; Khan, M.U.; Islam, E.; Mobin, A.; Anwar, S.; Khan, Q.M. Enhanced degradation of textile effluent in constructed wetland system using *Typha domingensis* and textile effluent-degrading endophytic bacteria. *Water Res.* **2014**, *58*, 152–159. [CrossRef]
56. Skrzypiecbcef, K.; Gajewskaad, M.H. The use of constructed wetlands for the treatment of industrial wastewater. *J. Water Land Dev.* **2017**, *34*, 233–240. [CrossRef]
57. Tara, N.; Arslan, M.; Hussain, Z.; Iqbal, M.; Khan, Q.M.; Afzal, M. On-site performance of floating treatment wetland macrocosms augmented with dye-degrading bacteria for the remediation of textile industry wastewater. *J. Clean Prod.* **2019**, *217*, 541–548. [CrossRef]
58. Chaturvedi, V.; Kumar, A. Diversity of culturable sodium dodecyl sulfate (SDS) degrading bacteria isolated from detergent contaminated ponds situated in Varanasi city, India. *Int. Biodeterior. Biodegrad.* **2011**, *65*, 961–971. [CrossRef]
59. Usharani, J.; Rajasekar, T.; Deivasigamani, B. Isolation and characterization of SDS (sodium dodecyl sulfate) degrading organisms from SDS contaminated areas. *Open Access Sci. Rep.* **2012**, *1*, 1–4. [CrossRef]
60. Effendi, I.; Nedi, S.; Pakpahan, R. Detergent disposal into our environmentand its impact on marine microbes. In *IOP Conference Series: Earth and Environmental Science*; IOP Publishing: Bristol, UK, 2017; Volume 97, p. 012030. [CrossRef]

Article

Microbial Communities Associated with Acetaminophen Biodegradation from Mangrove Sediment

Chu-Wen Yang, Yi-En Chen and Bea-Ven Chang *

Department of Microbiology, Soochow University, Taipei 11101, Taiwan; ycw6861@scu.edu.tw (C.-W.Y.); xx0521xx@gmail.com (Y.-E.C.)
* Correspondence: bvchang@scu.edu.tw; Tel.: +886-228-819-471 (ext. 6859)

Received: 10 May 2020; Accepted: 2 July 2020; Published: 4 July 2020

Abstract: Acetaminophen (ACE) is a widely used medicine. Currently, concerns regarding its potential adverse effects on the environments are raised. The aim of this study was to evaluate ACE biodegradation in mangrove sediments under aerobic and anaerobic conditions. Three ACE biodegradation strategies in mangrove sediments were tested. The degradation half-lives ($t_{1/2}$) of ACE in the sediments with spent mushroom compost under aerobic conditions ranged from 3.24 ± 0.16 to 6.25 ± 0.31 d. The degradation half-lives ($t_{1/2}$) of ACE in sediments with isolated bacterial strains ranged from 2.54 ± 0.13 to 3.30 ± 0.17 d and from 2.62 ± 0.13 to 3.52 ± 0.17 d under aerobic and anaerobic conditions, respectively. The degradation half-lives ($t_{1/2}$) of ACE in sediments amended with $NaNO_3$, Na_2SO_4 and $NaHCO_3$ under anaerobic conditions ranged from 1.16 ± 0.06 to 3.05 ± 0.15 d, 2.39 ± 0.12 to 3.84 ± 0.19 d and 2.79 ± 0.14 to 10.75 ± 0.53 d, respectively. The addition of the three electron acceptors enhanced ACE degradation in mangrove sediments, where $NaNO_3$ yielded the best effects. Sixteen microbial genera were identified as the major members of microbial communities associated in anaerobic ACE degradation in mangrove sediments with addition of $NaNO_3$ and Na_2SO_4. Three (*Arthrobacter*, *Enterobacter* and *Bacillus*) of the sixteen microbial genera were identified in the isolated ACE-degrading bacterial strains.

Keywords: acetaminophen; mangrove sediments; biodegradation; aerobic conditions; anaerobic conditions

---

1. Introduction

Acetaminophen (N-acetyl-*p*-aminophenol; ACE) is a widely used representative medicine. As a nonsteroidal anti-inflammatory drugs (NSAIDs), ACE exhibits analgesic and antipyretic properties and acts via inhibition of cyclooxygenase enzymes [1]. The reported concentrations of ACE range from 0.003 to 30 mg L^{-1} in stream water, sewage treatment plant influents and effluents [2–4]. The frequent detection of ACE in aquatic environments has raised concerns regarding its potential deleterious effects on the environments [5–8].

Mangrove ecosystems along the coastlines of tropical and subtropical regions are important intertidal estuary wetlands that considered to be significant sinks of pollutants from contaminated tidal water and discharges from freshwater [9]. The mangroves of Guandu and Bali are located on the banks of the Danshui River, one of the most polluted rivers in northern Taiwan. The concentrations and degradation of nonylphenol, sulfonamides and polycyclic aromatic hydrocarbons in the mangrove sediments have been reported [10–13].

Physicochemical methods such as primarily advanced oxidation processes (AOPs), homogeneous and heterogeneous photocatalysis, Fenton and Fenton like reactions, ozonation, and methods involving ultrasound and microwave treatments, or electrochemical processes are not appropriate for treating

ACE in mangrove sediments [14]. Biodegradation is an effective strategy to remove organic pollutants from sediments. Most studies on pharmaceutical biodegradation are focused on their removal during wastewater treatment processes [15]. Some investigations evaluated the microbial degradation of drugs and artificial compounds in freshwater [16] as well as in ocean and estuary waters [17]. The microbial degradation of drugs depends on the prevailing oxygen availability and redox conditions in sediments. However, little is known regarding the aerobic and anaerobic biodegradation potential of ACE in mangrove sediments.

Biodegradation is believed to be an effective strategy for eliminating contaminating ACE in environments. To enhance the efficiency of degradation, three remedial strategies have been proposed: natural attenuation, bioaugmentation and biostimulation [18]. Microbial degradation of ACE has been observed in bacterial strains. *Pseudomonas aeruginosa* strain HJ1012 was isolated on ACE as a sole carbon [19]. *Pseudomonas moorei* KB4 can metabolize ACE, with *p*-aminophenol and hydroquinone identified as degradation products [14]. Two ACE-metabolizing strains, *Delftia tsuruhatensis* and *Pseudomonas aeruginosa*, were isolated from the membrane bioreactor [20]. White rot fungus *Pleurotus eryngii* is one of the most widespread mushrooms consumed in the world. spent mushroom compost (SMC) is a mushroom industry waste which contains extracellular enzymes with organic pollutant degradation ability [21]. The addition of enzyme containing microcapsules (MC) was found to be effective for aerobic degradation of organic pollutants [12] and the aerobic degradation of tetracyclines in the river sediments [22]. Addition of $NaHCO_3$, Na_2SO_4, and$NaNO_3$ could create methanogenic conditions, sulfate-reducing conditions and nitrate-reducing conditions, respectively, as well as enhance anaerobic degradation of organic pollutants [13]. Therefore, the addition of ACE-degrading bacteria, MC and electron acceptors such as $NaNO_3$, Na_2SO_4 and $NaHCO_3$ were used to promote ACE degradation in this study.

The aim of this study is to evaluate strategies for enhancing biodegradation of ACE in mangrove sediments under aerobic and anaerobic conditions. Three strategies, including addition of MC, addition of ACE-degrading bacterial strains isolated from sediments and addition of electron acceptors ($NaNO_3$, Na_2SO_4 or $NaHCO_3$). The microbial communities involved in aerobic and anaerobic degradation of ACE in the mangrove sediments were investigated.

2. Materials and Methods

2.1. Chemicals

The ACE (99.0% analytical standard) used in this study was purchased from the Aldrich Chemical Co. Solvents were purchased from Mallinckrodt, while all other chemicals were purchased from Sigma Chemical Co. (USA).

2.2. Sample Collection

Samples were taken from the Guandu sampling site (25.11°68.43′ N, 121.46°41.53′ E) and Bali sampling site (25.15°86.13′ N, 121.43°55.75′ E) at Tamsui, northern Taiwan. Figure S1 shows the locations of the two sample collection sites. Sediments (0–15 cm) were collected in spring (March 2015). All samples from each sampling site were randomly collected, in triplicate, from an area approximately 1 m^2. Adaptation was performed by adding 100 mg kg^{-1} ACE to 500 g of sediment at 14-d intervals at 30 °C for 6 months under aerobic or anaerobic conditions. The sediment samples were analyzed for salinity, temperature, TOC, bacterial count, and ACE concentration. The salinity and temperature were measured by salinity/temperature meter (model 30, YSl, USA). The sediment samples were mixed in a ratio of 1:1 with distilled water in a beaker before inserting the probe. Readings were taken after allowing the instrument to stabilize. The TOC was measured using a TOC Analyzer (OI Analytical 1030 W, USA). The bacterial counts were enumerated by pour plate method and grown on R2A agar. Sediments properties for the Guandu and Bali sampling sites were listed in Table 1.

Table 1. Sediments properties for the two sampling sites.

Parameters	Guandu Sampling Sites	Bali Sampling Sites
Bacterial counts	$(1.9 \pm 0.2) \times 10^6$ CFU g^{-1}	$(2.6 \pm 0.3) \times 10^5$ CFU g^{-1}
Temperature	18.1 ± 1.1 °C	20.5 ± 1.9 °C
Salinities	11.1‰ $\pm$ 0.3‰	16.5‰ $\pm$ 1.5‰
TOC	3.6 wt% $\pm$ 0.2 wt%	2.5 wt% $\pm$ 0.4 wt%
ACE concentrations	0.89 mg kg^{-1}	0.45 mg kg^{-1}

2.3. Medium

The aerobic medium contained the followings (mg L^{-1}): K_2HPO_4, 65.3; KH_2PO_4, 25.5; $Na_2HPO_4 \cdot 12\ H_2O$, 133.8; NH_4Cl, 5.1; $CaCl_2$, 82.5; $MgSO_4 \cdot 7H_2O$, 67.5; and $FeCl_3 \cdot 6H_2O$, 0.75. The anaerobic medium contained the followings (mg L^{-1}): NH_4Cl, 2.7; $MgCl_2 \cdot 6H_2O$, 0.1; $CaCl_2 \cdot 2H_2O$, 0.1; $FeCl_2 \cdot 4H_2O$, 0.02; K_2HPO_4, 0.27; KH_2PO_4, 0.35; and resazurin, 0.001. The pH of the medium was adjusted to 7.0 using 0.9-mM titanium citrate as a reducing reagent before autoclaving. The aerobic and anaerobic medium were used under aerobic and anaerobic conditions, respectively.

2.4. Preparation of Enzyme Extract-Containing Microcapsules (MC)

The SMC of *Pleurotus eryngii* was obtained from a mushroom cultivation farm in Chiayi, Taiwan. The enzyme extract was extracted from the 120 g SMC with 600 mL sodium acetate buffer (pH 5.0) for 3 h at 4 °C. Alginate solution was made by dissolving sodium alginate (4 wt%) in 0.9-wt% sodium chloride with stirring for 1 h. Enzyme extract solution was then added into the alginate solution. An electrostatic droplet generator was used to prepare the MC. The mixture was drawn into a 10-mL syringe fitted with a needle and attached to a syringe pump that provided a steady solution flow rate of 25.2 mL/h and fixed voltage (12 kV) into a gently agitated aqueous solution of calcium chloride 1.5 wt% to form MC of 250 nm in diameter for experiments [12].

2.5. Experimental Setting

Experiments were performed under aerobic and anaerobic conditions. Aerobic experiments were performed using 125 mL serum bottles that contained 40 mL of aerobic medium, 5 g of sediment, 5 mL of the MC and 50 mg kg^{-1} of ACE, which were incubated on a rotary shaker (120 rpm) at 30 °C in the dark. Anaerobic experiments were performed using serum bottles containing 45 mL of anaerobic medium, 5 g of sediment, 20 mM of electron acceptors ($NaNO_3$, Na_2SO_4 or $NaHCO_3$) and 50 mg kg^{-1} of ACE. All experiments were conducted in an anaerobic glove box (Forma Scientific, model 1025 S/N, USA) filled with N_2 (85%), H_2 (10%) and CO_2 (5%) gases. The bottles were capped with butyl rubber stoppers and crimp seals, wrapped in aluminum foil, and then incubated without shaking at 30 °C. Inoculated controls containing 45 mL of aerobic or anaerobic medium, 5 g of sediment and 50 mg kg^{-1} of ACE (without the addition of MC or electron acceptors) were incubated at 30 °C. Sterile controls containing 45 mL of aerobic or anaerobic medium and 5 g of sediment were autoclaved at 121 °C for 30 min. The duration of aerobic or anaerobic experiments was 24 d. Each experiment was repeated 3 times.

The ACE concentrations in the sterile controls were examined for a 24-d incubation period. The remaining amount of ACE in the sediment ranged from 98.5% to 96.8%, indicating that the aerobic and anaerobic ACE degradation in all of the experiments were due to microbial activity.

2.6. Isolation, Identification and Tests of the ACE-Degrading Bacteria

ACE-degrading bacteria were isolated from sediments under aerobic or anaerobic conditions. The enrichment procedure was performed using aerobic or anaerobic medium agar plates containing 10 mg L^{-1} of ACE to isolate bacterial clones. To confirm the ACE-degrading ability of isolated bacterial strains, aerobic degradation experiments were performed using 5 mL cultures of the aerobic bacterial strains, 45 mL of aerobic medium and 2 mg L^{-1} of ACE on a shaker (120 rpm) at 30 °C in the dark. Anaerobic experiments were performed using serum bottles containing 45 mL of anaerobic medium, 5 g of sediment and 2 mg L^{-1} of ACE and were conducted in an anaerobic glove box. The bottles were capped with butyl rubber stoppers and crimp seals, wrapped in aluminum foil, and then incubated without shaking at 30 °C. Samples were periodically taken to analyze residual ACE. Each individual colony was purified and then identified by 16S rRNA gene sequencing after PCR amplification with the primers F8 and R1510. The PCR products were sequenced using an ABI Prism automatic sequencer. The 16S rRNA gene database was searched using BLAST of the National Center for Biotechnology Information (NCBI). Phylogenetic analysis was performed using ClustalX 2.0 with 1000 bootstraps [23].

Isolated bacterial strains were added to Guandu and Bali sediments under aerobic and anaerobic conditions to test their ACE-degrading ability. ACE degradation was assessed following the addition of aerobic and anaerobic ACE-degrading bacteria into the sediment. The experiment was performed using 1 mL of ACE-degrading bacteria, 5 g sediment, 45 mL of medium and 2 mg L^{-1} of ACE.

2.7. Analytical Methods

ACE was extracted from sediment samples twice by water (with 0.1% formic acid): acetonitrile: methanol (10:3:1), and then extracted using a C18 solid phase extraction cartridge. The SPE cartridges were Chromabond®HR-X (500 mg, 6 mL), and SPE was conducted at a sample pH of 3.0.

Extracts were analyzed using an Agilent 1260 HPLC system equipped with a 4.6 × 250 mm column (Zorbax Eclipse Plus C18, Agilent) and a photodiode array detector with monitoring at 270 nm. The mobile phase consisted of acetonitrile and water (containing 0.1% formic acid) at a ratio of 30%:70%. The recovery percentage was 96.4%, and the detection limit (LOD) was 0.05 mg L^{-1}, respectively. The remaining percentage of ACE (Rp) was calculated using the formula:

$$Rp = (RC_{ACE}/IC_{ACE}) \times 100\% \tag{1}$$

where RC_{ACE} is the residual ACE concentration and IC_{ACE} is the initial ACE concentration. The ACE degradation data collected in this study well fit first-order kinetics

$$t = -\ln(C/C_0)/k \tag{2}$$

where C_0 is the initial ACE concentration, C is the ACE concentration, t is the time period, $t_{1/2}$ is the half-live, and k is the degradation rate constant).

2.8. Microbial Community Analysis

Total DNA was extracted from experimental samples using a PowerSoil DNA Isolation kit (QIAGEN). DNA from three experiments (bottles) of each treatment were extracted and pooled together to perform NGS. The V5–V8 variable region of the 16S rRNA gene was amplified as described previously [12,13]. Next-generation sequencing (NGS) was performed at the Genome Center of the National Yang-Ming University, Taiwan using a MiSeq platform (Illumina, Inc.). Chimeric sequences in the 16S rRNA gene sequence data were removed using Chimera Check. The classifier from the RPD pipeline was used to assign taxonomic groups with a 95% sequence similarity. Cluster analysis of the microbial community compositions in the experimental samples was performed using the heatmap3 package of R. The differences in microbial composition between experimental samples were identified using the prop.test (two proportion test) function in the package MASS of R (https://www.r-project.org/). A p-value of less than 0.05 was considered to be significant. Specific microbial communities (such as those involved in the nitrogen cycle and aromatic compound degradation) were identified by integration of NGS data and the microbial list in the KEGG modules [24].

3. Results and Discussion

3.1. Aerobic ACE Degradation

The aerobic degradation of the ACE in the Guandu and Bali sediments are shown in Figure 1. We first found that the remaining percentage of ACE observed after 24 d in the Guandu and Bali non-ACE-adapted sediments without the addition of MC was 65.1% ± 4.1% and 80.8% ± 3.4%, respectively. The remaining ACE in the Guandu and Bali ACE-adapted sediments without added MC was 15.4% ± 3.7% and 38.4% ± 4.1%, respectively. The remaining ACE after 24 d in the Guandu and Bali ACE-adapted sediments without added MC was 15.4% ± 3.7% and 38.4% ± 4.1%, respectively. For the Guandu and Bali ACE-adapted sediment supplemented with MC, ACE was completely degraded after 12 and 18 d, respectively.

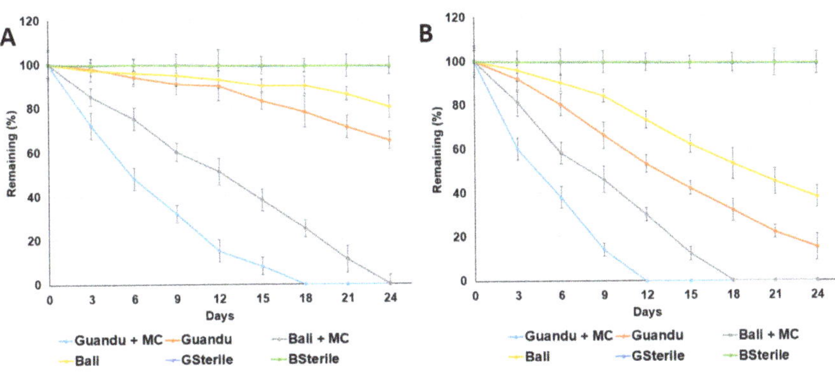

Figure 1. Aerobic degradation of acetaminophen with MC in non-acetaminophen (ACE)-adapted sediment (**A**) and ACE-adapted sediment; (**B**) Guandu: Guandu sediment. Bali: Bali sediment. GSterile: sterilized Guandu sediment. BSterile: sterilized Bali sediment. MC: enzyme extract-containing microcapsules. Data from three independent experiments are presented as the means ± SE.

The ACE degradation data in Figure 1 were fitted to first-order kinetics. For the Guandu and Bali non-ACE-adapted sediments supplemented with or without MC, the observed ACE half-lives were 3.24 ± 0.06 and 33.55 ± 0.16 d and 6.25 ± 0.31 and 66.66 ± 3.33 d, respectively. For the Guandu and Bali ACE-adapted sediments supplemented with or without MC, the observed ACE half-lives were 2.52 ± 0.12 and 7.42 ± 0.37 d and 4.33 ± 0.31 and 14.20 ± 0.71 d, respectively.

The results indicate that the rate of ACE degradation was higher in the Guandu sediments than in the Bali sediments. As shown in Table 1, the Guandu sediments exhibit higher TOC, bacterial counts and lower salinity and ACE concentrations than those of the Bali sediments. The salinities at the Guandu sediment and Bali sediment were 11.1‰ ± 0.3‰ and 16.5‰ ± 1.5‰, respectively. Salinity may affect organic pollutants degradation in the environment [25]. This result is consistent with biodegradation of phenanthrene by bacteria isolated from mangrove sediments [26].

A comparison of ACE degradation in sediments with or without ACE adaptation indicated that the adaptation process enhanced aerobic ACE degradation. The adaptation of microbial populations occurred by the induction of enzymes necessary for degradation followed by an increase in the population of degrading organisms [18]. The results also revealed the degradation of ACE was enhanced with the addition of MC in the sediments. Similar results were observed in a previous study which reported that the addition of MC enhanced sulfonamide degradation in sediments [12].

3.2. ACE Degradation by ACE-Degrading Bacteria

Ten bacterial strains with the capability to using ACE as a carbon source were isolated from the ACE-adapted sediments under aerobic conditions. Four of the ten strains (PF1, PF3, SC and SEC) exhibited higher ACE-degrading activities than the other strains (Figure 2A). The ACE degradation data collected in this study well fit first-order kinetics. The observed half-lives of ACE in the presence of strains SC, SEC, PF1 and PF3 were 0.32 ± 0.01, 0.48 ± 0.02, 0.55 ± 0.03 and 1.31 ± 0.06 d, respectively. The ACE-degrading ability of the strains exhibited the following order: strain SC > strain SEC > strain PF1 > strain PF3.

Nine bacterial strains with the ability to use ACE as a carbon source were isolated from the ACE-adapted sediments under anaerobic conditions. Three of the nine strains (E, G and J) exhibited a greater ACE-degrading capability than the other strains (Figure 2B). The ACE degradation data collected in this study well fit first-order kinetics. The observed degradation half-lives of ACE in the presence of strains E, G and J were 0.42 ± 0.02, 0.56 ± 0.03 and 0.67 ± 0.03 d, respectively. The ACE-degrading ability of the strains exhibited the following order: strain E > strain G > strain J.

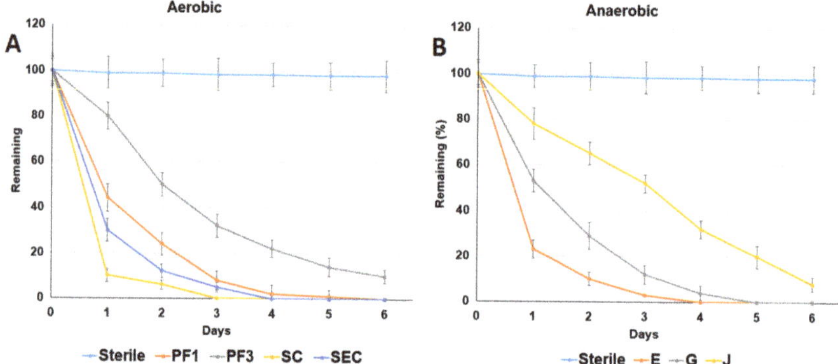

Figure 2. Comparison of the acetaminophen degradation abilities of the isolated bacterial strains. (**A**) Bacterial strains PF1, PF3, SC and SEC (without sediments) under aerobic conditions; (**B**) bacterial strains E, G and J (without sediments) under anaerobic conditions. Sterile: sterilized medium without the addition of bacteria. GS: Guandu sediment. BS: Bali sediment. GSterile: sterilized Guandu sediment. BSterile: sterilized Bali sediment. Data from three independent experiments are presented as the means ± SE.

A phylogenetic analysis of the 16S rRNA gene sequences of the isolated bacterial strains is shown in Figure 3. The aerobic strains PF1, PF3, SC and SEC were observed to be closely associated with *Bacillus pumilus*, *Bacillus aerius*, *Arthrobacter ginkgonis* and *Bacillus aryabhattai*, with similarities of 99%, 100%, 99% and 100%, respectively. In addition, the anaerobic strains E, G and J were observed be closely related to *Enterobacter hormechei*, *Clostridium sphenoides* and *Lysinibacillus* sp., with similarities of 99%, 99% and 99%, respectively (Table 2).

Table 2. The 16S rRNA gene-sequence blast results of the seven bacterial strains.

Strain	Name	Identity	Accession Number
PF1	*Bacillus pumilus* strain NBRC 12092	1251/1255 (99%)	NR_112637
PF3	*Bacillus aerius* strain 24K	1326/1326 (100%)	NR_118439
SC	*Arthrobacter ginkgonis* strain SYP-A7299	1283/1302 (99%)	NR_156061
SEC	*Bacillus aryabhattai* B8W22	1330/1330 (100%)	NR_115953
E	*Enterobacter hormechei* subsp. xiangfangensis strain 10–17	1292/1299 (99%)	NR_126208
G	*Clostridium sphenoides* JCM 1415 strain ATCC 19403	1260/1268 (99%)	NR_026409
J	*Lysinibacillus fusiformis* strain NBRC15717	1301/1305 (99%)	NR_112569

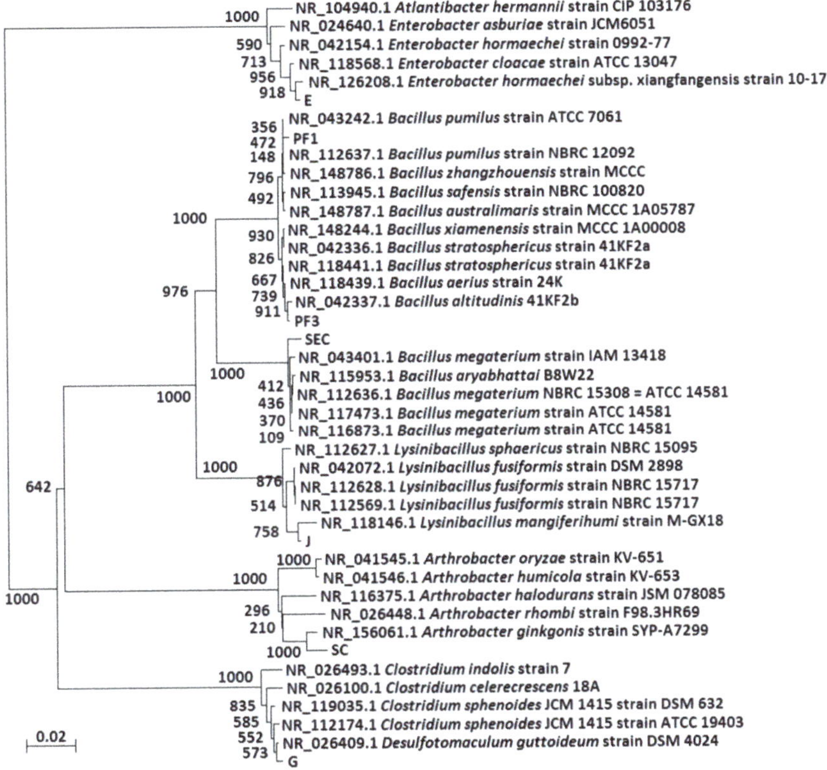

Figure 3. Phylogenetic analysis of 16S rRNA sequences of the seven bacterial strains (PF1, PF3, SEC, SC, E, G and J). Seven isolated bacterial strains: PF1, PF3, SC, SEC, E, G and J.

As shown in Figure 4, ACE degradation in the sediments was enhanced with the addition of the bacterial strains SC and E under aerobic and anaerobic conditions, respectively. The ACE degradation data in Figure 4 were fitted to first-order kinetics. The observed ACE half-lives (d) were 2.54 and 3.30 d in the Guandu and Bali sediments with the additions of bacterial strain SC, while values of 2.62 and 3.52 d were obtained in the Guandu and Bali sediments with the addition of bacterial strain E. These results indicated that ACE degradation in the Guandu and Bali sediments was enhanced by the addition of bacterial strains SC and E under aerobic and anaerobic conditions, respectively.

Bacillus pumilus has been previously shown to degrade total petroleum hydrocarbons in contaminated soil [27], while *Bacillus aerius* has been isolated, characterized and identified as a potential diuron-degrading bacterium [28]. *Bacillus aryabhattai* was previously isolated from pulp and paper mill wastewater and characterized as having lignin-degrading potential [29]. *Arthrobacter ginkgonis* was a Gram-positive, aerobic strain which displayed a rod–coccus growth lifecycle, was isolated from the rhizosphere of *Ginkgo biloba* L [30]. *Enterobacter hormechei* has been isolated from activated sludge and shown to transform diclofenac [31] as well as to bioconvert lutein into a new compound, 8-methyl-alpha-ionone [32]. *Clostridium sphenoides* is able to metabolize citrate as a sole carbon source, and a *Clostridium* sp. was previously shown to be capable of fermenting glucose, but not citric acid [33]. *Lysinibacillus fusiformis* has been shown to decolorize a selected azo dye [34]. These reports showed that *Bacillus pumilus, Bacillus aerius, Bacillus aryabhattai, Enterobacter hormechei, Clostridium sphenoides* and *Lysinibacillus* sp. had the degradation ability. However, there is no information about the degradation properties of *Arthrobacter ginkgo*.

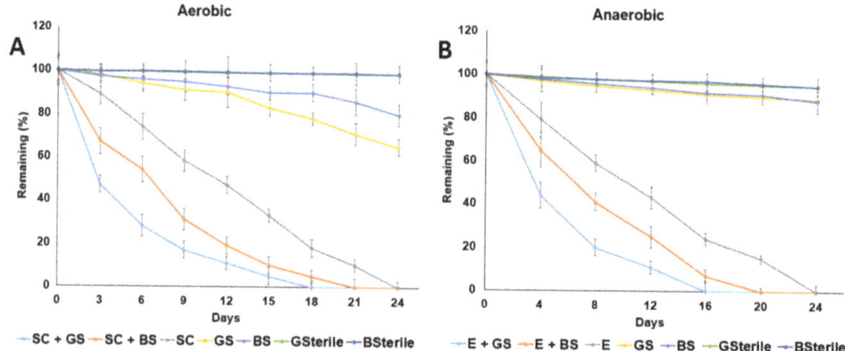

Figure 4. Comparison of the acetaminophen degradation abilities of the isolated bacterial strains. (**A**) Bacterial strain SC (*Arthrobacter ginkgonis*) with sediments under aerobic conditions; (**B**) bacterial strain E (*Enterobacter hormechei*) with sediments under anaerobic conditions. Sterile: sterilized medium without the addition of bacteria. GS: Guandu sediment. BS: Bali sediment. GSterile: sterilized Guandu sediment. BSterile: sterilized Bali sediment. Data from three independent experiments are presented as the means ± SE.

3.3. Addition of Electron Acceptors Improved Anaerobic ACE Degradation

As shown in Figure 5A, ACE was completely degraded after 12, 20 and 24 d in the Guandu non-ACE-adapted sediment supplemented with $NaNO_3$, Na_2SO_4 and $NaHCO_3$, respectively. For the Bali ACE-adapted sediments supplemented with $NaNO_3$, Na_2SO_4 or $NaHCO_3$, the remaining amount of ACE detected after 24 d was ND (non-detected), 4.1% ± 0.6% and 35.9% ± 1.8%, respectively (Figure 5B). The ACE anaerobic degradation data in Figure 5 were fitted to first-order kinetics. For the Guandu ACE-adapted sediments supplemented with $NaNO_3$, Na_2SO_4 or $NaHCO_3$, the observed ACE degradation half-lives were 1.16 ± 0.06, 2.39 ± 0.12 and 2.79 ± 0.14 d, respectively. For the Bali non-ACE-adapted sediments supplemented with $NaNO_3$, Na_2SO_4 or $NaHCO_3$, the observed ACE half-lives were 3.05 ± 0.15, 3.84 ± 0.19 and 10.75 ± 0.53 d, respectively. In contrast, for the Guandu non-ACE-adapted sediments supplemented with $NaNO_3$, Na_2SO_4 or $NaHCO_3$, the remaining amount of ACE detected after 6 d was ND, ND and 20.8% ± 1.1%, respectively (Figure 5C). For the Bali ACE-adapted sediments supplemented with $NaNO_3$, Na_2SO_4 or $NaHCO_3$, the remaining amount of ACE detected after 6 d was ND, 3.3% ± 0.6% and 35.9% ± 1.8%, respectively (Figure 5D). For the Guandu non-ACE-adapted sediments supplemented with $NaNO_3$, Na_2SO_4 or $NaHCO_3$, the observed ACE half-lives were 0.48 ± 0.02, 0.64 ± 0.3 and 1.79 ± 0.09 d, respectively. For the Bali ACE-adapted sediments supplemented with $NaNO_3$, Na_2SO_4 or $NaHCO_3$, the observed ACE half-lives were 0.55 ± 0.03, 0.93 ± 0.05 and 2.85 ± 0.14 d, respectively.

These results indicate that the rate of ACE degradation was higher in the Guandu sediments than in the Bali sediments with or without ACE adaptation. A comparison of ACE degradation in the sediments with or without ACE adaptation revealed that the ACE adaptation process enhanced ACE anaerobic degradation. Microorganisms with ACE-degrading activity may increase in response to ACE adaptation, and these results are similar to those of a previous study regarding the effects of nonylphenol adaptation on nonylphenol anaerobic degradation in sediments [11]. Compared with the control treatment (sediments without $NaHCO_3$, Na_2SO_4 or $NaNO_3$), the anaerobic degradation of ACE was enhanced under methanogenic, sulfate-reducing and nitrate-reducing conditions in the sediments supplemented with $NaNO_3$, Na_2SO_4 or $NaHCO_3$. The ACE degradation rates observed under the three conditions exhibited the following order: nitrate reducing conditions > sulfate-reducing conditions > methanogenic conditions. The results indicate that nitrate-reducing bacteria and sulfate-reducing bacteria, and methane may be the major contributors to anaerobic ACE degradation in all of the tested anaerobic mesocosms.

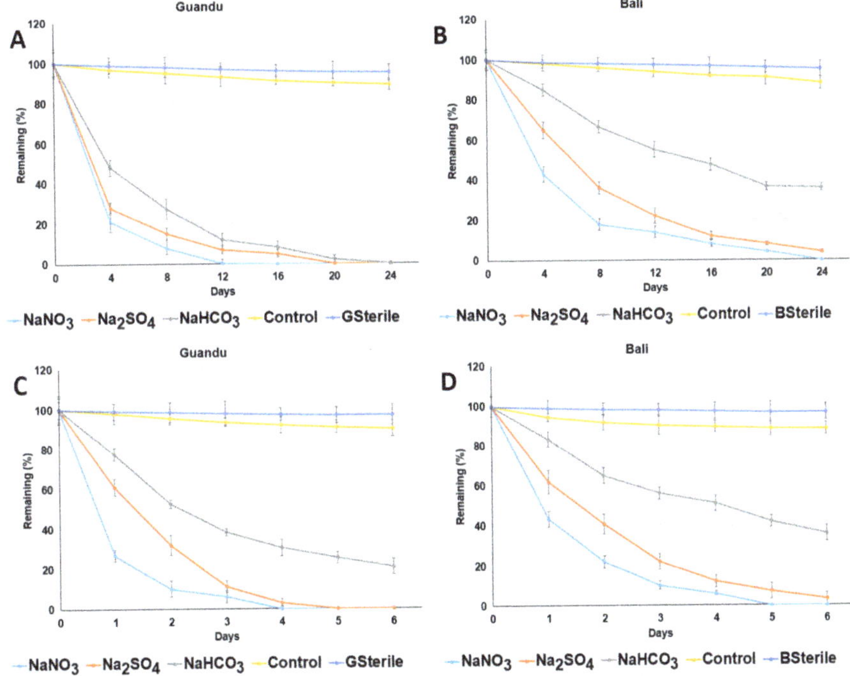

Figure 5. Anaerobic degradation of acetaminophen in (**A**) Guandu and (**B**) Bali non-ACE-adapted sediments and in (**C**) Guandu and (**D**) Bali ACE-adapted sediments. GSterile: sterilized Guandu sediment. BSterile: sterilized Bali sediment. Control: sediment of the sampling site. Data from three independent experiments are presented as the means ± SE. $NaNO_3$: addition of $NaNO_3$ in sediment. Na_2SO_4: addition of Na_2SO_4 in sediment. $NaHCO_3$: addition of $NaHCO_3$ in sediment.

Results of acetaminophen degradation half-lives ($t_{1/2}$, day) using different treatments (addition of ACE-degrading bacterial strains, MC and electron acceptors) were summarized in Table 3. All of the treatments can enhance the ACE biodegradation in mangrove sediments. Addition of $NaNO_3$ exhibited the best effects for ACE biodegradation in mangrove sediments.

Table 3. Comparison of acetaminophen degradation half-lives ($t_{1/2}$, day) using different treatments.

Treatment	Guandu Sediment		Bali Sediment	
	Without Additives	**With Additives**	**Without Additives**	**With Additives**
Aerobic conditions				
MC	33.55 ± 1.67	3.24 ± 0.16	66.66 ± 3.33	6.25 ± 0.31
Arthrobacter sp.	33.55 ± 1.67	2.54 ± 0.13	66.66 ± 3.33	3.30 ± 0.17
Anaerobic conditions				
Enterobacter sp.	95.59 ± 4.63	2.62 ± 0.13	98.04 ± 4.90	3.52 ± 0.17
$NaNO_3$	95.59 ± 4.63	1.16 ± 0.06	98.04 ± 4.90	3.05 ± 0.15
Na_2SO_4	95.59 ± 4.63	2.39 ± 0.12	98.04 ± 4.90	3.84 ± 0.19
$NaHCO_3$	95.59 ± 4.63	2.79 ± 0.14	98.04 ± 4.90	10.75 ± 0.53

3.4. Analysis of Microbial Communities Associated with ACE Degradation

To gain deeper insights into ACE degradation, NGS was performed to analyze the microbial community compositions of aerobic sediments supplemented with MC and anaerobic sediments supplemented with $NaNO_3$, Na_2SO_4 or $NaHCO_3$. As shown in Figure 6, the microbial community compositions of the aerobic sediments supplemented with MC were highly different from those of anaerobic sediments supplemented with NaNO$_3$, Na$_2$SO$_4$ or NaHCO$_3$.

The effects of the addition of NaNO$_3$, Na$_2$SO$_4$ or NaHCO$_3$ are shown in Figure 7. Under anaerobic conditions and with the addition of the electron acceptors, a large increase in the proportion of detected bacteria in the Guandu and Bali sediments (from 60% to 80% and from 47% to 71%, respectively) was observed, with a simultaneous reduction in the proportion of archaea (from 32% to 19% and from 37% to 27%, respectively).

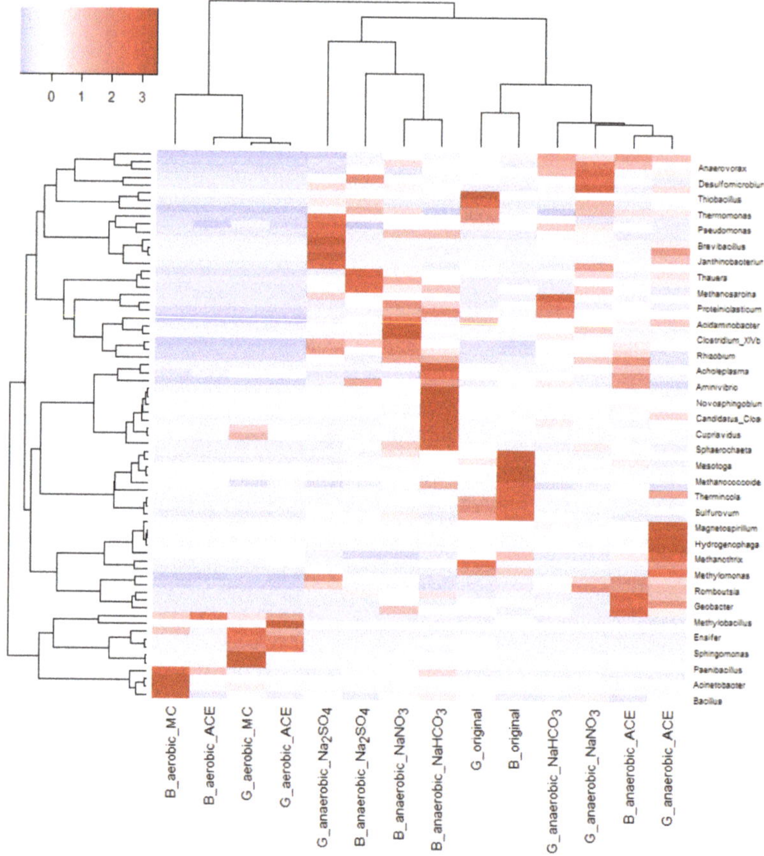

Figure 6. Cluster analysis of microbial communities (at genus level) involved in aerobic and anaerobic degradation of acetaminophen in mangrove sediments. G: Guandu. B: Bali. Original: sediments without additives. aerobic_ACE: sediments with ACE under aerobic conditions. anaerobic_ACE: sediments with ACE under anaerobic conditions. aerobic_MC: aerobic with addition of spent mushroom compost. anaerobic_Na$_2$SO$_4$: anaerobic with addition of Na$_2$SO$_4$. anaerobic_NaNO$_3$: anaerobic with addition of NaNO$_3$. anaerobic_NaHCO$_3$: anaerobic with addition of NaHCO$_3$.

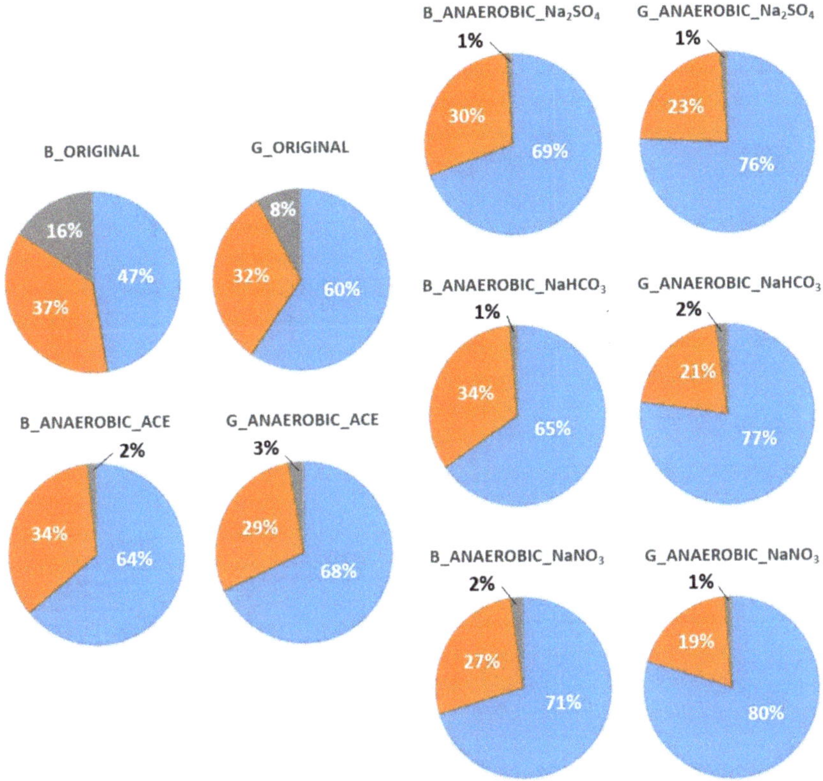

Figure 7. Compositions of Bacteria and Archaea in anaerobic experiments. ORIGINAL: original sediment. ACE: sediments with ACE. Na_2SO_4: sediments with ACE + Na_2SO_4. $NaHCO_3$: sediments with ACE + $NaHCO_3$. $NaNO_3$: sediments with ACE + $NaNO_3$.

The microbial communities associated with the sediments supplemented with $NaNO_3$, Na_2SO_4 or $NaHCO_3$ are shown in Figure 8. For the Bali sediments, the overall proportions of nitrogen metabolism bacteria (Figure 8A), sulfate–sulfur metabolism bacteria (Figure 8B) and methanogens (Figure 8C) observed under anaerobic conditions were greater than those detected under aerobic conditions (two proportion test, $p < 0.05$). Similar results were observed for the Guandu sediments.

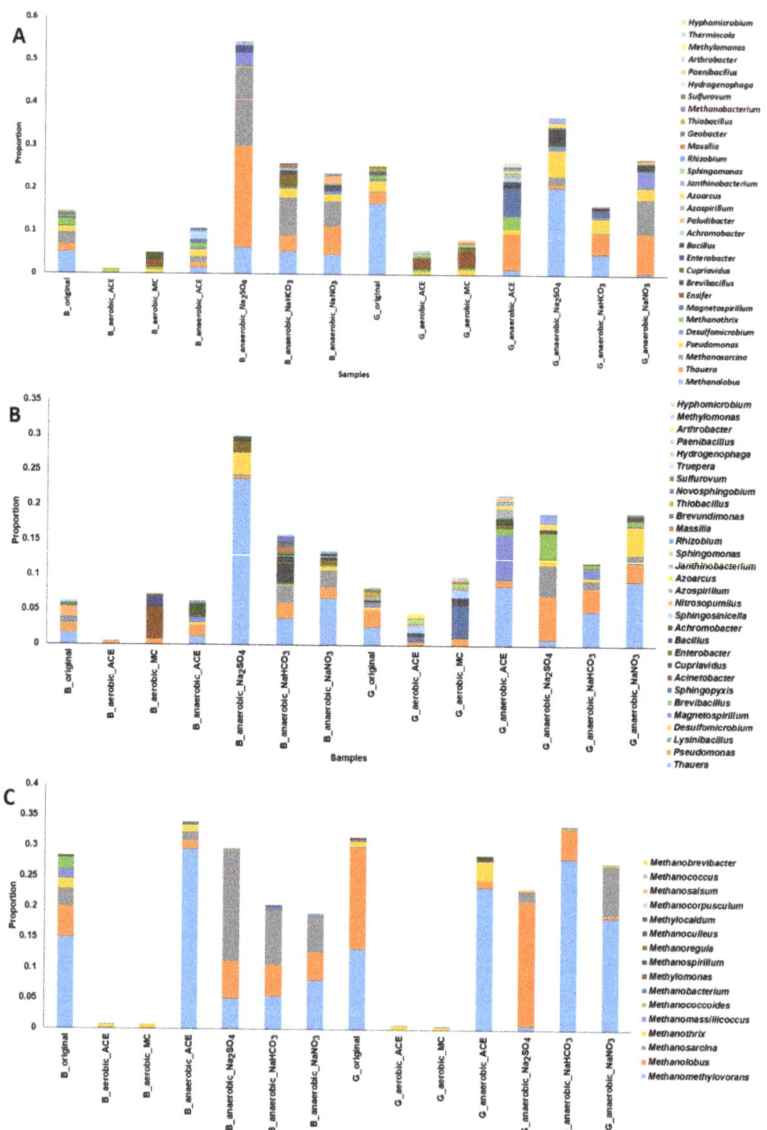

Figure 8. Microbial communities involved in the anaerobic degradation of acetaminophen in mangrove sediments supplemented with $NaNO_3$, Na_2SO_4 and $NaHCO_3$. (**A**) Major 30 microbial genera associated with nitrogen cycle; (**B**) major 30 microbial genera associated with sulfate cycle; (**C**) methanogens. The differences in microbial compositions between the aerobic and anaerobic experimental samples were assessed by the two-proportion test. The p-values of all of these tests are less than 0.05. G: Guandu. B: Bali. Original: sediments without additives. aerobic_ACE: sediments with ACE under aerobic conditions. anaerobic_ACE: sediments with ACE under anaerobic conditions. aerobic_MC: aerobic with addition of spent mushroom compost. anaerobic_Na_2SO_4: anaerobic with addition of Na_2SO_4. anaerobic_$NaNO_3$: anaerobic with addition of $NaNO_3$. anaerobic_$NaHCO_3$: anaerobic with addition of $NaHCO_3$.

Furthermore, for the Bali sediments, the overall proportions of microbes associated with aromatic compound degradation under anaerobic conditions were greater than observed under aerobic conditions (multiproportion test, $p < 0.05$) (Figure 9). Similar results were observed for the Guandu sediments.

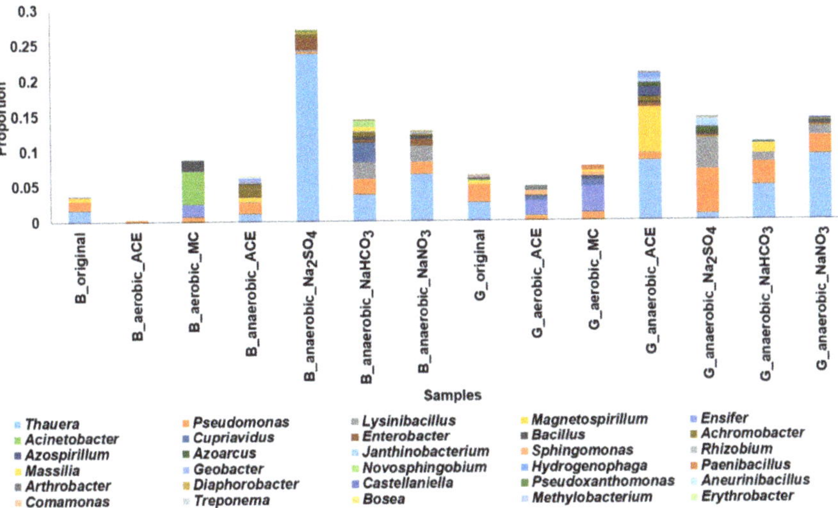

Figure 9. Major 30 microbial genera that have been reported to be involved in the degradation of aromatic compounds. The differences in the microbial compositions between the aerobic and anaerobic experimental samples were identified using the two-proportion test. The *p*-values of all of these tests are less than 0.05. G: Guandu. B: Bali. Original: sediments without additives. aerobic_ACE: sediments with ACE under aerobic conditions. anaerobic_ACE: sediments with ACE under anaerobic conditions. aerobic_MC: aerobic with addition of spent mushroom compost. anaerobic_Na_2SO_4: anaerobic with addition of Na_2SO_4. anaerobic_$NaNO_3$: anaerobic with addition of $NaNO_3$. anaerobic_$NaHCO_3$: anaerobic with addition of $NaHCO_3$.

The results presented in Figure 5 indicate that the addition of $NaNO_3$ or Na_2SO_4 had better ACE degradation effects than the addition of $NaHCO_3$ in both the Bali and Guandu sediments. To identify common and differential microbial genera associated with ACE degradation in the sediments with different additives, a Venn diagram analysis was performed. In Figure 10, "Aro_Deg" indicates the aromatic compound-degrading bacteria, while "Nitrogen" or "Sulfate" indicate the nitrogen or sulfate–sulfur metabolism bacteria in the sediments supplemented with $NaNO_3$ or Na_2SO_4, respectively.

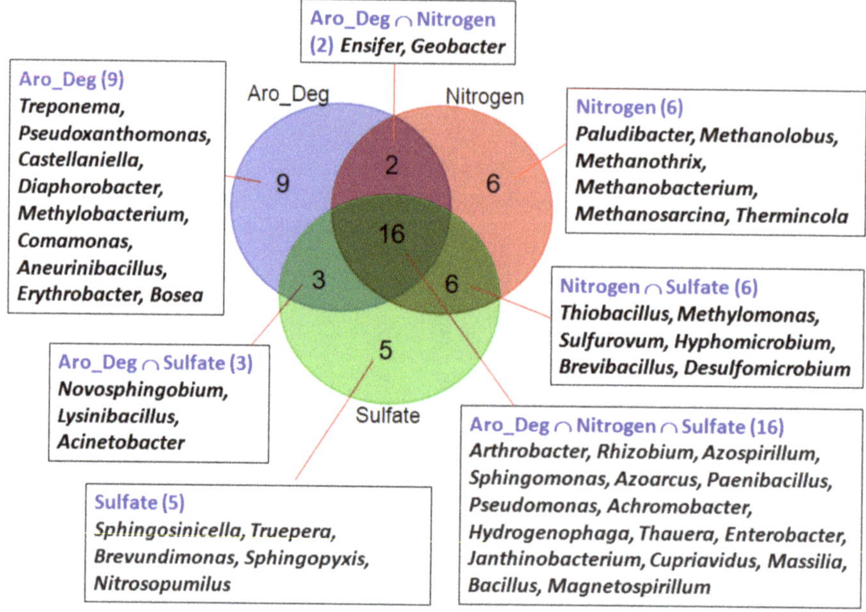

Figure 10. Identification of common and differential microbial genera detected between the different experiments by Venn diagram analysis. Aro_Deg: Major 30 microbial genera identified in the experiments using electron acceptors, which are associated with aromatic compound degradation. Nitrogen: Major 30 microbial genera identified in the experiments using electron acceptors that are associated with nitrogen metabolism. Sulfate: Major 30 microbial genera identified in the experiments using electron acceptors that are associated with sulfate–sulfur metabolism.

Sixteen microbial genera (*Arthrobacter, Rhizobium, Azospirillum, Sphingomonas, Azoarcus, Paenibacillus, Pseudomonas, Achromobacter, Hydrogenophaga, Thauera, Enterobacter, Janthinobacterium, Cupriavidus, Massilia, Bacillus* and *Magnetospirillum*) were detected in all of the sediments supplemented with $NaNO_3$ or Na_2SO_4. Four of the sixteen microbial genera (*Bacillus, Hydrogenophaga, Pseudomonas* and *Sphingomonas*) have been reported to be associated with ACE (and paracetamol compounds) degradation [14,35–37]. These results suggested that sixteen microbial genera may be involved in anaerobic ACE degradation in the sediments supplemented with $NaNO_3$ or Na_2SO_4. Three of the sixteen microbial genera (*Arthrobacter, Enterobacter* and *Bacillus*) were identified among the ACE-degrading bacterial isolates. Another two bacterial genera (*Ensifer* and *Geobacter*) represent additional aromatic compound-degrading bacteria that were not sulfate–sulfur metabolism bacteria. Three bacterial genera (*Novosphingobium, Lysinibacillus* and *Acinetobacter*) represent additional aromatic compound-degrading bacteria that were not nitrogen metabolism bacteria. These five bacterial genera can explain the different effects of the addition of $NaNO_3$ or Na_2SO_4 on ACE degradation in sediments observed in Figure 5. The distributions of number of major microbial genera with different aromatic compound degradation pathways are shown in Figure 11. Most of the microbial genera associated with six reaction modules, M00551 Benzoate degradation, benzoate => catechol/methyl benzoate => methyl catechol, M00568 Catechol ortho-cleavage, catechol => 3-oxoadipate, M00569 Catechol meta-cleavage, catechol => acetyl-CoA/4-methylcatechol => propionyl-CoA, M00548 Benzene degradation, benzene => catechol, M00541 Benzoyl-CoA degradation, benzoyl-CoA => 3-hydroxypimeloyl-CoA and M00638 Salicylate degradation, salicylate => gentisate. Moreover, more microbial genera associated with aromatic compound degradation were found in the sediment supplemented with $NaNO_3$ than that of the sediment supplemented with Na_2SO_4. These results can explain the different effects of the addition of $NaNO_3$ or Na_2SO_4 on ACE degradation in sediments observed in Figure 5.

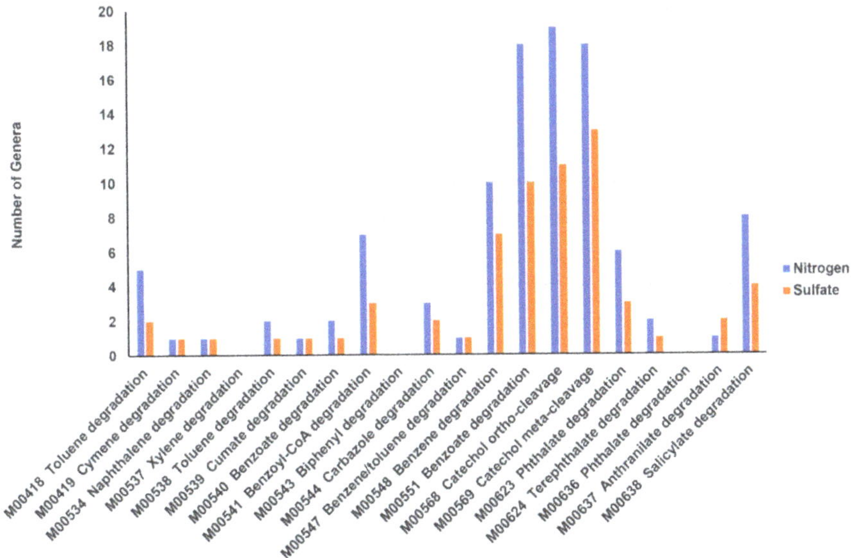

Figure 11. Distributions of number of major microbial genera with different aromatic compound degradation pathways. Mxxxxx are KEGG database module ID. Nitrogen: 30 major microbial genera identified in the experiments using electron acceptors that are associated with nitrogen metabolism. Sulfate: Major 30 microbial genera identified in the experiments using electron acceptors that are associated with sulfate–sulfur metabolism.

The sampling sites of mangrove sediments in this study are mangrove (*Kandelia obovata*) swamps. The study of Weng et al. investigated nitrogen cycle in rhizosphere of *Kandelia obovata* under nitrogen input shown that the potential nitrification intensity increased 200%–1500% compared with control under ammonium addition. The potential denitrification intensity increased more than 200% under nitrate addition. The abundance of ammonia-oxidizing archaea, ammonia-oxidizing bacteria, *nirK* and *nifH* containing microbes increased around the area of *Kandelia obovata* rhizosphere. The *nirS* containing microbes were decreased. Total nitrogen in tissues of *Kandelia obovata* increased more than 200% [38]. Results of another study indicated that addition of NH_4^+ significantly increased the growth rate of *Kandelia obovata* in mesocosm experiments [39]. A study of a 10-year-old constructed mangrove wetland planted with *Kandelia obovata* shown that the abundance of nitrogen cycle related bacteria, including ammonifiers, nitrifiers and denitrifiers, in planted belts were higher than that in unplanted control [40]. A study by Jian et al. demonstrated that addition of Na_2SO_4 enhanced the acid-volatile sulfide, crystalline Fe and exchangeable P contents, prompted iron plaque formation on root surfaces of *Kandelia obovata* and increased P accumulation in plant tissues. Sulfate and *Kandelia obovata* root activities had a confederate influence on the spatial distribution and geochemical cycling of Fe and P in the sediments. They enhanced *Kandelia obovata* resistance to the rugged surroundings in the mangrove environment and can prompt *Kandelia obovata* growth [41]. Therefore, additions of nitrate and sulfate to enhance ACE degradation (proposed in this study) are suitable to be applied at mangrove wetlands composed of *Kandelia obovata*. Moreover, these methods may be useful in a constructed wetland composed of *Kandelia obovata*.

4. Conclusions

In this study, the feasibility of bioremediation strategies for ACE removal in mangrove sediments were tested. All of the treatments, addition of ACE-degrading bacterial strains, MC and electron acceptors (NaNO$_3$, Na$_2$SO$_4$ or NaHCO$_3$) can enhance the ACE biodegradation in mangrove sediments. Addition of NaNO$_3$ exhibited the best effects for ACE biodegradation in mangrove sediments. Different treatments result in different bacterial communities and exhibited different ACE-degrading effectiveness. Sixteen microbial genera identified may be major microbes involved in anaerobic ACE degradation in mangrove sediments with electron acceptor addition. These results provide useful solutions for removal of ACE in mangrove sediments.

Supplementary Materials: The following are available online at http://www.mdpi.com/2071-1050/12/13/5410/s1. Figure S1: Mangrove sediment sampling sites Bali (25°09′48″ N 121°25′04″ E) and Guandu (25°07′21″ N 121°27′40″ E) in northern Taiwan.

Author Contributions: Data curation, C.-W.Y.; formal analysis, Y.-E.C.; funding acquisition, B.-V.C.; investigation, C.-W.Y.; methodology, Y.-E.C.; project administration, B.-V.C.; visualization, C.-W.Y.; writing—original draft, C.-W.Y. and B.-V.C.; writin—review & editing, C.-W.Y. and B.-V.C. All authors have read and agreed to the published version of the manuscript.

Funding: This research was supported by the Ministry of Science and Technology, Republic of China, Taiwan (grant no. MOST 107-2313-B-031-002).

Conflicts of Interest: The authors declare that they have no conflicts of interest.

References

1. Ouellet, M.; Percival, M.D. Acetaminophen inhibition of cyclooxygenase isoforms. *Arch. Biochem. Biophys.* **2001**, *387*, 273–280. [CrossRef] [PubMed]
2. Scott, P.D.; Bartkow, M.; Blockwell, S.J.; Coleman, H.M.; Khan, S.J.; Lim, R.; McDonald, J.A.; Nice, H.; Nugegoda, D.; Pettigrove, V.; et al. A national survey of trace organic contaminants in Australian Rivers. *J. Environ. Qual.* **2014**, *43*, 1702–1712. [CrossRef] [PubMed]
3. Campanha, M.B.; Awan, A.T.; De Sousa, D.N.R.; Grosseli, G.M.; Mozeto, A.A.; Fadini, P.S. A 3-year study on occurrence of emerging contaminants in an urban stream of So Paulo State of Southeast Brazil. *Environ. Sci. Pollut. Res.* **2015**, *22*, 7936–7947. [CrossRef]
4. Tran, N.H.; Reinhard, M.; Khan, E.; Chen, H.; Nguyen, V.T.; Li, Y.; Goh, S.G.; Nguyen, Q.B.; Saeidi, N.; Gin, K.Y.H. Emerging contaminants in wastewater, stormwater runoff, and surface water: Application as chemical markers for diffuse sources. *Sci. Total Environ.* **2019**, *676*, 252–267. [CrossRef] [PubMed]
5. Cao, F.; Zhang, M.; Yuan, S.; Feng, J.; Wang, Q.; Wang, W.; Hu, Z. Transformation of acetaminophen during water chlorination treatment: Kinetics and transformation products identification. *Environ. Sci. Pollut. Res. Int.* **2016**, *23*, 12303–12311. [CrossRef]
6. Sun, C.; Dudley, S.; McGinnis, M.; Trumble, J.; Gan, J. Acetaminophen detoxification in cucumber plants via induction of glutathione S-transferases. *Sci. Total Environ.* **2019**, *649*, 431–439. [CrossRef]
7. Philips, P.J.; Smith, S.G.; Koplin, D.W.; Zaugg, S.D.; Buxton, H.T.; Furlong, E.T.; Esposito, K.; Stinson, B. Pharmaceutical formulation facilities as sources of opioids and other pharmaceuticals to wastewater treatment plant effluents. *Environ. Sci. Technol.* **2010**, *44*, 4910–4916. [CrossRef]
8. Escher, B.I.; Baumgartner, R.; Koller, M.; Treyer, K.; Lienert, J.; McArdell, C.S. Environmental toxicology and risk assessment of pharmaceuticals from hospital wastewater. *Water Res.* **2011**, *45*, 75–92. [CrossRef]
9. Bernard, D.; Pascaline, H.; Jeremie, J.J. Distribution and origin of hydrocarbons in sediments from lagoons with fringing mangrove communities. *Mar. Pollut. Bull.* **1996**, *32*, 734–739. [CrossRef]
10. Chang, B.V.; Chang, I.T.; Yuan, S.Y. Anaerobic degradation of phenanthrene and pyrene in mangrove sediment. *Bull. Environ. Contam. Toxicol.* **2008**, *80*, 145–149. [CrossRef]
11. Chang, B.V.; Lu, Z.J.; Yuan, S.Y. Anaerobic degradation of nonylphenol in subtropical mangrove sediments. *J. Hazard. Mater.* **2009**, *165*, 162–167. [CrossRef] [PubMed]
12. Yang, C.W.; Tsai, L.L.; Chang, B.V. Fungi extracellular enzyme-containing microcapsules enhance degradation of sulfonamide antibiotics in mangrove sediments. *Environ. Sci. Pollut. Res.* **2017**, *25*, 10069–10079. [CrossRef] [PubMed]

13. Yang, C.W.; Tsai, L.L.; Chang, B.V. Anaerobic degradation of sulfamethoxazole in mangrove sediments. *Sci. Total Environ.* **2018**, *634*, 922–933. [CrossRef] [PubMed]

14. Żur, J.; Wojcieszyńska, D.; Hupert-Kocurek, K.; Marchlewicz, A.; Guzik, U. Paracetamol—Toxicity and microbial utilization. *Pseudomonas moorei* KB4 as a case study for exploring degradation pathway. *Chemosphere* **2018**, *20*, 192–202.

15. Joss, A.; Keller, E.; Alder, A.C.; Gobel, A.; McArdell, C.S.; Ternes, T.; Siegrist, H. Removal of pharmaceuticals and fragrances in biological wastewater treatment. *Water Res.* **2005**, *39*, 3139–3152. [CrossRef]

16. Groning, J.; Held, C.; Garten, C.; Claussnitzer, U.; Kaschabek, S.R.; Schlomann, M. Transformation of diclofenac by the indigenous microflora of river sediments and identification of a major intermediate. *Chemosphere* **2007**, *69*, 509–516. [CrossRef]

17. Ying, G.G.; Kookana, R.S. Degradation of five selected endocrine-disrupting chemicals in seawater and marine sediment. *Environ. Sci. Technol.* **2003**, *37*, 1256–1260. [CrossRef]

18. Pepper, L.L.; Gerba, C.P.; Gentry, T.J. *Environmental Microbiology*, 3rd ed.; Elsevier BV: Amsterdam, The Netherlands, 2015.

19. Hu, J.; Zhang, L.L.; Chen, J.M.; Liu, Y. Degradation of paracetamol by *Pseudomonas aeruginosa* strain HJ1012. *J. Environ. Sci. Health A Tox. Hazard Subst. Environ. Eng.* **2013**, *48*, 791–799. [CrossRef]

20. De Gusseme, B.; Vanhaecke, L.; Verstraete, W.; Boon, N. Degradation of acetaminophen by Delftia tsuruhatensis and Pseudomonas aeruginosa in a membrane bioreactor. *Water Res.* **2011**, *45*, 1829–1837. [CrossRef]

21. Li, X.; Lin, X.; Zhang, J.; Wu, Y.; Yin, R.; Feng, Y.; Wang, Y. Degradation of polycyclic aromatic hydrocarbons by crude extracts from spent mushroom substrate and its possible mechanisms. *Curr. Microbiol.* **2010**, *60*, 336–342. [CrossRef]

22. Chang, B.V.; Ren, Y.L. Biodegradation of three tetracyclines in river sediment. *Ecol. Eng.* **2015**, *75*, 272–277. [CrossRef]

23. Larkin, M.A.; Blackshields, G.; Brown, N.P.; Chenna, R.; McGettigan, P.A.; McWilliam, H.; Valentin, F.; Wallace, I.M.; Wilm, A.; Lopez, R.; et al. Clustal W and Clustal X version 2.0. *Bioinformatics* **2007**, *23*, 2947–2948. [CrossRef] [PubMed]

24. Kanehisa, M.; Furumichi, M.; Tanabe, M.; Sato, Y.; Morishima, K. KEGG: New perspectives on genomes, pathways, diseases and drugs. *Nucleic Acids Res.* **2017**, *45*, D353–D361. [CrossRef] [PubMed]

25. Castillo-Carvajal, L.C.; Sanz-Martín, J.L.; Barragán-Huerta, B.E. Biodegradation of organic pollutants in saline wastewater by halophilic microorganisms: A review. *Environ. Sci. Pollut. Res.* **2014**, *21*, 9578–9588. [CrossRef]

26. Tam, N.F.; Guo, C.L.; Yau, W.Y.; Wong, Y.S. Preliminary study on biodegradation of phenanthrene by bacteria isolated from mangrove sediments in Hong Kong. *Mar. Pollut. Bull.* **2002**, *45*, 316–324. [CrossRef]

27. Varma, S.S.; Lakshmi, M.B.; Rajagopal, P.; Velan, M. Degradation of total petroleum hydrocarbon (TPH) in contaminated soil using *Bacillus pumilus* MVSV3. *Biocontrol. Sci.* **2017**, *22*, 17–23. [CrossRef] [PubMed]

28. Hanapiah, M.; Zulkifli, S.Z.; Mustafa, M.; Mohamat-Yusuff, F.; Ismail, A. Isolation, characterization, and identification of potential diuron-degrading bacteria from surface sediments of Port Klang, Malaysia. *Mar. Pollu. Bull.* **2018**, *127*, 453–457. [CrossRef]

29. Zainith, S.; Purchase, D.; Saratale, G.D.; Ferreira, L.F.R.; Bilal, M.; Bharagava, R.N. Isolation and characterization of lignin-degrading bacterium *Bacillus aryabhattai* from pulp and paper mill wastewater and evaluation of its lignin-degrading potential. *3 Biotech.* **2019**, *9*, 92. [CrossRef]

30. Cheng, J.; Zhang, M.Y.; Zhao, J.C.; Xu, J.; Zhang, Y.; Zhang, T.Y.; Wu, Y.Y.; Zhang, Y.X. *Arthrobacter ginkgonis* sp. nov., an actinomycete isolated from rhizosphere of *Ginkgo biloba* L. *Int. J. Syst. Evol. Microbiol.* **2017**, *67*, 319–332. [CrossRef] [PubMed]

31. Aissaoui, S.; Ouled-Haddar, H.; Sifour, M.; Harrouche, K.; Sghaier, H. Metabolic and co-metabolic transformation of diclofenac by *Enterobacter hormaechei* D15 isolated from activated sludge. *Curr. Microbiol.* **2017**, *74*, 381–388. [CrossRef]

32. Zhong, G.F.; Wang, F.F.; Sun, J.H.; Ye, J.B.; Mao, D.B.; Ma, K.; Yang, X.P. Bioconversion of lutein by *Enterobacter hormaechei* to form a new compound, 8-methyl-alpha-ionone. *Biotech. Lett.* **2017**, *39*, 1019–1024. [CrossRef] [PubMed]

33. Francis, A.J.; Dodge, C.J.; Meinken, G.E. Biotransformation of pertechnetate by *Clostridia*. *Radiochim. Acta* **2009**, *90*, 9–11. [CrossRef]

34. Sari, I.P.; Simarani, K. Decolorization of selected azo dye by *Lysinibacillus fusiformis* W1B6: Biodegradation optimization, isotherm, and kinetic study biosorption mechanism. *Adsorp. Sci. Technol.* **2019**, *37*, 492–508. [CrossRef]
35. Chang, T.J.; Chang, Y.H.; Chao, W.L.; Jane, W.N.; Chang, Y.T. Effect of hydraulic retention time on electricity generation using a solid plain-graphite plate microbial fuel cell anoxic/oxic process for treating pharmaceutical sewage. *J. Environ. Sci. Health Pert A* **2018**, *53*, 1185–1197. [CrossRef]
36. Chang, Y.T.; Yang, C.W.; Chang, Y.J.; Chang, T.C.; Wei, D.J. The treatment of PPCP-containing sewage in an anoxic/aerobic reactor coupled with a novel design of solid plain graphite-plates microbial fuel cell. *Biomed. Res. Int.* **2014**, *2014*, 765652. [CrossRef]
37. Zhang, L.; Hu, J.; Zhu, R.; Zhou, Q.; Chen, J. Degradation of paracetamol by pure bacterial cultures and their microbial consortium. *Appl. Microbiol. Biotechnol.* **2013**, *97*, 3687–3698. [CrossRef]
38. Weng, B.; Xie, X.; Yang, J.; Liu, J.; Lu, H.; Yan, C. Research on the nitrogen cycle in rhizosphere of Kandelia obovata under ammonium and nitrate addition. *Mar. Pollut. Bull.* **2013**, *76*, 227–240. [CrossRef]
39. Cui, X.; Song, W.; Feng, J.; Jia, D.; Guo, J.; Wang, Z.; Wu, H.; Qi, F.; Liang, J.; Lin, G. Increased nitrogen input enhances *Kandelia obovata* seedling growth in the presence of invasive *Spartina alterniflora* in subtropical regions of China. *Biol. Lett.* **2017**, *13*, 20160760. [CrossRef]
40. Tian, T.; Tam, N.F.Y.; Zan, Q.; Cheung, S.G.; Shin, P.K.S.; Wong, Y.S.; Zhang, L.; Chen, Z. Performance and bacterial community structure of a 10-years old constructed mangrove wetland. *Mar. Pollut. Bull.* **2017**, *124*, 1096–1105. [CrossRef]
41. Jian, L.; Junyi, Y.; Jingchun, L.; Chongling, Y.; Haoliang, L.; Spencer, K.L. The effects of sulfur amendments on the geochemistry of sulfur, phosphorus and iron in the mangrove plant (*Kandelia obovata* (S. L.)) rhizosphere. *Mar. Pollut. Bull.* **2017**, *114*, 733–741. [CrossRef]

MDPI

St. Alban-Anlage 66

4052 Basel

Switzerland

Tel. +41 61 683 77 34

Fax +41 61 302 89 18

www.mdpi.com

Sustainability Editorial Office

E-mail: sustainability@mdpi.com

www.mdpi.com/journal/sustainability